Modern Land Drainage

Modern Land Drainage

Planning, Design and Management
of Agricultural Drainage Systems

2nd Edition

Willem F. Vlotman
PhD, MSc, Ing

Lambert K. Smedema
PhD, MSc

David W. Rycroft
PhD, CEng, MICE

CRC Press
Taylor & Francis Group
Boca Raton London New York Leiden

CRC Press is an imprint of the
Taylor & Francis Group, an **informa** business

A BALKEMA BOOK

Cover photo is by Willem F. Vlotman, Drainage control structure, Binnenveld, Wageningen, The Netherlands, taken March 2001

CRC Press/Balkema is an imprint of the Taylor & Francis Group, an informa business

Typeset by Apex CoVantage, LLC

Printed in the Netherlands by Printforce

Second Edition: Revised and updated 2020
Copyright © 2004 Taylor & Francis Group plc, London, UK, The Netherlands
Copyright © 1983 Lambert K Smedema and David W. Rycroft
First published 1983 by Batsford Ltd. UK.
ISBN 90-5809-554-1
Paperback edition published 1988

Library of Congress Cataloging-in-Publication Data
Applied for

Published by: CRC Press/Balkema
                Schipholweg 107c, 2316 XC Leiden, The Netherlands
                e-mail: Pub.NL@taylorandfrancis.com
                www.crcpress.com – www.taylorandfrancis.com

ISBN: 978-0-367-45866-9 (Hbk)
ISBN: 978-1-003-02590-0 (eBook)

DOI: https://doi.org/10.1201/9781003025900

Contents

Tables

Figures

Boxes

Preface

'A day in the field is worth six days in the office'
C.C. Inglis, Superintending Drainage Engineer,
Public Works Department, Bombay, India 1937.

This book is the 2nd edition of Modern Land Drainage that was first published in 2004. It is the third version of 'Land drainage: planning and design of agricultural drainage systems' authored by Lambert K. Smedema and David W. Rycroft and first published in 1983. Willem F. Vlotman was added as a co-author to provide modern and new perspectives of development in drainage in 2004 and became the lead contributor of the 2020 version.

This new edition presents a comprehensive and current description of land drainage and includes the use of remote and drone investigations. It recommends a thorough cause and effect analysis of waterlogging-, drainage- and salinity problems before planning of a drainage system; the problems are the effect of a cause and treating the cause may alleviate the need for drainage; a new paradigm of drainage altogether. The book covers new theories, concepts, technologies, methodologies, findings and experiences developed and gained since the previous publications. It also familiarises the younger generations of drainage professionals with the large body of still highly relevant knowledge of "drainage" accumulated in the past. This edition has retained its focus on analysis and understanding of the underlying field processes that are key to arriving at a correct diagnosis and finding the best solution for the drainage problems at hand.

The geographical scope of the book is global. The story line is based on the classic drainage of the rainfed agricultural land of the humid temperate zone. A major part is dedicated to the drainage for waterlogging and salinity control of irrigated land in the (semi) arid zone, which is a zone of considerable drainage needs in which the authors have worked extensively. The new frontier of drainage of the humid tropics, with emphasis on drainage of rice land, has also been covered.

The thematic scope has been further expanded by a much more extensive coverage of the institutional and management aspects of drainage development. Subjects like public drainage organisations, participatory development, stakeholder involvement, financing and cost recovery, maintenance and performance assessments using remote sensing have been addressed in several new chapters.

The discussion of the possible adverse impacts of drainage interventions on the environment has been enhanced in scope and depth. The role and the procedures of the environmental impact assessment in the preparation of drainage projects are described. The hazards of irrigation-induced salt mobilisation and other irrigation-induced land and river salinisation processes have been addressed as well as the related issue of the disposal of saline drainage water.

Computer applications for drainage planning, design and management are described where appropriate but no specific programs are provided. Some of the most widely used computer programs are described to illustrate some of their generic characteristics and features. Website addresses and other references where programs can be reviewed and procured have been verified and correct up to September 2019.

As with the first edition, this book targets professionals in drainage engineering and management practice in both developed and developing countries. The book is also intended as a university level textbook.

References have been used sparingly in the text. It would have required an unwieldy number of references to do justice to all those researchers and professionals who have contributed to building up the present body of drainage knowledge as compiled herein. Suggestions for further reading and perusal have been given with the references and are marked with an asterisk after the first name.

All three authors have had the privilege and benefit of having worked with distinguished colleagues and been associated with reputed institutions. Books like these are the fruits of a lifetime of exchanges and interactions rather than the brainchild of merely three individuals. The authors gratefully acknowledge the contributions made by former colleagues and by the various members of the international drainage community met at International Commission on Irrigation and Drainage (ICID) meetings and other gatherings over the past few decades. Specific mention is made of the contributions by Robert Broughton (Canada), Mohammed Bybordi (Iran) and Mohammed Nawaz Bhutta (Pakistan), who by commenting on the first edition helped shape this one, and Yoshihiko Ogino (Japan) and Walter Ochs (USA) who readily responded to requests for review of major parts of the book.

It is our firm belief that modern land drainage is a key element for continued agricultural and rural development worldwide and has a crucial role to fulfil in the management of our environment, now and in the future. We sincerely hope that this book will make a significant contribution towards achieving this.

Arnhem/Wageningen/Winchester 2004   *LKS/WFV/DWR*
Canberra 2019   *WFV*

"Drainage design and assessment is best done in the field when it rains."
                                                                          Willem F. Vlotman, 2019

About the authors

Willem F. Vlotman has worked both with consultants and research organisations through-out his professional career. He received his BSc (Civil Engineering) in 1975 at the Haar-lem Technological College, the Netherlands, his MSc (Agricultural Engineering) in 1982 at the University of Arizona, Tucson, USA, his PhD in Irrigation and Agricultural Engi-neering at Utah State University (Logan, Utah, USA) in 1985 and a Master of Business Administration (MBA) from the Wageningen Business School in 2005. From 1976 till 1989 he worked with Dutch and American consultants on agricultural drainage, urban drainage and irrigation projects in the Netherlands, Jamaica, USA, Bangladesh, Nepal and Sri Lanka. From 1988–2004 he worked with the International Institute for Land Reclama-tion and Improvement, ILRI in Pakistan, Egypt and the Netherlands. In 2004 he moved to Australia working with consultancy firm SKM and for the Murray-Darling Basin Author-ity (MDBA) from which he retired in October 2017. Dr. Vlotman has written numerous publications and co-authored a book on drain envelope design (Envelope design for sub-surface drains, ILRI, 2000). He was chairman of the Working Group on Drainage of the International Commission on Irrigation and Drainage (ICID) from 1995–2013.

Lambert K. Smedema is a graduate of Wageningen University, Netherlands (MSc Irriga-tion, in 1960) and Cornell University, USA (PhD Drainage, in 1964). For most of his career, he worked as an engineer with Euroconsult overseas consultants (1965–1991), ini-tially on irrigation in Africa and South America, later on drainage in the Middle East and South and Central Asia. He was a lecturer/researcher at the University of Nairobi (Soil Physics/Drainage, 1978–79) and at the Delft University of Technology (Water Manage-ment, 1980–84). During the nineties, he was a long-term consultant with the World Bank (1992–1998) and FAO (1999–2000) and he is currently an independent consultant. Dr Smedema's professional career has spanned a wide range of countries and environmental conditions. He is an international consultant, guest lecturer and speaker at meetings of the International Commission on Irrigation and Drainage (ICID) and other international drainage conferences. He has published widely on drainage issues.

David W. Rycroft is a graduate of St Andrews University (in Civil Engineering) and Dundee University (PhD in 1971). He began his professional career as a Civil Engineer work-ing for the UK Ministry of Agriculture at the Field Drainage Experimental Unit based in Cambridge. The unit had been established in order to introduce more science and objectivity into the practice of Field Drainage. Subsequently he joined consultants Sir M MacDonald and Partners, working as their Land Drainage specialist on large irrigation projects in Iraq, Somalia and Egypt. In 1979 he joined the Institute of Irrigation Studies, part of the Department of Civil Engineering at Southampton University. He taught Soil Physics and Land Drainage to Postgraduates studying Irrigation Engineering. During this time, he undertook many assignments in Land Drainage and Salinity control, mainly in the Middle East, North Africa, Turkey and Pakistan. His research interests centred around the drainage, reclamation and control of salinity in heavy clay soils. Thereafter he became responsible for developing a master's research training programme directed towards identifying the impacts and the responses to rising sea levels on low lying coasts.

Part I

Introduction

Chapter 1

Land drainage for agriculture

Agricultural drainage as a practice to improve the prevailing natural drainage conditions of the land is now more than two centuries old. Although even some ancient farmers were probably already aware of the benefits of improved drainage and some early applications are on record, they were few in number and often more motivated by military than by agricultural considerations. The agricultural use of drainage remained very limited until the second half of the 18th century when as part of the birth of modern agriculture, improved drainage started to attract wider interest and application.

Most land has periods during which excess water occurs. However, these need not be too harmful provided the quantities are small, the periods of occurrence are of short duration, or the excess occurs during a non-critical part of the growing season. Most land also has some natural drainage, which assists in the removal of a certain amount of the excess water (Box 1.1, Figure 1.1). It is only when large quantities occur for prolonged durations at critical periods, that its removal by artificial means may be feasible and desirable.

Land drainage is applied for the following two quite different purposes:

- reclamation of naturally waterlogged land for agricultural use (*drainage for horizontal expansion*)
- improvement of the drainage conditions of existing agricultural land (*drainage for vertical expansion*).

Drainage for horizontal expansion was till the end of the 19th century generally the main drainage activity but has now become relatively less important as in most countries this form of land reclamation has come to standstill, partly due the non-availability of suitable land, partly due to the societal wish to preserve remaining natural wetland resources. Although drainage for horizontal expansion is still an important activity in some countries in SE Asia and South America, most drainage investment is for the purposes of vertical expansion.

FAO assessments put the world's area of potentially suitable cropland at some 3 200 million ha. Almost all of the best land has already been developed (1 500 million ha) while the remaining not yet developed land (1 700 million ha) includes a large percentage of marginally suitable and environmentally sensitive land. Of the developed cropland, about 1 200–1 250 million ha are used for rainfed cropping while about 270 million ha (about 18% of the total cropland) are provided with irrigation facilities. The area provided with improved drainage is estimated to be of the order of 150–200 million ha (10–14% of the total cropland) of which 100–150 million ha are rainfed land and 25–50 million ha are irrigated land (see Table 1.1).

Box 1.1   The hydrologic cycle

The precipitation reaching the soil surface will partly enter the soil (*infiltration*) where it may be retained in the upper layers (*soil moisture storage*) or percolate through to the deeper layers (*deep percolation*). The deep percolation eventually reaching the ground water is termed the *groundwater recharge*.

The rainfall that initially fails to infiltrate remains ponded on the land, mainly filling up the various (micro) depressions. Once the storage potential on the soil surface (*surface retention*) has been fully occupied, the water will start to move down the slope as *overland flow*. This water will eventually collect in the (natural or man-made) field drains and then discharge through the main drains to the outlet. The water in transit as overland flow and in the field-drains constitutes the *surface detention* but once this water reaches the main drainage system it is referred to as *surface runoff*.

The groundwater recharge will cause the watertable to rise. When the watertable rises above the local subsurface *drainage base*, as formed by the water level in the deep drainage system, a hydraulic gradient is established which causes *groundwater flow* to this system. Part of the water that has infiltrated may also find its way towards the drains as lateral flow above the watertable (*interflow*), especially in the case of impeded percolation.

Transpiration by the vegetation and evaporation directly from the soil surface (jointly referred to as *evapotranspiration*) cause a water loss from the soil into the atmosphere. The evapotranspiration and continued deep percolation deplete the soil moisture storage; rainfall replenishes this storage.

Irrigation water essentially goes through the same cycle as rainfall.

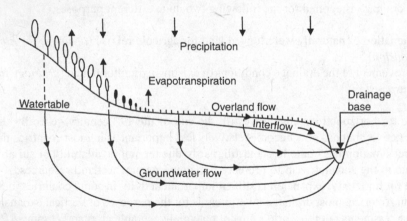

Figure 1.1 Main components of the hydrological cycle

The available statistics are very rough and the defined drained land and provided drainage facilities cover a wide and diverse range, from extensive to very intensive drainage and from area-wide to only incidental coverage. It also includes land which is only provided with some elementary regional flood control facilities (coastal and river plains protected against flooding from the sea or by the rivers but with no regular land drainage). Most of this drained area is found in Europe and in North America where 30–40 percent of the cropland is provided with improved drainage. In most developing countries, this percentage is less

Table 1.1 Drainage status of the world's cropland in 2002 (million ha)

	In need of improved drainage		Adequate natural drainage	Total
	already drained	not yet drained		
Irrigated land	25–50	100–150	50–100	275–310
Rainfed land	100–150	250–300	800	1150–1250
Total	125–200	350–450	850–900	1325–1550

than 5–10 percent. The provided drainage is estimated to account for some 10–15 percent of the world food production.

The need for drainage is a dynamic attribute of the land which generally may be expected to increase with the level of agricultural development (see also section 1.9). Available statistics suggest that at a global scale, about two-thirds of the rainfed cropland is natural sufficiently drained to allow it to be used for regular cropping without or with only minimal provision for artificial drainage, implying that only one-third of this land is in need of improved drainage (Smedema et al., 2000). According to this rule of thumb, some 400 million ha of rainfed cropland needs improved drainage while only 100–150 million ha is provided. For irrigated land, the relative drainage needs are conceived to be higher than for rainfed land and the drained area of 25–50 million ha (Table 1.1, only 10–20% of the total) is almost certainly far below the requirements.

Drainage has become a much more diverse instrument over the years. Instead of focusing almost exclusively on the removal of excess water, modern drainage is very much part of integrated water management, removing or conserving water as required and also being much concerned about water quality and environmental values. Drainage development is not only applied to raise crop yields and crop production but also to lower production costs, to broaden the range of land use that can be applied and crops that can be grown, to overcome farm management constraints, to protect the environment and to advance rural welfare and rural well-being. The wider purposes of drainage mostly evolved in parallel with the changing views of the role of agriculture in society.

1.1 Drainage objectives; Scope of the book

Within the context of the above-described horizontal and vertical expansion of drainage, the following specific and additional objectives may generally be distinguished.

Waterlogging control: in poorly drained waterlogged land, much of the pore space in the rootzone is occupied by water and as a consequence air is in short supply. Crop growth can generally be improved by draining out the excess water. This has traditionally been the first and foremost purpose of drainage. Waterlogging also restricts the use of farm machinery and the efficiency of the involved farm operations and imposes constraints on the crop choice and farm calendar. The latter rationales for drainage have become one of the main driving forces for drainage investment in Europe and North America and are gaining in importance in the developing countries.

Salinity control: irrigated land in the arid zone risks becoming salinised due to the accumulation of excess salts in the rootzone. This accumulation is usually due to a combination of insufficient leaching of the salts imported with the applied irrigation water and capillary rise of salts from the underlying saline groundwater. Although crops differ in their salt tolerance, the growth and yields of many common crops become severely affected once this accumulation of salts in the rootzone rises above certain threshold levels. To prevent this salt accumulation, it is essential to maintain a downward drainage flow in the soil by which salts are removed from the rootzone and the level of the watertable can be controlled.

Erosion control: loss of topsoil poses a serious threat to the productivity and sustainability of much agricultural land used in the semi-arid zone as much land has too little vegetation at the beginning of the rainy season to protect it against the erosive forces of the intense rain. Improved drainage may help to reduce this loss of valuable topsoil and the flooding of lower areas by checking and controlling the runoff of rain or other types of excess water and by preventing the runoff flow from reaching erosive velocities.

Flood control: drainage problems may not only be caused by local water sources (rain, seepage, irrigation, etc.) but also by invasion and inundation by water coming from an outside source. While most of the flood control is considered to be outside the domain of land drainage, some forms of local flooding may be controlled by regular land drainage.

Environmental protection: drainage designs should always strive to achieve the above-mentioned agricultural objectives while protecting (preferably enhancing) the environmental values of the drained area. The relationships between drainage and the environment are described in section 1.9.

Public health and rural sanitation: improved drainage by restricting the breeding opportunities for the insect vectors, can significantly contribute to the control of malaria, filariasis, schistosomiasis (bilharziasis) and other vector borne water related diseases. Drainage can also help to improve sanitary conditions in areas with stagnating and polluted water and provide improved drainage opportunities for villages without an adequate drainage outlet.

Protection of infrastructure: flooding often disrupts many economic activities but can also cause considerable direct damage to roads and other infrastructure. The rising of the watertables in many irrigated areas affects trafficability of roads, raises the road maintenance costs, undermines the foundations of buildings and has led to the collapse of many (mud-based) houses.

Rural development and food security: the nature and wide range of its objectives and impacts make drainage a suitable instrument for rural development. Drainage has been extensively used for this purpose in the developed countries during the second half of the 20th century. It has greatly contributed to the agricultural development and rural welfare of backward regions. As such it has also contributed to the national and global food security. Drainage could play a similar role in many of the developing countries.

Integrated water resources management; controlled drainage

Integrated Water Resources Management (IWRM) is a process that promotes the co-ordinated development and management of water, land and related resources, to maximise economic and social welfare in an equitable manner without compromising the sustainability of vital ecosystems. Drainage is a vital part of this process. While the traditional role of drainage i.e., the control of the incidences of excess water, is still very valid and needed, it is also true that cases have emerged where classic drainage designs have resulted in 'over-drainage' of the land. In such cases, a balance needs to be struck between waterlogging control and water/moisture conservation. Designs may need to be modified to be able to exercise more control over the drainage process and to avoid undesirable drawdown of the watertable (for further description, see Chapter 20).

Modern land drainage may in future also be expected to serve wider societal objectives and purposes than just agriculture, some of which have already been indicated in the previous sub-section. In particular, modern drainage design and management may be expected to give more specific and dedicated recognition to the non-agricultural functions of the different land and water resources and to the different landscapes in a drainage basin. Values attached by various stakeholders to biological aspects, residential, recreational and other environmental functions have shifted with local, national and global developments. While in most cases, the non-agricultural functions and values can readily be accommodated as 'add-ons' to regular agricultural drainage projects, in other cases it may require a fully new approach to project design and management (i.e., stakeholder engagement in section 3.4).

European Union Water Framework Directive

The EU Water Framework Directive (WFD) was established in 1999 and provides information about European river basin districts, river basin district sub-units, surface water bodies, groundwater bodies and monitoring sites used in the first and second River Basin Management Plans (RBMP, in 2010 and 2015 respectively). The data sets are part of the Water Information System for Europe (WISE), and compile information reported by the EU Member States and Norway to the European Commission (EC) and the European Environment Agency (EEA) since 2010. The WFD harmonises water policy in member states in the domain of ground and surface waters. The river basin approach of WFD accommodates trans-boundary agreements for better water quality. WFD affects primarily main drainage systems. The main implementing agency is the European Environmental Agency in Stockholm and their website, databank and reports provide the cutting edge for integrated water management (https://www.eea.europa.eu).

Scope of the book

The core and story line of this book is based on classic agricultural drainage i.e., the control of the rainfall-induced waterlogging. This form of drainage developed in the humid parts of the temperate zones of Europe and North America but has since spread to other areas with similar drainage problems. Drainage for salinity control of irrigated land is treated separately although making full use of the common methodological and technical base of classic drainage. Full attention has been given to the new drainage paradigm of *controlled drainage*

The control of erosion is covered within the context of surface drainage but full coverage is considered to be outside the scope of this book (belongs to the field of *soil and water conservation*). Flooding may occur at different scales. Flood control which requires major civil works (upstream reservoirs, embankments, diversion channels, etc.) is technically (but often also institutionally) outside the domain of land drainage. The control of flooding is often preconditional for undertaking effective land drainage. Local flood control, caused by processes within typical drainage catchments, is part and parcel of land drainage and covered as such.

Drainage is treated as a technical discipline and full attention has also been given to the non-technical context in which drainage systems are developed and operate. The non-agricultural applications and the broader societal and developmental aspects of drainage are described where appropriate but have generally not been treated to the same level as the core subjects.

1.2   Global drainage zones

The agricultural drainage conditions around the world can broadly be grouped into three zones: the temperate zone, the (semi) arid tropical zone and the (semi) humid tropical zone (Figure 1.2).

1.2.1   Temperate zone

By far most of the world's drainage assets exist in the developed countries of the temperate zone, especially in Europe and in North America. Drainage development in these regions started in earnest in the early parts of the 19th century and in most countries between 20–35% of the agricultural land has been provided with some form of improved drainage. This percentage appears to be close to saturation, the remaining land apparently being naturally well enough drained to suit the current land use or otherwise not warranting investment in drainage. This explains, jointly with the prevailing environmental concerns, the stagnation in drainage development in this zone.

Drainage has greatly contributed to agricultural development in this zone, both horizontally (expanding the agricultural area) and vertically (increasing per ha production). Much formerly waterlogged land has been upgraded to highly productive agricultural land. Improved drainage has also helped to create the conducive soil-water-air conditions for producing high yielding and high value crops at low cost.

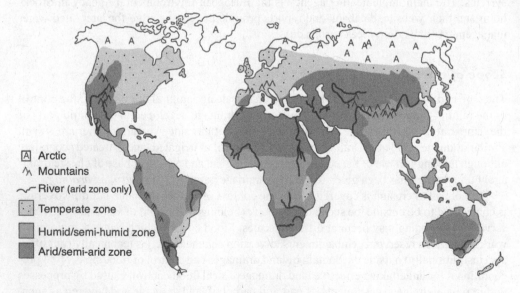

Arctic
Mountains
River (arid zone only)
Temperate zone
Humid/semi-humid zone
Arid/semi-arid zone

Figure 1.2 Global drainage zones

1.2.2 Arid and semi-arid zone

The principal land drainage need is for the control of waterlogging and salinisation in the 100–120 million ha of irrigated land (out of a world's total of 270 million ha) located in this zone. No reliable data exist on the extent of the problem, but the available data suggest that some 20–30 million ha of this land is already seriously affected and that this is growing by some 0.25–0.50 million ha per year.

Most of the irrigated land in this zone has some surface drainage, typically consisting of a combination of prior existing natural drains and a few additionally constructed drains. The drain densities are seldom more that 2–5 m per ha (average spacing of 2 to 5 km), which leaves much of the land without a nearby drainage outlet. Moreover, existing drains often do not function properly due to inadequate maintenance and flow blockages by infrastructure crossings and other man-made obstructions. On-farm surface drainage generally relies on unstructured migratory plot-to-plot flow and is often grossly inadequate.

Not all the land needs improved subsurface drainage as the natural drainage may suffice to provide sufficient leaching and keep the watertables at a safe low level. Watertable control is generally also not a problem in areas underlain by fresh groundwater as the ongoing tubewell pumping for irrigation typically provides more than enough drainage. The water-logging and salinisation problems are mostly confined to the saline groundwater zones and to the areas underlain by poorly permeable substrata, which are less suitable for tubewell development. For these areas, pipe drainage is usually the best solution. It is estimated that the area already provided with this type of drainage is in the order of 2.5 million ha to which some 100 000–200 000 ha is added each year. Clearly, considering the above estimates of the waterlogged/salinised area, this covers only a fraction of the requirements.

1.2.3 Humid and semi-humid zone

Agricultural development in this zone, both of the rainfed and the irrigated land, is severely constrained by the flooding and waterlogging caused by the high monsoon rainfall. It especially affects the extensive lowland plains. It is a typical feature of these plains that the land use is adapted to the prevailing hydrological regimes and that the opportunities for growing improved rice varieties and high value non-rice crops are very limited. The flooding and waterlogging also limits the introduction of mechanisation and of other modern farming techniques. Improved flood control and land drainage would increase these opportunities. The drainage would, however, have to done in steps, as full control of the excess water would technically be difficult to achieve and be quite costly. The drainage challenge for this zone is to define the next feasible steps, keeping in mind the economic and technical limitations.

The drainage development needs of this zone are easily in the order of 100–200 million ha while the benefiting population may well surpass the 0.5 billion mark. This is essentially a new drainage frontier as the drainage knowledge and technology of the other zones is of limited relevance. Most countries in this zone are lacking in technical capabilities and urgently need to strengthen their capacity building frameworks.

1.3 Agro-hydrological regimes

Direct rainfall constitutes by far the major and most common source of excess water. However, another major source of excess water in many cold and moderate climates is snowmelt in spring. Other sources such as irrigation, seepage, runoff and floodwater are mostly of local or minor

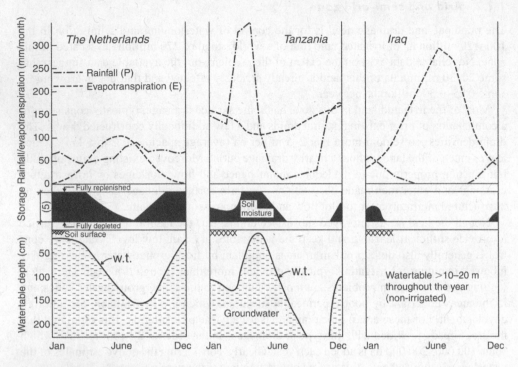

Figure 1.3 Water balances and agro-hydrological conditions

importance. The occurrence of excess rainfall applies especially to humid climates. However, it may also occur in (semi-) arid climates following the common type of intense, heavy storm or in general during the rainy season. The drainage load from rainfall not only depends on the amount of rainfall (P) but also on the storage capacity of the soil (S) and on the rate of evapotranspiration (ET). Part of the rainfall may be stored beneficially in the soil profile or be readily evaporated so that only the remaining excess needs to be removed from the land. Excess water and drainage problems for three quite different climatic conditions are illustrated in Figure 1.3. More generalised descriptions of drainage in various climate zones, are given in section 1.2.

The Netherlands: here P is in excess of ET throughout the winter (November to March) while the summer is characterised by a small deficit of rainfall. Watertables are high during the winter and the soil moisture content even in the topsoil is close to saturation during much of the time. In the early spring, as ET rises above P, the soil begins to dry out and watertables fall. Crops compensate for the lack of adequate summer rainfall by depleting the soil moisture storage. For most soils the soil moisture storage is sufficient to meet the water requirements of the crops. When in late autumn P once again exceeds ET, the excess rain first replenishes the depleted soil moisture storage and then any further excess percolates downwards to recharge the groundwater, causing watertables to rise (KNMI 2018).

Tanzania: during the early part of the rainy season P and ET are more or less in balance. The soil moisture storage, which was thoroughly depleted during the preceding dry season, is able to accommodate most of the short-term excess rain. Major drainage problems do not tend to occur during this period. However, when the rainfall increases in March, the soil moisture storage is rapidly replenished and watertables rise high, remaining so until June when the rainy season nears its end and ET again overtakes P.

Iraq: under non-irrigated conditions high watertable problems are not to be expected as P « ET almost all year round. Recharge to groundwater from rain is practically nil. It is only during short periods in the rainy season that excess rainfall may occur and soil moisture storage may then become partly replenished. Some surface drainage problems may occur as rain often falls in intense storms. Very little cropping can be done without irrigation. The introduction of irrigation, however, changes the hydrological balance. More water will percolate downwards to the groundwater as frequent irrigation keeps replenishing the soil moisture storage, and watertables may rise to critical levels, causing salinisation of the root zone.

1.4 Waterlogging control

In this section, the nature, the adverse impacts and the desired control of agricultural waterlogging are described in more detail. When waterlogging is prevalent in agricultural areas installing a field drainage system will allow control of the watertable at desirable depth in the agricultural fields provided the drainage water can be disposed-off effectively via a main drainage system (see Chapter 10). Alternatively, during water-short periods of the growing season it may be desirable to not drain; i.e., apply controlled drainage (see Chapter 20).

1.4.1 *Positive and adverse impacts*

Waterlogging may occur on the land surface (surface ponding, often combined with near saturation of the topsoil) or deeper down in the soil profile (excess water in the root zone due to impeded percolation or due to high watertables). The adverse impacts of such waterlogging on farming broadly fall into the following two categories: impaired crop growth and impaired farm operations.

Impaired crop growth

Most common crops are mesophytic[1] plants, which grow best when there is both sufficient water and air in the rootzone. They respire by gaseous exchange in the root zone, the process whereby roots absorb oxygen (O_2) from the soil atmosphere and release carbon dioxide (CO_2) back into it. Roots are also able to absorb O_2 dissolved in the soil water, but this capacity is very limited. Rice is unusual in being able to transfer oxygen taken in through the stomata internally to the roots (Chapter 19) and is therefore able to grow well even though its roots are submerged.

In waterlogged soils, the air content of the soil is low because most pores are filled with water. Moreover, the exchange between the remaining air in the soil and the air in the atmosphere above (O_2 moving into the soil, CO_2 moving out) is very restricted under these conditions. In consequence, respiration is restricted by the oxygen deficiency while at the same time the carbon dioxide accumulates to toxic levels, directly impairing the root growth and the roots ability to absorb nutrients. Anaerobic conditions in the soil may also lead to the formation of toxic concentrations of reduced iron and manganese compounds, sulphides and organic gases.

Rootzone aeration generally becomes inadequate when the effective air-filled pore volume in the main rootzone falls below 5–10%. However, the duration of waterlogging and its timing in relation to the activity and stage of development of the crop are also of importance. Waterlogging of the entire rootzone for a period of two to three days can be fatal when it occurs during the seedling stage, whereas a well-developed crop is likely to suffer relatively little damage from a similar incident. Furthermore, a vigorously growing healthy crop is able to withstand waterlogging better than a poor one. Most crops have a large proportion of their roots in the topsoil and so early restoration of aerated topsoil is of particular importance.

Crops suffer much more from waterlogging under warm than under cold weather conditions. The reasons for this are that plants are physiologically more active at higher temperatures, increasing the oxygen consumption (of the crops but also of the soil flora and fauna) and leading to earlier deficits. The lower solubility of oxygen in water at higher temperatures also plays a role.

In temperate climates, waterlogged soils often remain cold for too long in spring for good crop growth. Waterlogging also affects plant growth by its adverse effects on soil biological life and on the structure of the soil. Thus, waterlogging during the winter period in Northern Europe is known to impair mineralisation and nitrification by microbes, resulting in nitrogen deficiency to crops in the spring. This, however, may be partially offset by increasing the nitrogen fertilizer application. It may also cause the soil structure to disintegrate or prevent it being restored by the action of frost. Off-season drainage in temperate climates is therefore largely undertaken to overcome these adverse indirect effects. Similar effects may also be expected in the humid tropics from excessive and prolonged waterlogging during the rainy season. Common indices to assess the waterlogging problem are shown in Figure 1.4.

Impaired farm operations

Excess water on or in the soil adversely affects the accessibility of the land and the workability of the soil. There are fewer workable days on poorly drained land and essential farm operations i.e., seedbed preparation, planting, weeding, spraying and harvesting may be

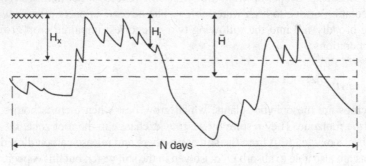

The average seasonal watertable depth: $\bar{H} = \sum_{i=1}^{N} H_i / N$

The number of daily exceedances: $W_x = \sum_{i=1}^{N}$ days (where $H_i < H_x = 1$; $H_i \geq H_x = 0$)

The sum of daily exceedances: $SEW_x = \sum_{i=1}^{N} (H_x - H_i)$ for $H_i < H_x$

where: H_i = watertable depth on day i,
  H_x = reference watertable depth,
  N = number of days per season.

Figure 1.4 Commonly used watertable depth indices

critically delayed. If, through necessity, these operations are not delayed but go ahead under unsuitably wet soil conditions, compaction, puddling and smearing of the soil is likely to occur, and the soil structure may seriously deteriorate. Besides affecting future yields, a poor soil structure also hampers the infiltration and percolation of rainwater into and through the soil, leading to further reductions in the number of workable days. Details are described in section 6.5. The economic significance of the effects of excess water on farm operations depends on the type of farming; modern mechanised farming, for example, being much more affected than traditional subsistence farming.

When waterlogging prevails at the beginning and/or end of season, improved drainage helps to assure early planting, resulting in more growth-days and higher yields and those crops can be timely and properly harvested. Improved drainage also allows farmers to grow a wider range of crops. Pasture can be converted into more productive grassland or even be converted into cropland. In the humid tropics, improved rice varieties and more upland crops can be grown. In most cases, the greater freedom of crop choice allows farmers to include more rewarding crops in their farming systems and raise their incomes.

1.4.2   Responses to improved drainage

Aeration conditions in the rootzone are inversely related to the soil water content of the upper soil layers and since the latter are closely dependent on the watertable depth, this readily determined depth is often used as a diagnostic index for the prevailing aeration conditions. Another commonly used index for the aeration conditions in the rootzone, is the duration of surface water ponding. These two indices may also be considered to be highly indicative of the prevailing farm operation conditions. Available data generally show an improvement in accessibility/ workability conditions with increasing watertable depths up to 50–100 cm for light soils and up to 100–150 cm for heavy soils, then remaining constant. Relevant further background and details on the nature and behaviour of water in the soil has been provided in Chapter 6.

Watertable depth indices

The dependence of crop yields on watertable depth has been documented for various countries with a range of soil and climatic conditions. A strong relationship between the average watertable depth and crop yield as shown in Figure 1.5 can usually only be found when the watertable varies within a narrow depth range. Even under more variable watertable conditions, fair relationships may be expected as the average depth often captures some response significant regime characteristics. In temperate climates, different curves are typically found for the summer and the winter seasons whereby the summer curves reflect the aeration conditions during the growing season and the winter curves the adverse impact of off-season waterlogging on the available nitrogen and on the soil structure.

Crop responses vary with the pattern of the watertable regime (the duration and timing of the high watertables are especially important), which patterns may to some extent be captured by the W_x and the SEW_x indices (Figure 1.4). Good correlations with a range of crop yields were obtained in the UK for the W_{40} index and for yields of sugar cane in Australia for the W_{50} index. Figure 1.6 shows two cases of similarly good correlations for the SEW_{30} index. The nature of the response shown by these indices depends on the period to which they apply (winter/off-season indices mostly showing the adverse impact of high watertables on soil structure while for the summer/growing season indices this is mostly the impact of

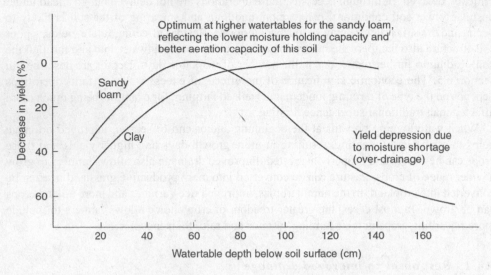

Figure 1.5 Crop yield - watertable depth curves for two different soils (The Netherlands)

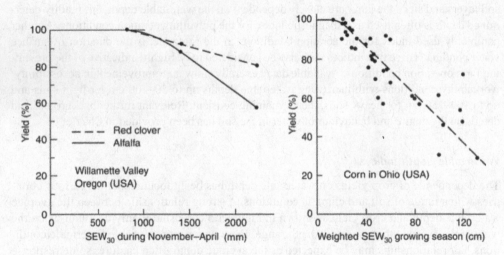

Figure 1.6 Relationships between crop yield and SEW values

oxygen deficiency). In some cases, weighted indices were used which took into account the occurrence of the waterlogging in relation to the development cycle of the crops, but these have not always proved to be superior to the non-weighted indices. For references and more detailed discussion, see Smedema (1988).

Ponding indices

The adverse impacts on crop yields of prolonged ponding of water on the surface of the land may generally be best captured by the length of the period during which this ponding prevails. Research results from a range of countries are shown in Table 1.2 and Table 1.3. In all

Table 1.2 Percentage yield losses due to surface ponding in Hungary (after Salamin, 1957)

Days of full ponding	J	F	M	A	M	J	J	A	S	O	N	D
1. Grassland 3												
7												
11				10	15	20	20	10				
15				20	30	30	20	20				
			10	30	50	50	50	30	10			
2. Fodder crop 3												
7				10	10	10	10	10	10			
11			10	25	30	40	40	30	30	10		
15	5	5	20	40	50	70	70	50	50	20	5	5
	10	10	30	60	100	100	100	80	70	30	10	10
3. Winter grains 3												
7			5	10	30	20						
11	5	5	15	25	40	50				5	5	5
15	10	10	30	40	70	80	10			10	10	10
	15	15	50	70	100	100	20			20	20	20
4. Summer grains 3												
7			10	15	15	20						
11			20	40	50	50						
15			40	75	75	75	10					
			100	100	100	100	20					
5. Maize 3												
7					20	10	10					
11					80	50	40	10	10	10		
15					100	80	75	50	40	20	10	
					100	100	100	80	60	30	10	
6. Sunflower 3												
7					10	10	10	10				
11					20	30	40	40	10			
15					40	60	80	60	30			
					80	100	100	80	50			
7. Fibre crops[1] 3												
7					20	20	10	10				
11					40	50	40	30				
15					60	75	60	50	20			
					100	100	80	70	20			
8. Sugar beets 3												
7				10	10	10	10	10	10	10		
11				50	50	50	40	40	40	40	10	
15				100	90	90	90	90	90	90	30	
				100	100	100	100	100	100	100	50	
10. Potatoes 3												
7					30	30	40	50	50	50	20	
11					80	80	90	100	100	100	40	
15					100	100	100	100	100	100	60	
					100	100	100	100	100	100	80	

1 hemp, flax

Table 1.3 Yield losses due to surface ponding (after Gupta et al. 1992)

Location	Crop	D100 (days)	Slope (%/day)	D50 (days)
India, Delhi	Pigeon pea	1.6	23.2	3.8
Hisar	Cow pea	0.8	6.6	8.4
	Pigeon pea	0.5	9.2	6.0
Karnal	Wheat	0.0	7.0	7.2
	Pearl millet	0.0	5.3	9.4
Ludhiana	Wheat	1.0–1.9	9.2	7.3
Madhipura	Maize	0.0	9.3–14.2	3.5–5.4
Pusa	Groundnuts	0.0	8.9–10.3	4.9–5.6
	Maize	0.0–1.2	9.4–9.6	3.8–6.4
Venezuela, Merida	Potato	0.0	9.0	5.5
	Beets	2.8	10.3	7.7
	Forage	2.1	7.9	8.4
	Sunflower	2.0	7.1	9.0
	Pasture	1.5	3.6	15.4
USA, Texas	Grain sorghum	0.0	14.6	3.4
	Green peas	0.1	18.3	2.8
	Cotton	2.3	14.9	4.6
	Corn/maize	0.0	9.9	5.0

cases the results were obtained on artificially submerged plots whereby the applied submergences closely simulated the prevailing natural flooding/ponding regimes (winter and summer flooding/ponding in Hungary, monsoon/summer flooding/ponding in India, Venezuela and Texas, Table 1.3).

The results shown in Table 1.2 clearly show that the impacts of ponding on yields is most severe in the spring and summer when the crops are physiologically most active and temperatures are higher. The damage, even from prolonged ponding, is relatively low when it occurs during the dormant/winter period. Grassland is clearly more tolerant to flooding while almost all crops are most tolerant at the end of the development cycle (ripening stage).

The results shown in Table 1.3 have been fitted to a broken-linear model (Figure 1.7), similar to the model also commonly used to describe the yield response to soil salinity (Figure 1.6 and Figure 14.7). It shows that yields of most crops start to be affected by only short periods of ponding (often <1 day) while the 50% yield reduction ($Y_{50\%}$) for most crops occurs after 5 to 8 days of ponding. Ponding appears to be more harmful in Texas than in the other two countries, but this may also be due to differences in the set-up of the research. The researchers noted that most crops are somewhat more sensitive to ponding at the seedling stage. The impact of ponding depth was not evaluated but might explain some of the higher sensitivity at the seedling stage (seedlings are more likely to become submerged, compare to sensitivity to flooding of rice as described in section 19.1 (Table 19.1).

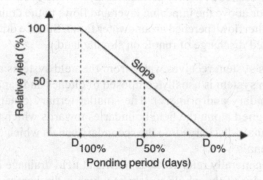

Figure 1.7 Yield losses due to surface ponding (see also Figure 14.7)

1.5   Salinity control

Salinity and the effects of salinity cannot be controlled by simply installing a drainage system. Salinity is managed by adding excess (fresh/low salinity) water to the agricultural land and this excess water needs to be drained. In other words, unless the salts are flushed down the rootzone and then disposed of, there will be no benefit of installing a drainage system unless the water for leaching the salts is provided. Generally, salinity can only be managed and only in rare cases can the problem be eradicated sustainably. Irrigation is critical and maintaining a net downward seasonal water movement through the rootzone is essential.

That is not always possible, for instance, the Warabundi (irrigation) system of the Indian sub-continent was designed to provide only 70–80 percent of the peak crop water requirement, occurring during a short period in the crop growth cycle. This eventually caused salinisation of the irrigated lands. With social pressure to take more land under cultivation the problem of salinity risk only increased resulting in wide-spread emergence of salinity problems. Yet, some farmers in the waterlogged and salinity-stricken areas manage to grow salt intolerant citrus successfully; they have access to the required amount of water to maintain a net-downward seasonal water movement through the rootzone. This was achieved by sacrificing some land, so the water entitlements of those lands can be used for the citrus groves.

1.6   Drainage systems

A typical agricultural drainage system (Figure 1.8) has the following main components:

Field system: this system gathers the excess water from the land by means of a network of field drains, and where necessary these are supplemented by measures, which promote the flow of excess water to these drains. The systems used depend mainly upon the drainage characteristics of the soil. Two principal types may be distinguished:

   a) subsurface drainage systems: these are used in soils in which the excess water is able to infiltrate and percolate through the main root zone to the watertable and then move as groundwater flow through the subsoil/substratum to the drains;

   b) surface/shallow drainage systems: these are used where the infiltration or percolation of excess water is impeded at the surface or at shallow depth in the root zone due to the presence of poorly permeable layers. The excess water either ponds on the surface or above the impeding layer and flows to the drains, either as overland flow or as interflow (perched groundwater flow). Surface drainage is also used for the controlled discharge of runoff on sloping land.

Main system: this system receives water from the field systems and conveys it to the outfall. A main system is usually composed of ditches and canals of differing order (tertiary, secondary and primary). The smaller tertiary canals (*collector ditches*) are usually aligned along the field boundaries towards which the field drains flow. The tertiary canals discharge into the secondary canals which discharge in turn into the primary canals.

Outfall: outfalls (generally referred to as outlets at field drainage level, Figure 1.8) are the terminal point of the whole system at which it discharges into a major element of the natural open water system of the region (river, lake or sea). The water level at the outfall constitutes the drainage base for the area concerned. This level, relative to the land level, governs the amount of hydraulic head available for the drainage flow. It determines how far the water levels/watertables may be lowered below the land surface. It also determines whether the area can be drained by gravity or requires pumping.

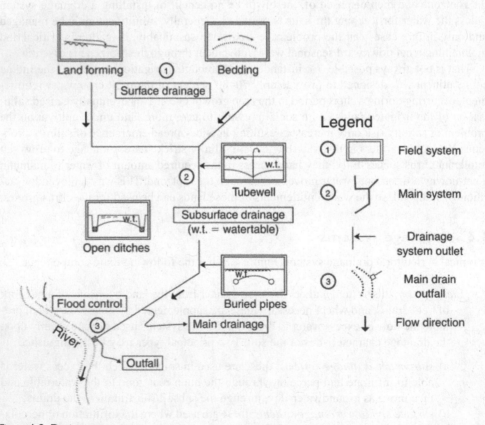

Figure 1.8 Drainage systems

1.7   Bio-drainage

Conventional land drainage is generally based on the laws of potential flow. A sink facility is installed (a ditch, a well or any other facility which establishes a line or point of low potential) which creates the hydraulic gradients by which excess water from the surrounding land is drawn to the sink. There it is captured and disposed to some ultimate sink outside the area. The provision of such drainage requires both capital investment and O&M expenditure.

In comparison, biological drainage would seem to be a highly appealing alternative: why not use the evapotranspirative capacity of the vegetation to remove the excess soil water or expressed more popularly: *let the vegetation drink itself out of the waterlogging problem.* This water removal mechanism would not only be of low cost as it would not require the installation of any physical field facilities and would not require a disposal system but, on the contrary, may actually yield some marketable products (fodder, fibre, wood). There may also be environmental benefits: enhanced biodiversity, water quality protection, enhanced scenery.

The cases shown in Figure 1.9 are fairly representative of situations where bio-drainage may be considered. The first case depicts a waterlogged landscape depression. The conventional drainage solution would involve the installation of a pumped well by means of which the watertable in and around the depression would be drawn down. The sketch suggests that this drainage improvement could also be achieved by the planting of trees, shrubs or other suitable vegetation in the depression (starting from the outer edges and gradually closing in on the most waterlogged areas). The second case depicts a canal seepage situation and it is suggested that instead of the conventional solution of installing an interceptor drain at the toe of the embankment, drainage conditions could also be improved by the planting of a strip of trees along the canal. The final case shows the biological drainage equivalent of the conventional parallel field drainage system with the function of the physical line sinks (ditches or underground pipe drains) being taken over by biological line sinks (planted tree/shrub strips). This layout of parallel tree/shrub strips is fairly similar to the typical windbreak

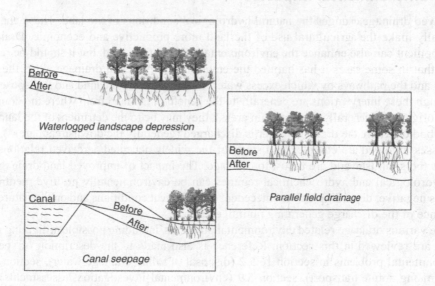

Figure 1.9 Examples of bio-drainage

layout and the creation of a more favourable micro-climatic environment may actually be an added asset.

Biological drainage is of course an inherently somewhat dubious concept as it is well-known that plants generally do not grow well when the rootzone is waterlogged and the root functions are impaired by lack of air. Therefore, biological drainage is generally not a realistic option when common crops are facing severely waterlogged conditions; in such situations there is no alternative to conventional physical drainage.

The concept of bio-drainage has proved to be effective in a variety of other situations. Planting of reed has effectively been used in the Netherlands to accelerate the dewatering of newly reclaimed polder land; the seed is broadcast from the air soon after the land has emerged from under the water. In the Sava Valley (Croatia), it was found that on poorly drained heavy clay soils, seedbed preparations in the spring could generally be started one-two weeks earlier when the land had a transpiring grass cover (ley[2] cover remaining from the previous year grain crop). On Kilombero Sugar Estate (Tanzania), actively growing/transpiring fully developed, full canopy sugar cane crops suffered fewer waterlogging problems during the rainy season than the less transpiring just planted or just ratooned crops.

The scope for bio-drainage would seem to be most favourable in the arid zones where the drainable surpluses are generally quite small in relation to the evapotranspiration rates (some 1–2 mm/d vs. some 10 mm/d) and limited evapotranspirative surfaces are able to cope. While some bio-drainage advocates claim that this type of drainage cannot only cope with waterlogging but also with the salinity menace of the irrigated lands in the arid zone, this is not supported by the available experimental evidence from Australia and California. Experience there indicates that bio-drainage, without any physical drainage, is likely to lead in the long-term to a harmful build-up of the rootzone salinity (Heuperman et al. 2002).

1.8 Environmental impacts

Improved drainage changes the natural hydrological conditions of the land. These changes generally make the agricultural use of the land more productive and economic. Drainage development can also enhance the environmental values of the land, but it should be recognised that in some cases it has harmed the environment. Artificial drainage alters the processes and the pathways by which excess water is removed from the land and is disposed of. Although these interventions are generally to the benefit of the land from where the drainage water originates (generally the upstream areas), they may be to the detriment of the land and water bodies where the drainage water is discharged (generally the downstream areas). The processes involved are complex and some of the widely accepted/perceived relationships are far too simplistic and are unjust to drainage. The impact of improved land drainage on the hydrological and hydro chemical regimes can be environmentally positive, neutral as well as negative, depending on the antecedent hydrological conditions and on the nature and sequence of the discharge generating rainfall events.

The various drainage related environmental problems, including possible mitigating measures, are reviewed in this section. Reference is also made to the description of specific environmental problems in section 16.5.2 (disposal of saline drainage water), section 20.2 (minimising solute transport), section 5.9 (environmental investigations/assessments) and section 24.3 (maintenance of open drains).

1.8.1   Stream flow regimes

Although improved drainage generally does not greatly alter the annual water balance of the land, it does remove the excess water more rapidly and by processes and routes which differ from the natural ones. As such, drainage may alter the hydrographs of the drainage discharge. Depending on the type of on-farm drainage, storage and residence times may increase or decrease and more or less water may be removed by surface and/or subsurface flow. Improved surface drainage will generally lead to an increase in the rate of field discharge while improved subsurface drainage, depending on prevailing conditions (rainfall intensity, antecedent rainfall, season, land use, etc.) may increase as well as decrease these rates. Almost all of these changes at the field level are largely attenuated as the field discharge is passed down the main system and they have hardly any impact on the downstream stream flows.

These downstream stream flows may be greatly affected by various main drainage system works such as clearance/enlargement of channel sections, straightening and short-cutting of alignments, construction of embankments, etc. Such works, designed to promote rapid discharge and/or to confine the flow within the designated flow sections, may significantly increase downstream peak flows. Complex interactions between the type and spatial pattern of the works, the actual and antecedent rainfall, the land use, etc. may occur which make the magnitude and direction of the changes in downstream stream flow often quite unpredictable.

The potentially harmful impacts of these drainage induced regime changes are partly economic, partly environmental. Downstream areas may become exposed to greater flood risks. Natural ecological conditions and terrestrial and aquatic life in the stream bed and in the riparian land will adapt to the new regimes and biodiversity may be lost. The river works may also lead to loss of scenic values. Because of these environmental/ecological impacts, and the loss of, or change in, biodiversity in the river basin systems, modern land drainage and modern integrated water management, advocates engineering approaches that maintain/restore many of these scenic and biological features.

1.8.2   Water quality

The disposal of nutrient rich drainage water, exceeding the natural assimilation capacities of the receiving waters, has led to some large scale eutrophication of rivers, lakes and estuaries (typically manifested by excessive growth of phytoplankton, algae and duckweed). Drainage induced disposal of (remnants of) pesticides and herbicides has also led to the toxification of the receiving waters but this is only likely to occur when the toxic elements can become concentrated (as may occur in lakes, bays, lagoons, marshes and other water bodies with limited through flow). Salinity originating from natural geological formations, or intrusion from the sea and salinity induced by irrigation practices, affects water quality (see Chapter 14). The pollutants are carried by regular drainage discharge, but seepage inflows may also contribute. The special case of acidification of water bodies due to the reclamation of acid sulphate soils is described in section 18.3. It should be recognised that most of the drainage induced pollution is primarily due to the increased use of agrochemicals in modern agriculture and disposal of manure from intensive livestock production and not due to the drainage as such (Figure 1.10).

Each arrow in Figure 1.10 indicates a pollution pathway to be controlled by national and EU environmental policies: control points. Nitrogen is a rather inert gas which makes

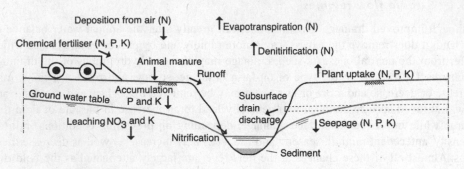

Figure 1.10 Solute transport in soil, groundwater and surface water (after Goewie and Duqqah, 2002)

up some 78 % of the atmosphere. In the soil, atmospheric nitrogen may convert by various physical, chemical and biological processes into various nitrogen compounds which processes largely drive the key soil-borne organic matter production and decomposition cycles. Nitrification (the conversion of atmospheric nitrogen to nitrate) and denitrification (the reduction of nitrate to atmospheric nitrogen) are important processes in these cycles. Many of these processes rely heavily on soil bacterial activities.

Improved surface drainage may even reduce the disposal of phosphorus (mostly carried by erosion material), especially when the remaining uncontrolled runoff is intercepted by riparian grass strips. Improved subsurface drainage may indeed enhance the nitrogen disposal although the increased rates in most cases are only slightly higher that the natural rates.

In addition to the open water pollution from diffuse (agricultural) sources, much of the pollution in the drainage system (including rivers) is from point sources. Point sources are generally uncontrolled disposal of untreated residential and industrial wastewater (typically disposed of via the drainage systems, these being the sink systems for the area). These can be mitigated by appropriate legislature and setting emission standards appropriate for intended downstream water use.

The assimilation capacity of the receiving water bodies is difficult to predict and was found to be highly inadequate around the 70's in Western Europe after three decades of intensive farming. The assimilation depends on the prevailing rates of biodegradation in the water body and on the interactions between the biomass and the bed sludge. Phosphates may be taken out by particle sedimentation and by biological and chemical fixation by the biomass and the bed sludge while nitrogen concentrations may be reduced by denitrification (Figure 1.10) and other transformation processes. Bed sludge may however also release pollutants when beds are disturbed by maintenance measures or are scoured during periods of high discharge. In most systems the pollutant levels will fall somewhat as the drainage water is transported down the system. Pollutants, which end up in the sludge, may also present a spoil disposal problem during maintenance (section 24.3).

The water quality requirement depends on its intended use (Table 1.4). The guidelines and directives for water quality assurance and control, as well as that for safety and human health (see section 1.8.4) are continuously under development as new insights and research results become available from various environmental protection agencies. These agencies consider the full water chain or hydrologic cycle and its multiple (sequential) uses of water. At (critical) control points, typically the entry and exit point of certain water usage, the

Table 1.4 Water quality assessment from the water user's point of view

Aquatic systems	Typologies of ecological systems are under development by various organisations		
Domestic	Human usage		
	Consumptive	Utilitarian (grey water)	
		Full contact Drinking water	Intermediate contact Washing water
Recreation	Full contact	Intermediate contact	
Industry	Cooling, steam generation, processing, product water, utility water, washing		
Agriculture (FAO, incl. WHO standards)	Livestock watering (FAO)	Irrigation (see also Table 15.1)	Aquaculture
		Direct contact with foliage	Indirect contact (sub-irrigation)

quantity, quality and economic value of water can be ascertained. Depending on the use of the water for ecology, agriculture, livestock, swimming, drinking, or industrial, indicator target values for water quality will be set differently. Indicator values suggested are given by FAO (1985, 2002a) and may be found on the US EPA, WHO, EEA and the South African Department of Water Affairs and Forestry websites. Literally, hundreds of contaminants are listed on those websites and the number is growing daily.

Major groups of these contaminants, which include the traditional salinity, sodicity, phosphorous and nitrogen indicators, are:

• *Physical parameters and general water quality indicators*, such as colour, taste odour, turbidity, temperature, BOD (Biological Oxygen Demand), COD (Chemical Oxygen Demand), pH, Total Suspended Solids (TSS), Total Dissolved Salts (TDS) or Total Dissolved Solid (also TDS), Total Volatile Solids (TVS), Dissolved Oxygen
• *Microorganism*, covering most of the human and animal faecal waste (pathogenic, bacteria, E-coli, viruses, giardia, etc.), as well as insect vectors (malaria, bilharzias, etc.)
• *Disinfectants and disinfection by-products* (DBPs i.e., chlorine and others) used for drinking water disinfection and additives used to control microbes. A term commonly used in this context is PCBs (Poly chlorinated biphenyls)
• *Inorganic contaminants*, ranging from the heavy metals to salts, nitrate, phosphorous, selenium, etc. expressed as indicator ratios e.g., EC_e, SAR, etc. Salinity, sodicity, green algal blooms (Harmful Algal Blooms (HAB) Stumpf et al. 2016 in IOCCG 2018) and eutrophication are some of the terms used to describe the environmental impacts. Note that N-P-K (nitrogen, phosphate and potassium) are also crop nutrients. Calcium and magnesium are crop nutrients and are major components of soluble salts as well (Goewie and Duqqah 2002)
• *Organic contaminants*, which include most of the pesticides, biocides, nutrients (ammonia). These are characterised, or further grouped, by abbreviations such as VOC (Volatile Organic Chemicals), SOC (Semi-volatile Organic Chemicals), POP (Persistent Organic Pollutants), and PAH (Poly Aromatic Hydrocarbons)
• *Radio nuclides*, such as alpha, and beta particles, Radium 226/8 and uranium that are all linked with cancer and uranium also with kidney toxicity.

Various agencies use different terminology and procedures with respect to the level of allowable concentration of substances in water. For instance, the US EPA forges ahead with the concept of Total Maximum Daily Load (TMDL), while in the Netherlands traditionally a term describing alert levels of contaminants in rain, agricultural and drinking water has been used. In South Africa, the concept of fitness of water for use is applied and their guidelines specify the Target Water Quality Range (TWQR), which includes some economic considerations. The FAO (2002a) simply uses the term Guideline Value. The latter approach is adopted by most nations: a simple target value below which the contaminant level has to stay. It is therefore very difficult to give a definite set of water quality guidelines, but those most appropriate for consideration on deciding what are acceptable water qualities for the receiving water bodies of agricultural drainage are shown in Table 1.5. When values given in Table 1.5 are based on guidelines/standards for non-agricultural uses this is indicated.

1.8.3 Wetlands and conservation drainage

Wetlands (naturally waterlogged lands) generally play an important regulating hydrological function as well as constituting valuable natural habitats. Depending on the local site conditions and on the amassed volume, the incoming water is retained, or it overflows to neighbouring wetlands and/or to nearby outfalls. In the process, it attenuates the discharge hydrographs of the regional streams and rivers. The storage/retention potential of the area is considerably reduced when wetlands are reclaimed for agricultural use. This has been confirmed by experiences in some basins in the USA where wetland reclamation has resulted in 30–50% increases in peak flows (USDA 1987). Wetlands add uniquely to the regional biodiversity and provide traditional refuge and foraging sites for migratory birds. Worldwide areas have been designated with special value from a natural point of view: the Ramsar sites, the European Union (EU) Habitat areas and the EU bird protection areas. Australia has migratory bird agreements with China (1986), Japan (1974) and Korea (2007) which provide an important mechanism for pursuing conservation outcomes for migratory birds, including migratory water birds.

Reclamation and drainage of wetland areas also affect the groundwater hydrology in the region. Wetlands set the regional drainage base and lowering of this base may be expected to lead to a general drawdown of the watertables in the surrounding areas. Wetlands also recharge watertables, feed local springs and augment the low flows of local streams. Overall, the hydrological regime in and around the area would become drier and the ecology of the regions would adjust accordingly. Wetlands also have an important filtering, water purification and sediment trapping function. Drainage and development of the wetlands for agriculture would largely eliminate these important functions, which may be expected to have an adverse impact on the water quality in the region.

Horizontal expansion type of drainage developments during the 18th, 19th and early part of the 20th century have converted large areas of formerly naturally waterlogged land into agricultural land. Much of the Mid-West of the USA which is now one of the world's most important grain belts was originally too wet for farming. Reclamation of naturally waterlogged land has now in almost all countries become subject to scrutiny and regulation. Wherever wetlands are still being lost, decisions are based on a careful analysis and weighing of the societal priorities, mostly nature conservation vs. food security/settlement needs.

Conservation drainage applies to cases where regular agricultural drainage would create too dry a hydrological regime. Regular drainage may e.g., lead to insufficient recharge

Table 1.5 Selected water quality parameters and indicator values for open waters

Pollutant type	Dutch Guidelines CV 2000	US EPA MCL or TT[1)7)]	South Africa[11)]	FAO (WHO)
Physical parameters				
Acidity/alkalinity (pH)	$6.5 \leq pH \leq 9.8$[4)]	$6.5 \leq pH \leq 8.5$[7)]	6–9 [4)6)8)9)]	
Algae [13]			[5)6)7)9)]	
Dissolved oxygen (O_2, BOD, COD)			[8)10)]	
Odour	no smell	threshold no 3	< 1 T.O.N.[9)]	
Temperature [14]	$\leq 25°C$		[6)10)]	
Total dissolved solids mg/l			0–1 000 [4)5)6)8)10)]	
Total suspended solids			< 50 [4)6)8)10)]	
Turbidity [2)]	≤ 750 mgSiO$_4$/l	≤ 5 NTU (TT)	0–1 NTU [9)]	
Micro-organism (only a few selected parameters are indicated)				
Faecal coliforms e-coli, count/100 ml, incl. pathogens	1–10		0–200 [4)5)7)]	
Heterotrophic Bacteria, count/100 ml			0–100 [7)]	
Total coliforms		5.0%	0–5 cnt/100 ml [7)]	
Disinfectants and disinfection by-products				
Chlorine Cl$_2$/Chloride Cl-	–	≤ 4.0 mg/l	0–3 000 mg/l [4)6)7)8)]	

etc. see references. Mostly health related problems occur. Values: Max. Residual Disinfectant Level

(Continued)

Table 1.5 (Continued)

Pollutant type	Dutch Guidelines CV 2000	US EPA MCL or TT[1) 7)]	South Africa[1)]	FAO (WHO)
Inorganic contaminants: N-P-K[12)], salts, (heavy) metals				
Nutrients:				
Ammonia NH_3 (N[12)])	≤ 20 µg/l - N		0–1[6) 7)]	5 mg/l[4)] - N[12)]
Nitrate NO_3 (form of N[12)])	≤ 10 mg/l -N	≤ 10 mg/l -N	0–6[5) 6) 7)]	50 mg/l acute[3) 7)]
Nitrite NO_2 (form of N[12)])	≤ 10 mg/l -N	≤ 1 mg/l -N	0–6[5) 6)]	3 mg/l acute[3) 7)]
Total Phosphate $H_2PO_4^-$	≤ 0.15 mg/l - P		[6)]	- FAO 1992a[4)]
Potassium KNO_3			0–50 mg/l[7)]	- FAO 1992a[4)]
Salts:				
EC, TDS (Total Dissolved Salts), SAR and, Chloride Cl, Sodium (Na), Boron B, etc. see Table 15.1 Bicarbonates HCO_3, Calcium Ca and Magnesium Mg (both also crop nutrients), Sodium Na, are not listed but used in SAR calculation (Eq. 14.3)				
Sulphate SO_4	≤ 100 mg/l		0–1 000[7) 8)]	
Fluoride Fl	-	≤ 4 mg/l	0–2 mg/l[4) 5) 7) 10)]	
Metals (NB Cobalt Co, and Iron Fe are mentioned but no values given):				
Arsenic As	≤ 50 µg/l	≤ 0.05 mg/l	0–1 mg/l[4) 5) 6) 7) 10)]	0.01–1 mg/l[3/5) 7)]
Cadmium Cd	≤ 2.5 µg/l	≤ 0.005 mg/l	< 0.01 mg/l, [4) 5) 10)], 0–5 mg/l[7)]	0.003–0.025 mg/l[7) 5)]
Chromium (VI) Cr	≤ 50 µg/l	≤ 0.1 mg/l	0–1 mg/l[4) 5) 6) 7) 10)]	0.05–0.1 mg/l[3/4) 7)]
Copper Cu	≤ 50 µg/l	≤ 1.3 mg/l (TT)	0–1 mg/l[4) 5) 6) 7) 10)]	0.2–2 mg/l[4) 7)]
Iron Fe			0–10 mg/l[4) 5) 6) 7) 8) 10)]	
Lead Pb	≤ 50 µg/l		0–0.2, 10 mg/l[4) 5) 6) 7) 10)]	0.01–5 mg/l[7) 4)]
Manganese Mn	≤ 0.2 – 0.5 µg/l		<0.02 mg/l[4)], 0–50 mg/l[5) 6) 7) 8) 10)]	0.2–0.5 mg/l[4) 3)]
Mercury Hg	≤ 0.5 µg/l	≤ 0.002 mg/l	0–1 µg/l[5) 6) 7) 10)]	-
Molybdenum Mb	≤ 0.02 mg/l[5)]		< 0.01 mg/l[4) 5)]	0.07 mg/l
Nickel Ni	≤ 50 µg/l		0–1 mg/l[4) 5)]	0.02 mg/l[3)]
Selenium Se	≤ 0.02 mg/l[4) 7)]	≤ 0.05 mg/l	< 2 µg/l[4)] . 0–50 mg/l[5) 6) 7) 10)]	0.01 mg/l[7)]
Zinc Zn	≤ 200 µg/l		<1 mg/l[4)], 0–20 mg/l[5) 6) 7) 10)]	2–24 mg/l[4) 5)]

Organic contaminants for details see FAO 2002, US EPA websites. Pesticides are in this group: insecticides, fungicides, herbicides, acaricides, rodenticides.

Approx. number listed:	> 7 groups	> 50	> 20 [5,6,7,10]	> 40
Radio nuclides				
Alpha particles	≤ 2.7 pCi/l	15 pCi/l	0–0.5 Bq/l [7]	
Beta particles	≤ 27 pCi/l	4 millirems/yr	0–1.38 Bq/l [7]	
Uranium		30 µg/l	0–0.5 Bq/l [7]; <0.01 mg/l [4]	0.002 mg/l [3]

Sources: CV 1988 & 2000, FAO 1992a, FAO 1992, FAO 1996a & FAO 2002, https://www.epa.gov/environmental-topics/water-topics, http://www.dwaf.gov.za/Documents/RSP.aspx https://www.who.int/publications/guidelines/en/. NB. I mg = 1 000 mg, pCi = pico Curies, Bq/l = Becquerels/litre of water, acaricides are substances lethal to ticks and mites. T.O.N. = threshold odour number as determined by a panel of judges.

1) MCL Maximum Contaminant Level. TT = Treatment Technique; a required process intended to reduce the level of contamination in drinking water
2) Turbidity (cloudiness of water) measured in nephelometric turbidity units (NTU)
3) Provisional guideline value FAO 2002
4) Agricultural use; irrigation
5) Agricultural use; livestock drinking water
6) Agricultural use; aquaculture
7) Domestic use; drinking water
8) Industrial water
9) Consumer complaints; recreational use
10) Aquatic ecosystems
11) Different TWQR values apply for different uses. Often ranges are classified in different severity classes; full range for all uses is indicated in this column; see www. dwaf.gov.za for details
12) Nitrogen (N) occurs for 78% (by volume) in the air and moves in and out of various constituents in life; the nitrogen cycle (Figure 1.10), which includes nitrification and denitrification processes
13) Different measures apply. Aesthetic effects are measured by chlorophyll in mg/l, while human health effects are measured by cell counts of blue-green algae
14) Together with pH level determines many of the biochemical processes.

of the groundwater in areas of potable water abstraction. In areas where lower lying agricultural land occurs interspersed with higher lying forest land, regular drainage would cause the agricultural land to become sinks and watertables in the forest areas to fall too deep for the trees to survive. In all these cases controlled drainage should be applied which ensures that the installed drainage will not provide over-drainage i.e., does not remove too much water and does not cause watertables to fall unnecessarily deep (for details see Chapter 20). Conservation drainage also applies to cases where agricultural land adjoins wet type nature conservation areas and where special measures (usually in the form of buffer zones) need to be taken to maintain the desired wet regimes in the latter area.

1.8.4   Public health

Drainage can greatly help to control the major water related vector-borne diseases like malaria and filariasis (transmitted by mosquitoes) and bilharziasis (transmitted by freshwater snails). Especially the drainage of stagnating shallow water bodies deprives the vectors and/ or intermediate hosts of some of their favourite habitats and breeding grounds. Such water bodies may occur naturally (landscape depressions with stagnating rain water) but in many cases they are also created by poor water management (irrigation waste water, well waste water, household waste water, etc.) or by incomplete finishing of construction works (borrow pits along roads, canals, and other construction sites, blockage of natural drainage ways, non-judicious spoil disposal, etc.).

To be effective, the drainage systems should be well maintained which in the developing countries is often not the case (applies also to much of the road drainage, dam site drainage and urban drainage). Overgrown and silted up open agricultural drains with slow moving water and stagnant pools provide good breeding conditions for mosquitoes and snails.

1.9   Drainage development considerations

In general, soil and climatic conditions together with the farm system, determine the economic scope for drainage improvements. On many soils, notably on the so-called heavy lands (described in Chapter 9), a high degree of excess water control is technically difficult or even impossible to achieve, at any event only at very high cost.

On such soils a combination of improved drainage, achieving a moderate but significant excess water control, and adapted farming, imposing some moderate constraints on the use of the land, often provides a satisfactory solution (tolerant crops, farm calendars such that critical farm operations (planting and harvesting) take place in drier periods, etc.). The design of such integrated drainage solutions clearly requires a thorough understanding of the agricultural conditions of the area. Decision charts as shown in Figure 1.11 may help to arrive at sound decisions. If the source of excess water is not local (i.e., inefficient irrigation upstream, upland runoff) then investigating upstream of the waterlogged area may provide insights into the source of excess water that may be resolved upstream, thus dissipating the waterlogging in the downstream area.

Investment in drainage generally becomes more opportune with rising levels of agricultural productivity. Drainage development in the developing countries may generally only be expected to take off after the more easily overcome productivity constraints have been addressed, the productivity of the land has risen to reasonable levels and drainage has become the next critical constraint. This relationship between agricultural and drainage

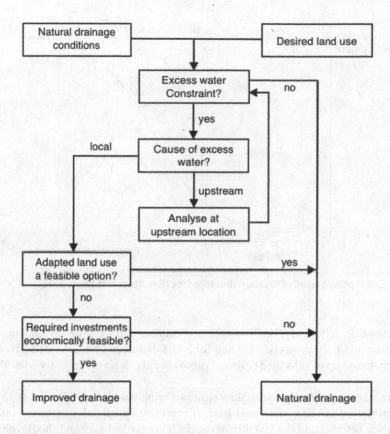

Figure 1.11 To drain or not to drain?

development is schematically indicated in Figure 1.12. When the area is still in a development stage, only a low level of drainage development is warranted (only some elementary flood control and arterial drainage). Opportunities for profitable drainage investment increase when the threshold level of development is passed. Drainage development may even be a pre-condition for agricultural development to reach advanced levels.

Eventually a saturation stage will be reached when most of the land benefiting from improved drainage will have been covered and opportunities of drainage investment become more restricted and selective (e.g., some upgrading of existing drainage system to meet the higher requirements of the changing land use). Although, the relationship depicted in Figure 1.12 in principle also applies to drainage for salinity control of irrigated land, this does not hold true in cases where without drainage the land would go out of production. In such cases, the choice is between saving and losing the land and the socio-economics of the two alternatives should dictate which choice is made.

It needs to be recognized that although drainage development should in principle be governed by the economic opportunities and that these should be seized when the time is ripe, these opportunities and developments will generally not occur automatically but only in a suitable environment. The latter would generally include the prevalence of a conducive

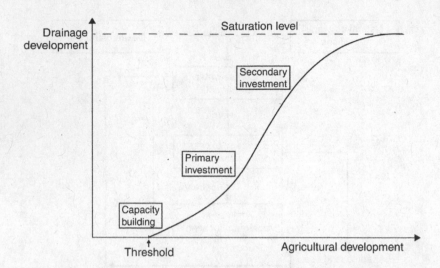

Figure 1.12 Conception stages of drainage development (Smedema and Shiati 2002)

policy framework, certain levels of research and training and the prior undertaking of some demonstration and pilot projects. Also helpful is the existence of a dedicated advocacy and lobby group, made up of informed drainage professionals, drainage industry and other interest group.

Land drainage is generally undertaken either to bring land into production or to increase the productivity of existing cultivated land. It represents a capital investment intended to result in future benefits and the viability of the drainage project may and should, be assessed like any other investment on the basis of sound economic analysis.

Notes

1 Mesophytes are terrestrial plants which are neither adapted to particularly dry nor particularly wet environments. Unlike hydrophytic plants, such as water lily or pondweed, that grow in saturated soil or water, or xerophytic plants, such as cactus, that grow in extremely dry soil, mesophytes are ordinary plants that exist between the two extremes.
2 A piece of land put down to grass, clover, etc., for a single season or a limited number of years, in contrast to permanent pasture.

Chapter 2

Planning and design considerations

The preparation of a drainage plan involves the determination of the optimal combination of the following groups of plan and design variables:

- system variables: types of drains, structures and outfalls; alignments, spacing, depths, capacities; materials and construction methods; matching irrigation system variables if applicable, etc.
- land use variables: crops and crop rotations, farming systems, farming practices, etc.
- environmental variables: water quality standards to be maintained, wetlands to be protected; permissible impacts on the downstream flow regime, etc.
- management variables: institutional organisation and procedures, operation and maintenance (O&M), financial arrangements, stakeholder involvement, etc.

Most of these variables are mutually dependent (drain type depends on land use; drain spacing depends on drain depth; etc.) or are subject to outside constraints (dictating e.g., crops to be grown, location of the outlet, acceptable effluent water quality, etc.). The planner must endeavour to fix, within the limits imposed by the above constraints, the variables to optimise the objectives of the project (*optimal plan*, see Figure 2.1). In practice, optimisation using straightforward planning methods such as systems analysis and operations research[1] can seldom be applied because many variables of the multi-disciplinary relationships, and relationships between objectives and the variables are inadequately understood. Early stakeholder involvement should be very broad based i.e., from the general public to the end-user of the planned system and it is very important that in the planning and design process this is incorporated (see section 3.4).

Drainage planning methods have developed historically and are essentially a loosely structured set of methods and procedures by which the variables are determined partly on the basis of a-priori assumptions and ad-hoc decisions, partly on the basis of empirically or analytically established relationships between some of the main variables. Variables are generally not considered in an integrated manner, nor simultaneously, nor in all their mutual relationships. These procedures may be tested and refined to the extent that, applied by experienced staff to routine situations, they result in reasonably integrated and optimised plans. No such guarantee, however, exists when applied to new areas and new situations.

Methods all start with a formulation in technical terms of the desired excess water control, this being referred to as the *basic design criterion*. It is implicitly assumed that the drainage plan which best meets this criterion is the optimal plan. The formulation of the basic drainage criterion is a short cut for the present methodological inability to consider all the relationships in an integrated manner (Figure 2.1), simultaneously and explicitly.

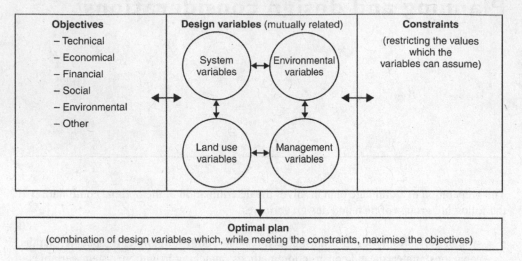

Figure 2.1 Relationships to be considered in drainage design

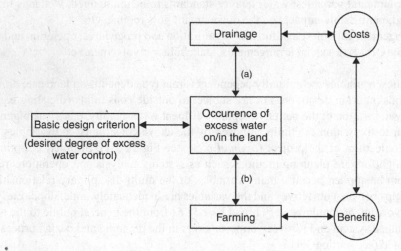

Figure 2.2 The role of the basic drainage criteria in drainage design

The physical characteristics of the drainage system can be worked out using the basic design criteria. Some land use characteristics and operation rules are considered beforehand and incorporated into the basic design criterion, whilst others are defined alongside in the process of defining the physical works. Various subsidiary design criteria are used as well, partly derived from the basic criterion, partly in addition to the latter. These are, implicitly or explicitly, described in the text where applicable.

Figure 2.2 illustrates the central role of the basic drainage criterion in the drainage planning. In this figure, the relationships between drainage and farming have been broken down into:

(a) the relationships between the provided drainage and the occurrence of excess water on or in the land

(b) the impacts of these excess water occurrences on the farming of the land.

The relationships of (a) are fairly well established (Chapter 10) and having formulated the desired degree of excess water control (i.e., the basic design criterion), the required drainage system can be readily established. The formulation of the desired degree of excess water control based on the relationships according to (b) is more difficult since these are generally far less well understood. Criteria, therefore, are mostly based on experience rather than firm scientific knowledge.

Subsurface drainage

The basic design criterion for subsurface drainage in humid regions prescribes the required watertable control during and after the occurrence of a specified high rainfall event (design rainfall). This of course applies where groundwater is recharged by rainfall. Elsewhere, recharge due to irrigation, seepage, etc., may need consideration. The required watertable control may be formulated in a steady state or a non-steady state form. Under steady state conditions, the recharge and discharge rates are assumed to be equal and steady (= constant in time) whilst the watertable level remains constant (static watertable criterion). Under non-steady state conditions, the criterion may specify how fast the watertable should fall after rising to an unacceptably high level (falling watertable criterion), or limits may be imposed on the frequency and duration of occurrence of unacceptably high watertable levels (fluctuating watertable criterion).

Surface/shallow drainage

The basic design criterion for this type of drainage prescribes within which span of time the excess water on the land, resulting from the design rainfall, must be evacuated. Other sources of excess water, such as snowmelt, irrigation losses, etc., may also be important, individually or in combination with rainfall, and should where relevant be incorporated in the design criterion.

Main drainage

As main drainage is subservient to field drainage, criteria for the design of main drainage systems may in principle be derived from the field discharges they collect. Other requirements, however, also enter into the design of main systems and although the relevant criteria will always be closely related, they are seldom identical to those used for field drainage.

2.1   Design rainfall

As the rainfall depth increases, its frequency of occurrence decreases. This holds true for all types of rainfall (hourly, daily, monthly, even annual rainfall). The nature of this relationship may be determined by noting how often different rainfall depths occurred during say the last 20–30 years. The results of such an analysis are commonly expressed *as frequency curves*. The curve in Figure 2.3 shows, for example, that a 48 hours rainfall higher than 40 mm is

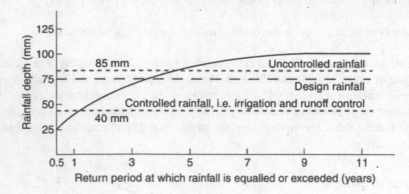

Figure 2.3 Rainfall frequency curve for 48 hours rainfall periods

to be expected to occur on average once every year (1 × 1 year) while depths higher than 85 mm occur on average only once every five years (1 × 5 years).

The design rainfall is the most critical rainfall event that the drainage system should be able to cope with; it generates the most serious of all excess rainwater incidences, which the system is designed to control. Even higher rainfalls may occur, but which are not fully controlled by the designed system and therefore these will cause damage (Figure 2.3). Such high rainfalls, however, only occur rarely and it is generally economic to accept occasional damage rather than constructing a failproof, expensive drainage system, capable of coping with even the highest rainfalls.

Damage and benefits

A drainage system designed for a 48 hours rainfall of 75mm will be able to cope with all 48 hours rainfalls up to 75mm whilst rainfalls in excess of 75mm will result in there being an uncontrolled element of rainfall which may cause damage. The relationship between the uncontrolled rainfall depth and the damage incurred is complex and not always fully understood. Generally, however, the damage is proportional to (although usually not linearly) the depth of the uncontrolled rainfall (for example as in Figure 2.4).

The benefits of a drainage system may be increased by increasing the design rainfall but this involves an increase in cost. The design rainfall is thus an economic parameter and its selection essentially involves the optimisation of the expected benefits in relation to the costs. The procedure for evaluating this optimum is set out in Table 2.1 and Table 2.2 whilst the results of the calculations are illustrated in Figure 2.5. This figure shows that the total cost of the drainage system is composed of two elements, the actual cost of the installation and the cost of the damage that occurs in spite of the drainage system. The total costs decrease to a minimum at the optimum point A.

For agricultural drainage it is generally assumed that this optimum is reached when systems are designed on the basis of the 1 × 2 to 5 years rainfall event. Such systems may be expected to discharge safely and adequately all storms of the critical type occurring more frequently (these storms inflict little or no damage) but cannot fully control heavier storms

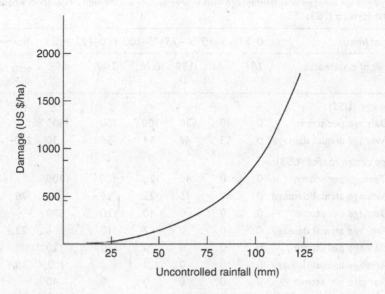

Figure 2.4 Relationship between uncontrolled rainfall depth and damage incurred, cost levels ±1983

Table 2.1 Analysis of frequency of occurrence of 48 hours rainfalls

$R = 48$ hours rainfall (mm)[1]	Return period of rainfall $\geq R$[2] (years)	Annual number of rainfalls $\geq R$	Rainfall interval (mm)	Annual number of rainfalls per interval[3]
25	0.5	2.00	0–25	180.5
50	1.5	0.67	25–50	1.33
75	3.5	0.28	50 75	0.39
100	10.0	0.10	75–100	0.18
125	50.0	0.02	100–125	0.08
			>125	0.02

1 48h rainfall is typical for humid regions

2 return periods based on Figure 2.3

3 there are 365/2 = 182.5 non-overlapping 48 hours rainfall periods per year. In two
   of these periods, the rainfall ≥ 25 mm, so there are 180.5 occurrences in the 0–25 mm
   interval. Similarly, it follows that there are 2.00 − 0.67 = 1.33 occurrences in the 25–50 mm
   interval, etc.

(occurring less frequently). However, no hard and fast rules exist. In situations where uncontrolled rainfall can do considerable damage (e.g., in the case of a high value crop) and the cost of a greater degree of control is not excessive, it may well pay to extend the control to less frequent events and base the design on, for example, the 1 × 10 year event.

On the other hand, a lower standard is often economically most feasible for drainage of land of low agricultural value. In the UK for example, the 1 × 2 years frequency storms are often taken as the design storm for the design of field drainage systems for grassland, the 1 × 5

Table 2.2 Analysis of average annual damage and marginal-average benefits due to drainage (damage cost levels ±1983)

Rainfall interval (mm)	0–25	25–50	50–75	75–100	100–125	>125	Average annual costs of damage	Marginal average annual benefits[4]
Annual number of occurrences	181	1.33	0.39	0.18	0.08	0.02		
Without drainage (US$)								
Damage per storm[1]	0	40	120	300	700	1500		
Average annual damage[2]	0	53	47	54	56	30	240	
With drainage design rainfall (US$)								
25 mm/d Damage per storm[1]	0	0	40	120	300	700		164
Average annual damage	0	0	16	22	24	14	76	
50 mm/d Damage per storm[1,3]	0	0	0	40	120	300		53
Average annual damage	0	0	0	7	10	6	23	
75 mm/d Damage per storm[1]	0	0	0	0	40	120		17.6
Average annual damage	0	0	0	0	3	2.4	5.4	
100 mm/d Damage per storm[1]	0	0	0	0	0	40		4.6
Average annual damage	0	0	0	0	0	0.8	0.8	

1 The damage per storm US$/ha based on Figure 2.4 and Table 2.1 Note that because of many housing and industrial developments since the mid-eighties the current average costs of damage have risen substantially.
2 The annual damage due to 1.33 storms in interval 25–50 mm is 1.33 × 40 = US$ 53/ha.
3 With a design rainfall of 50 mm/d the uncontrolled element of 48 hours rainfalls of 25, 50, 75, 100, 125 mm is 0, 0, 25, 50, 75 mm for which the damage per storm is US$ 0, 0, 40, 120 and 300 per/ha
4 Equals the benefit (avoided damage) by increasing the design rainfall one step.

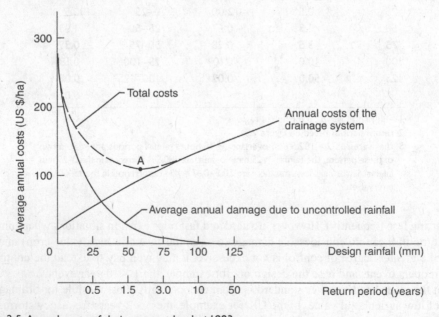

Figure 2.5 Annual costs of drainage, cost levels ±1983

years storms for arable land and the 1 × 10 years for horticultural land. Lower frequencies are generally used for shallow field drainage systems as compared to subsurface drainage systems (1 × 2 to 5 years vs. 1 × 1 to 2 years) and for main systems as compared to field systems (1 × 5 to 25 years vs. 1 × 1 to 5 years), reflecting the differences in risk of damage for each of these systems.

Critical rainfall types

The most harmful type of rainfall varies with each drainage situation. For subsurface drainage (watertable control), prolonged periods of moderately intensive rainfall, lasting some three to five days and occurring during the rainy season, are usually most critical. Much of this rain can infiltrate into the soil and since during the rainy season soil profiles are at or near field capacity, these rains easily lead to saturated conditions in the soil and to high watertables. For shallow drainage, the more intensive shorter duration storms are usually more critical (12–48 hours storms).

For main systems fed predominantly by subsurface discharge, the critical rainfall types are the same or similar to those considered for the field drainage. This also applies to basins with a large proportion of shallow field drainage provided the land slopes remain minimal. In sloping basins, shallow field drainage and especially overland flow will respond directly to rainfall while in addition a concentration of the field drainage discharge is liable to occur. For these types of basins, storms with a duration equal to the time of concentration of the basin are most critical. Depending on the area and characteristics of the basin, critical durations vary from 0.5–6 hours (Chapter 12).

Of course, the season of occurrence of the rainfall in relation to the farm calendar for the area is also of great importance. A high rainfall during the off-season would not normally be a good basis for design. Rather, design should be based on the rainfall occurring during periods when the crop is at a sensitive stage (planting and seedling stage) or during periods in which important farm operations have to be done (seedbed preparation, harvesting, etc.). Interactions between rainfall frequency, critical rainfall type and critical season should also be considered.

2.2   Percolation of excess irrigation water

Although traditionally drainage design evolves around rainfall intensities, in particularly in the temperate climate zones, this is not always the case. In irrigated areas the efficiency of the water delivery system determines the need for drainage (Table 2.3). In drier climate

Table 2.3 Efficiency of various irrigation methods

Method	Efficiency (%)	Remarks
Flood irrigation	50–85	New water management control technologies
Sprinkler irrigation	65–90	From high pressure to low pressure application
Trickle irrigation	75–95	Reliability, durability and water management
Sub-surface irrigation	50–95	Shallow soil management
Controlled drainage	50–85	Maintain and manage high watertable as appropriate
State of the art water management	85–100	Soil moisture management and water delivery system management combined

zones flooding is caused by runoff from upstream areas congregating in the lower reaches of the catchment. The fact that the "problem" is caused upstream suggests a closer look at what is happening upstream. Should we re-forest certain areas, should we change land use (Baoa et al. 2017), should we build water and salt interception schemes?

When the intensity of the irrigation exceeds the infiltration rate of the soil, runoff from the irrigated fields will occur, but also when the infiltration rate is greater than the irrigation intensity (deep) percolation may occur. This, eventually, may lead to waterlogging and salinity problems. Details of the processes involved are described in detail in Chapter 15.

2.3   Design of field drainage systems

The procedures for the design of the field drainage systems are briefly outlined below, further details being given in Chapters 11 and 12.

Subsurface drainage systems: the following main variables have to be defined in the design:

- type and layout of the system
- discharge capacity of the system (q)
- watertable depths to be maintained in the field relative to the soil surface (H)
- the field drainage base depth (W) i.e., the installation depth of the pipe drains or the water level to be maintained in the ditches
- spacing of the field drains (L).

Various types and layouts of pipe drainage systems are described in Chapter 8, together with the relevant selection criteria. The determination of the remaining variables is dealt with in Chapter 11. Although the discussion focuses on pipe drainage, most of it applies equally well to ditch drainage.

Surface/shallow drainage systems: here the main variables to be defined are:

a)   type and layout of the system
b)   discharge capacity of the system (q)
c)   field drainage base depth (W) i.e., water level to be maintained in the field ditches.

Aspects (a) and (c) are described in Chapter 9. By comparing the pros and cons of the relevant alternatives available, choices can be made. Very little design is involved since the various parameters are either standard or are defined based on experience.

The determination of the required discharge capacities are described in Chapter 12 and having also defined the levels to be maintained, the dimensions of the drains and of the structures can be readily determined by means of standard hydraulic calculations.

Controlled drainage systems: the design of controlled drainage systems in principle follows the same sequential steps as enumerated above for the conventional systems. It, however, includes a facility for the control of the discharge and potentially controls water levels in upstream and downstream sections of the drainage system by automated control gates and adjustable weirs. Further details can be found in Chapter 20. Controlled drainage systems may include reuse facilities with options to mix fresh and saline or polluted water.

2.4   Determination of design criteria

Design criteria are generally established partly based on sound scientific theory and analysis and partly based on experience. The role of prior experience is reflected mainly in the use of (semi-) empirical formulae and engineering rules of thumb. In addition, the design of a new system in an area usually relies heavily on experience obtained on similar work in the area or under comparable conditions elsewhere. All this should be judiciously applied, especially when the conditions to which the experiences apply are not well-defined.

Often locally tested criteria will be unavailable, while results from test plots and/or pilot projects often can't be waited for (easily takes three to five years). In such cases estimated criteria may have to be used, at least for preliminary planning/early implementation. When a testing/monitoring programme is started early on in parallel with the progress of the work, preliminary adopted criteria may be verified and/or revised as work proceeds. However, by a judicious combination of theory, analysis and relevant experience, reasonable estimates can in fact be made in most cases.

The planner should of course be cost-conscious throughout the planning and design process. This applies especially to the selection of the design discharges which largely determine the required canal dimensions, pump capacities, etc., making it a main cost-determining factor. Other design parameters/criteria, however, also have a considerable influence on drainage costs, both on investments and on operational costs. With respect to construction, the local availability of skills and materials should always be an important consideration. It should also be stressed that design criteria should be established with a view to the future operation of the project. The future functioning of the project should be thoroughly analysed and defined prior to and alongside the design and the planner should make sure that the technical aspects of the drainage system and its envisaged operation are mutually compatible. Operational considerations should play an important and fully integrated role in the planning and design of a drainage project and include stakeholder involvement.

Note

1 System analysis techniques, including (non-)linear programming, dynamic programming, stochastic techniques, and multi-objective analysis have been applied extensively to solve the water resources management problems but these are not yet widely used in drainage (Cai et al. 2001, Hall 2001, Itoh et al. 2003, McKinney et al. 1999).

Chapter 3

Remote sensing and field reconnaissance

Planning and design require a great deal of information to be collected by means of field investigations. Recent advances in aerial photography with drones and advances in the use of satellite remote sensing applications will allow to determine the need for drainage in more holistic ways than before. This information will in particular be used:

- to diagnose the drainage problem and conceive possible solutions
- to prepare plans and designs.

3.1 Need for drainage and problem diagnosis

A correct diagnosis of the drainage problem is essential for the preparation of a sound drainage plan, and as such it is the first thing to be done. It involves establishing:

- the nature and cause of the problem (impeded percolation, high watertables, sources of excess water, bottlenecks in the collection/disposal of excess water, etc.)
- the harmful effects (affected area, frequency and seasonal period of excess water, effects on crops and farm operations, etc.)
- the potential environmental impacts both upstream and downstream of the area under consideration for possible drainage enhancements.

A preliminary diagnosis can often be made based on available information and field inspections, the latter preferably including actual observations of the occurrence of excess water and of the resulting damage. It should generally also involve some interviewing of the affected farmers and other stakeholders. A programme of further investigations can then be planned based upon the results of this preliminary diagnosis. This process is essentially the same as described under performance assessment in Chapter 25.

Although traditionally drainage design evolves around rainfall intensities, in particularly in the temperate climate zones, this is not always the case and it is changing rapidly. In irrigated areas the efficiency of the water delivery system determines the need for drainage (Table 2.3). In drier climate zones flooding is caused by runoff from upstream areas congregating in the lower reaches of the catchment. The fact that the "problem" is caused upstream suggests a closer look at what is happening upstream. Should we re-forest certain areas, should we change land use (Baoa et al. 2017), should we have built water and salt interception schemes upstream (Cause and Effect)?

Diagnosing is a skill which is partly based on theoretical understanding of the involved processes and relationships, partly on experiences gained through field work and practice. This mixing and interaction of understanding and experience greatly adds to the quality of the diagnosis. As most of the diagnosis is non-formalized and non-explicit, it would obviously be of great value if it could be captured in a logical structured knowledge decision-making system. Efforts have been made (Haie and Irwin 1987, Strzepek and Garcia, 1987, Durnford et al. 1984 & 1987, Ehteshami et al. 1988) but these remain primarily in the early stage of (model) development (Amlan and Das 2000, Itoh et al. 2003) without gaining much practical use. Tuohy et al. 2016a experimented with a low-cost visual drainage assessment methodology for landowners with limited knowledge of, and access to, land drainage know-how and means to estimate soil permeability. For six sites in Ireland, three design methodologies (1 Visual Drainage Assessment, 2 Ideal design and 3 Standard design) were tested and modelled with resulting drain discharge and minimum watertable depth used as indicators of achievement (see section 5.8.3 for more description).

Various types of maps are useful during this stage: geographical, hydrological, soils, administrative, etc., showing relief, characteristic elevations, the present (main) drainage systems, soil-patterns, land use, main infrastructure features and other relevant information. Aerial photos (incl. drone imagery) and remote sensing imagery are also particularly useful. The assessment of the outlet conditions and of foreign water involvement, may require the investigation of a wider area than the project area.

Plan preparation/elaboration

Once a satisfactory diagnosis of the problem has been made, plans for its solution can be conceived. Parallel with the planning and design process, further investigations are made to establish the environmental, technical, socio-economic and institutional conditions under which the planned measures and works are to be implemented and to operate. These provide the basis for the formulation of the boundary conditions and design criteria for the planning. This is also the time to start stakeholder engagement (see section 3.4).

3.2   Remote sensing and aerial survey

Since the early seventies the availability of Landsat imagery and Google Earth from 2010 onwards provides a source of potentially useful information to assess, plan and design drainage systems (Vlotman, 2017). Similarly, to involving stakeholders from the beginning to the end, a thorough technical analysis of the condition at, up and downstream of the intended drainage system is essential. The advances in remote sensing techniques and the availability of these services as well as the skills of stakeholders allow a sophisticated process to be included in the reconnaissance stage of the design process. These processes may actually lead to the conclusion that drainage is not necessary if other, potential cheaper solutions, upstream or downstream in the water management system show promise that will negate the need for drainage, or, show means of controlling water quality at downstream locations; a thorough assessment of *Cause and Effect* is in order, noting that high watertables and salinity problems are the effect of a cause most likely upstream or downstream in coastal areas.

The continued advances in remote sensing in the last couple of decades (Figure 3.1) have been significant and will continue to evolve at a rapid pace when more satellites are

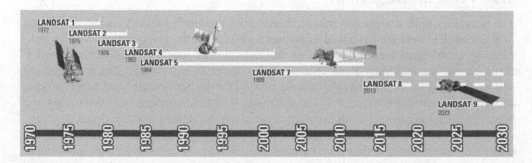

Figure 3.1 Overview of Landsat satellites past, present and future

launched (Landsat, IKONOS, MODIS, SPOT, QuickBird, WorldView, RapidEye, GRACE[1], etc.). Access to the raw outputs of Landsat imagery is easy via the web. More advanced outputs are commercially available and can include those from other platforms such as aerial photography and drones. Many government agencies and private companies are developing tools to help with accessing and assessing the data available via the web and internal computer network systems. This is a far cry from the good old days when draftsmen prepared drawings on tracing paper; then to be printed; the smells of ammonia filling your nose; something un-imaginable with today's attention to Occupation Health and Safety procedures (OHS).

Agencies in the US and Australia now make water observation from space (WOfS) maps available (US Landsat web, http://landsat.usgs.gov, and WOfS web from Geoscience Australia, www.ga.gov.au). These maps are used to inform flood inundation modelling and mapping which allows us to assess the extent and duration of flooding at certain flow rates from low overbank flows that occur on a regular basis (several times a year) to events that occur only 1 in 50 or 1 in 100 years. As Figure 3.1 shows the remote sensing information is available from 1972 and one will find events that occur 1 in every 50 or more years are likely captured.

The emergence of drones with cameras at retail outlets, rather than the sophisticated multi-million-dollar drones used by the military has opened a whole new avenue of reconnaissance. Drones equipped with sophisticated cameras of the type that are used on satellites with various band widths can be acquired or specialised survey companies can provide these. With appropriate algorithms the drone imagery can provide information on water use, plant health, open water in the landscape, surface moisture content, surface salinity (see also section 14.5.3), etc. These can then be used with further algorithm development as indicators of drainage problems.

Many earth observation sensors have been designed, built and launched with primary objectives of either terrestrial or ocean remote sensing applications (CEOS 2018). Often the data from these sensors are also used for freshwater, estuarine and coastal water quality observations, bathymetric and bentic mapping. As they can only map open surface water the usefulness for drainage design is limited to assessing waterlogged areas. Concurrent and complementary on-site surveys are necessary for ground-truthing and to meet the requirements for detailed design.

3.3   Field investigations

The necessary field surveys and related investigations required for the planning and design of a drainage system are described in this section. Most of the investigations and the indicated levels of detail are for the detailed design phase and would generally also be undertaken during the feasibility stage at a somewhat lower level of detail. The investigations and/or evaluations are comparable to those needed for the performance and benchmarking activities described in Chapter 25. The difference is that for performance assessment and benchmarking long-term records are needed but at a lower density than for detailed design.

Traditionally field surveys are executed by governments, with the aid of consultants, but rarely are the end stakeholders (i.e., farmers and general public) involved from inception to planning to execution of a drainage scheme. Yet, when we reach the Operation, Management and Maintenance (OMM) phase of a water management system, they are expected to takeover, run, bear and be responsible for the costs of OMM. This has not worked and will not work ever; therefore, in this chapter successful methodologies and processes to be followed for engagement of stakeholders at all levels, from government to farm, from minister to farmer, are described.

Land drainage for agriculture includes various approaches to assessment and prevention of waterlogging and salinity problems. It also should consider the water-food-energy nexus approach (Vlotman and Ballard, 2013, 2014 & 2016) and give due attention to ecological and economic considerations, referred to as the Triple Bottom Line, TBL, i.e.:

1   Social aspects; stakeholder engagement
2   Environmental/ecological aspects, and
3   Financial/economic aspects.

The most important consideration to be included in drainage system design is the consideration of Cause and Effect. Waterlogging and salinity problems are the Effect of something that is occurring most likely upstream; the Cause.

3.4   Planning stakeholder engagement

To achieve active stakeholder involvement a planned process will need to be executed (MDBA 2015):

* Assessment of state of institutional development at all levels;
* Needs assessment;
* Plan development reflecting:

    o   Who you will engage with;
    o   Why you will engage them;
    o   Why they will want to engage with you;
    o   How you will engage them;
    o   When you will engage them; and
    o   How you will monitor and evaluate your engagement approach?

The key for involvement of stakeholders in irrigation and drainage system operation, management and maintenance (OMM) is the central question: what is in it for me? Incentives do not necessarily need to be economic in nature. They can be improvement in lifestyle, improvements in physical environment and in general improvement in social well-being. Hence, in order to involve stakeholders in water management, incl. irrigation, drainage and environmental watering, it is essential to find out first in what type of TBL environment they operate and what their needs are. It is not just involvement in water management but consideration of all aspects of being successful (i.e., all TBL elements and all water-food-energy nexus considerations).

All stakeholders from farmer to system operator to top level regional and national government staff need to have a clear understanding of the potential benefits of being involved and they need assurance that those benefits are sustainable. Stakeholder engagement is a planned process with the specific purpose of working across organisations, stakeholders and communities to shape the decisions and actions of the members of the community, the stakeholders and the organisations involved. Typical questions to be asked in planning for the involvement of water managers at all levels, including foremost farmers, are:

- What issues do you face in being successful in your (water operation, management and maintenance) enterprise/organisation?
- Do you consider all TBL aspects for the design and future operations?
- Are you willing to share water with the environment/ecology?
- What additional knowledge, skills and information do you need to make an informed decision?

These questions will generate discussions, anger, trepidation, excitement and raise a range of socio-economic issues that should not be ignored and are essential to consider for successful involvement of stakeholders in the design process of the drainage system and the eventual successful OMM of the system.

3.5   Stages of project preparation

Investigations and planning for a drainage project normally proceed in stages. These form a sequence in which the intensity of the investigations and the degree of detail of the plans increase progressively. Thus, by a process of successive approximations, increasingly more refined and optimised plans are made.

In general, the following four stages of project preparation may be distinguished. Depending on the project (available information, size, urgency, etc.) certain stages may be omitted or combined (combination of the first two stages is quite common).

Identification stage

- first formulation of the project, mostly on the basis of available information; hardly any analysis and/or appraisal. This is generally performed by engineers in the office.

Reconnaissance/pre-feasibility stage

- information collected through reconnaissance type field investigations (broad and general), including assessing stakeholders and engaging them early on in the planning and design process

- preliminary diagnosis of the drainage problem
- rough outline of possible solution(s); delineation of the project area and its subdivisions
- evidence that the proposed project is promising, desirable and shown interest of stake-holders from farmer to minister.

Feasibility stage

- information collected through semi-detailed type of field investigations (map scale 1:10 000/50 000)
- presentation of the proposed plans in sufficient detail to demonstrate convincingly that they are technically sound and to enable costs to be estimated within some 10% accuracy
- proof without doubt that the proposed project is the best available solution, is administratively workable, economically/financially viable and socially acceptable.

Detailed design stage

- information collected through detailed type of field investigations and detailed stake-holder surveys
- elaboration of all plans to the extent that they can serve as working documents for the implementation (detailed plans and designs, construction drawings and specifi-cations, etc.).

The *Feasibility Report*, containing the results of the work done during this stage, represents the basic document upon which the decision to proceed with the project will be based. Most banks require such a report before they will consider a loan application. Once the go-ahead decision is made, the final stage can be entered to make the project ready for implementation.

Depending on the size and nature of the project, a feasibility study may be carried out by one drainage engineer or require a small multi-disciplinary team (composed for example of an engineer, a soil scientist, an agronomist, an (socio-/agro-) economist, an institutional expert and an environmentalist). This study may take, for a large complex project, almost a year to complete. The identification and reconnaissance/pre-feasibility studies often take the form of short missions by highly qualified and experienced drainage specialists. The final design stage usually takes the longest and would normally be carried out by a group of senior and junior engineers employed by the public drainage organisation or by a consulting engi-neering firm. The required standards for the reconnaissance, semi-detailed and detailed type of field investigations are implicitly formulated by the objectives of the stage to which they apply. For some investigations the required standards have been more explicitly formulated (see Chapter 5).

3.6   Operation, management and maintenance

Upon completion of the feasibility stage and the various investigations one should also con-sider the operation/management and maintenance stage of any project, and its transition from project to a system fully transferred to the organisation responsible for operating, managing and maintaining (OMM) it. In many cases, the organisation for planning, designing and con-structing the project is not going to be the same as the one for OMM of the system. More than ever, the involvement of (future) stakeholders is essential (see Section 3.4). Often, this is not considered during the various stages of project planning and design, but it is mandatory if the project is to be successful when construction is completed and OMM starts. Timely involvement of all potential stakeholders as described in section 3.4 is the key to sustainability of the drainage system.

Note

1 The Gravity and Climate Recovery Experiment (GRACE) satellite measures changes in the total water storage of river systems, providing a unique opportunity for better understanding connections between stream salinity and changes in catchment water storages at the large river basin scale (Heimhuber et al. 2019). GRACE is a constellation of two satellites that measure the relative position between each other using microwave-ranging from 2002 to 2017. Its successor, GRACE-FO, was successfully launched on 22 May 2018.

Chapter 4

Assessment of costs

The scope of this Chapter is restricted to the evaluation of the costs and benefits directly related to the improved drainage as would for example be relevant to a farmer who is considering installing a drainage system. In larger public drainage projects, other objectives (besides improved farming) are often served by the project, while the costs and the benefits are often less straightforward economic or financial quantities. In the evaluation of these projects the contribution of the project to the national economy and welfare needs to be considered while for farmer's projects a narrower view can be taken. The cost and benefits are based on those prevalent of the 1980–2000 period.

4.1   Required information

In some cases, the benefits of land drainage are so clearly in excess of the costs as to make detailed analysis unnecessary. In most cases, however, calculations are required to enable a decision to be reached about its worthiness and to obtain the required financing. The information, which must be established has been briefly reviewed below. Commonly used terminology is also explained.

Costs

The construction costs of the project may be estimated based on the recommended design or from contractors estimates. In the case of phased development, the amount and timing of the capital inputs must be established.

Operation costs arise on most projects for maintenance, re-moling (see section 9.3), reconstruction of field ditches, and in the case of pumped drainage, for power. In some circumstances, e.g., shallow drainage with a bedding system, increased farm costs also arise because the farm operations take extra time and care. A detailed cost calculation for a pipe system has been given in section 4.5.

The time period

For the purposes of cost evaluation, a number of different "lifetimes" may be identified. Firstly, the drainage project will have an *actual life* which for a well-maintained pipe drainage scheme may be anywhere between 50 and 100 years. Secondly, the project has an *economic life* corresponding to a notional lifetime at the end of which the project will be renewed. The inherent value of the project is at a maximum immediately after construction declining steadily thereafter until, at the end of the economic life, it has a zero-terminal value. The economic lifetime could coincide with the actual lifetime but in economic terms future costs occurring after say, 30 years, are insignificant. For this reason, the economic life of the project is generally taken to be between 20 and 30 years. Thirdly, the project has a *financial life*, often imposed by the lending agency. For example, a farmer having obtained a loan from the bank to finance the project may be required to pay back the loan over a ten-year period. In this case, the financial life is very much shorter than either the economic or the actual life. Also, the project has a terminal value at the end of its financial life which should be considered. The farmer's main concern is that the cash flow benefits equal or preferably exceed the costs arising during the financial life, the existence of a terminal value in this case being of little significance.

Financing and annuity

The cost of borrowed capital can be readily estimated knowing the period over which the loan has to be repaid and the interest rate imposed upon the loan. The cost of using available own capital may be determined from the return likely to have been obtained from the alternative use to which the capital might have been put.

Capital cost (interest and depreciation) can be expressed in terms of an annual cost by the following annuity formula:

$$R = \frac{i(1+i)^T}{(1+i)^T - 1}$$ Eq. 4.1

where i = interest (in decimals) and T = depreciation period (yrs). The annuity R is also referred as the *cost recovery factor*. Example: i = 0.10 and T = 25 → R = 0.11 (11%).

Benefits

The principal benefit of land drainage is usually a yield increase but improved cropping intensity and growing of more remunerative crops may also be important. The benefits of drainage vary in response to climatic factors, particularly rainfall which is also highly variable. The reduction of the crop yield due to poor drainage may vary from zero in dry years to perhaps 50% in very wet years, a reduction that does not of course occur on well-drained land. In fact the real timing of the benefits has a major effect on the overall cash flow but since these cannot be predicted in advance the financial evaluation has to be based upon an average benefit due to an average increase in crop yield (see section 4.4, example 1).

Important benefits may also arise from the improved workability and trafficability of drained land resulting, for example, in timely planting in spring (important for obtaining good yields) and more workable days (reducing the machinery costs). The same problems apply to evaluating these benefits as mentioned for the benefits due to yield increases.

With good drainage, the farmer is more assured of a reliable yield. This assurance can provide a firm foundation for initiating a whole train of consequential investments leading to benefits due to their mutual dependency.

At the project level (especially for public projects) various non-agricultural/off-site benefits (see section 1.1) should also be considered. This applies also to the costs.

4.2 Discounting

The costs and the benefits of a land drainage project occur at different times during the project period (the main costs occurring mostly during the construction phase, the benefits after the project has reached maturity). Economically, the timing of the costs and benefits is important as different interest periods apply. The *discounted cash flow* (DCF) method of analysis takes this into account. It is based upon the simple observation that a sum of money available now is worth more than the same sum of money in ten years time. This is because the sum available now could, for example, be placed in a savings account in a bank and increase in value due to accrued interest over the intervening period.

Numerically, discounting is the inverse of charging compound interest. For example, the sum of £100 now will with an interest rate of 10% increase to £110 in one year. Conversely, £110 a year later is equivalent to £100 today (*the present value*) the discount factor being 0.909. Discount factors for different interest rates and varying periods of time are presented in Table 4.1, enabling any future sum to be converted to a present value. Thus, a benefit of £90/ha arising in year 15 of the project has a present value at a 10% interest of 90 × 0.239 = £21.51 (the sum of £21.51 invested now at a compound interest of 10% will increase in value to £90 in 15 years time).

Interest rates in a free economy are closely linked to the rate of inflation and the prevailing interest rate tends to be 3–4% above the general level of inflation. Two strategies may be adopted in relation to the choice of interest rate in an inflationary economy. In the first, inflation may be neglected in which case an appropriate but rather low interest rate should be adopted. In the second strategy, the future costs and benefits are increased in value to take account of an assumed rate of inflation and the interest rate is then based upon the prevailing interest rates on opportunity capital.

Table 4.1 Discount factors for calculating present values of future sums

yr	Rate of interest				
	3%	*5%*	*10%*	*15%*	*20%*
1	0.971	0.952	0.909	0.870	0.833
2	0.943	0.907	0.826	0.756	0.694
3	0.915	0.864	0.751	0.658	0.579
4	0.888	0.823	0.683	0.572	0.482
5	0.863	0.784	0.621	0.497	0.402
6	0.837	0.746	0.564	0.432	0.335
7	0.813	0.711	0.513	0.376	0.279
8	0.789	0.677	0.467	0.327	0.233
9	0.766	0.645	0.424	0.284	0.194
10	0.744	0.614	0.386	0.247	0.162
11	0.722	0.585	0.350	0.215	0.135
12	0.701	0.557	0.319	0.187	0.112
13	0.681	0.530	0.290	0.163	0.093
14	0.661	0.505	0.263	0.141	0.078
15	0.642	0.481	0.239	0.123	0.065
20	0.554	0.377	0.149	0.061	0.026

4.3   Evaluation indices

The costs and the benefits may be evaluated based on the following indices:

1.   *Net Present Value* (NPV) being the difference between the present value of benefits and costs; a positive value clearly being desirable.
2.   *Benefit/Cost ratio* (B/C) being the present value of benefits divided by the present value of costs; a value in excess of unity being desirable.

In many situations the benefits of land drainage are very difficult to quantify. In these cases, it is more meaningful to approach the problem from the opposite end to establish what the benefit would have to be in order to exactly cover the costs. The farmer or engineer can then consider whether this break-even benefit, is likely to be obtained or even exceeded. In view of the relatively small timelapse between costs and benefits and the long depreciation period, sophisticated economic analyses are often not required (see example 1).

For large public projects, the *internal rate of return* (IRR) is often used as the evaluation index. The IRR reflects the interest made by the project on the investments in the project. This index is not relevant to most projects initiated by farmers who are mainly interested in financial cash flow considerations in order to meet predetermined interest rates on loans.

4.4   Cost evaluation of open and pipe drainage systems incl. O&M

Example 1

A farmer plans to make a US$ 500/ha investment in improved on farm drainage. The annual costs are estimated to be in the order of US$ 60/ha/year at levels of the early nineteen eighties (7% annuity + O&M). To break even, the net productive value (NPV) would have to increase by at least this amount. For an NPV of US$ 600/ha, the increase would have to be at least 10% while it would have to be at least 6% in case of a NPV of US$ 1000/ha.

Example 2

In this example, the costs of two drainage systems for the Sava Valley (Croatia) are compared. The first system is a regular pipe drainage scheme whilst the second is a parallel passable ditch system as described in section 9.2.1.

	Costs Dinar/ha
System A (regular pipe drainage)	
cost of construction (plastic pipe spacing 15m)	16 200
pipe flushing 2 years after installation, thereafter every 7 years at Dinar 0.5/m	650
annual replacement of some end pipes	20
System B (parallel passable ditch system)	
cost of construction	
first year-(approx. 50%)	3 000
second year (approx. 25%)	1 500
third year (approx. 25%)	1 500
maintenance of field ditches:	
twice a year rotary ditching	100
part reconstruction every 5 years	1 000
extra farm costs:	
extra time and care in tillage, transport, etc.	400

Table 4.2 Comparison of costs of two drainage systems

Year	Undiscounted costs in Dinar/ha		Discounted costs in Dinar/ha	
	System A	System B	System A	System B
1	16 200	3 000	15 066	2 790
2	20	1 500	18	1 335
3	670	1 500	563	1 260
4	20	500	16	395
5	20	500	15	375
6	20	500	14	350
7	20	500	13	330
8	20	1 500	13	945
9	20	500	12	295
10	670	500	375	280
11	20	500	11	265
12	20	500	10	250
13	20	1 500	9	705
14	20	500	9	220
15	20	500	8	210
16	20	500	8	200
17	670	500	248	185
18	20	1 500	7	525
19	20	500	7	165
20	20	500	6	155
21	20	500	6	145
22	20	500	6	140
23	20	1 500	5	390
24	670	500	168	125
25	20	500	5	115
26	20	500	4	110
27	20	500	4	105
28	20	1 500	4	300
29	20	500	4	92
30	20	500	4	88
			16 638	12 845

Notes (the results of the calculations are shown in Table 4.2):

1 Only those costs that differ between the two systems have been taken into account. This is justified in this case since the objective is simply to evaluate the additional benefits needed from the more expensive system.
2 Inflation may be ignored since it is likely to have a similar effect on both streams of costs considered.
3 The discount rate of 6% reflects the prevailing opportunity cost of capital in the country.
4 The costs are evaluated over a 30-year period since the present values for periods in excess of 30 years are relatively insignificant.

System A costs Dinar 3 793/ha more than System B over thirty years which amounts to an annual average undiscounted cost of Dinar 276/ha (R = 0.073 according to Eq. 4.1). The net profit on wheat or maize is approximately Dinar 2/kg so that System A would have to produce 138 kg/ha more yield than System B in order to be as financially attractive.

4.5   Cost calculations for pipe drainage systems

Costs of pipe drainage systems vary from project to project and are valid for a limited period only. Therefore, emphasis is upon methods of cost calculation and on the structure of the total and component costs rather than on absolute costs. For some items the range of comparative costs is indicated.

4.5.1   Cost structure for pipe drainage construction

The breakdown of the total costs of pipe drain construction given below shows that the machinery cost component is rather small in spite of their high purchase costs. This is due to the high efficiency and high capacity of these machines. The materials represent well over half of the costs when a good envelope is needed (for example a gravel envelope; see section 8.4):

	% of total costs
• machinery costs	10 to 15
• costs of material (pipe and envelope)	40 to 65
• labour costs (W. Europe)	10 to 15
• overheads (including design and supervision)	15 to 25

This costs breakdown applies to singular drainage. The example of cost calculation presented further on gives an idea of the cost structure for a composite pipe drainage system.

4.5.2   Guidelines for cost calculations

General guidelines are presented together with some standards which enable rough cost estimates to be made. Final estimates should always be based on locally applicable standards.

Drainage machines

- *annual running time*: 1 000–1 500 hours in humid climates, 2 000 hours in arid climates; annual output for a light machine 250–400 km (humid climate) to 500–600 km (arid climate), for trenchless machines 500–700 km (humid) to 1 000 km (arid); output of heavy machines more variable, very much depending on drain depth
- *fuel consumption*: engines assumed to be running on average at half their rated power, consuming then about 0.3 litre diesel oil per HP-hour
- *oil and lubrication*: about 20% of the fuel costs
- *depreciation*: five- and seven-years periods respectively for trench and trenchless type (10 000–15 000 hrs)
- *maintenance and repairs*: about 15% of the purchase price.

At 10% interest, the total of depreciation, interest, maintenance and repair costs add up to an annual cost factor of 35% of the purchase cost for a trench type machine and 25% for a trenchless machine.

Labour costs (four to five men crew)

Taking total employer's costs as 150% of the salary costs and assuming the salaries of respectively the operator, assistant-operator and labourers to relate as 3:2:1, the total annual labour costs comes to roughly 10 to 12 × S where S = annual labourer's salary.

Materials

The *relative costs* of different types of drainpipe in Table 4.3 are indicative only, actual costs depending very much on the locally available materials and production facilities.

For a gravel envelope, some 0.05 to 0.10 m^3 granular material per meter drain length is needed (for a 20–30 cm wide trench including 25% wastage). Costs depend very much on the availability and quality of the local gravel material.

For the on-site transport of materials and for the backfilling of the trench a tractor with a trailer and a blade are required, costing per annum about 50% of the purchase price of this equipment.

4.5.3   Example cost calculation

- drainage of irrigated land, Iraq, 1980 (Abu Ghraib project)
- composite pipe drainage system with laterals 220 cm deep, 40 m spacing and 250 m length, consisting of 8 cm corrugated plastic pipes with gravel envelope; collectors 250 cm deep and 1 000 m length consisting of 20 and 30 cm concrete pipe (these days mostly replaced by plastic perforated pipes)
- outfall by gravity (this is an exception as most of these systems would be pumped systems)

Table 4.3 Relative costs of drainpipes with respect to 100 mm plain corrugated plastic pipe; i.e., plain is without envelope

Diameter[1] (mm)	Clay pipe	Corrugated plastic pipe	
		Plain	Coconut fibre envelope[2]
50–60	30–35	30–35	60–65
80	40–50	50–60	110–120
100	70–80	**100 standard**	150–165
150	100–125	250	
200		400–500	

1) Internal diameter for clay pipe, Outside diameter for plastic pipe
2) thick synthetic envelope 50–100% more costly than the organic equivalent

- laterals laid by machine (trench type 175 HP machine costing ID 40 000 cif[1] Basra; annual output 2 000 hours × 150 m = 300 km; fuel consumption 30l/hr at Iraqi Dinar (ID) 0.02 per litre[2]
- labourer's salary ID 600 per annum.

Material and equipment costs:

Laterals: pipe	ID 0.250/m
gravel, supply to site	ID 0.100 ⎱
installation	ID 0.050 ⎰
drainage machine (depreciation, interest maintenance and repairs)	ID 0.047 ⎫
fuel, oil and lubrication	ID 0.005 ⎬
labour	ID 0.024 ⎭
other costs	<u>ID 0.024/m</u>
Total:	ID 0.500/m

Material costs continued:

Collector: pipe	ID 0.500 ⎱
installation	ID 1.000 ⎰
Junction:	ID 50/piece
Gravity outlet:	ID 35/piece

Construction costs:

	Units	Requirements per ha	Unit cost ID	Total costs ID/ha
Laterals	m	250	0.5	125
Collectors	m	20	1.5	30
Junctions	piece	0.50	50	25
Outlets	piece	0.02	35	1
				181
Overhead (15%) and engineering (20%)				63
				ID 244

Annual costs:

capital annuity (recovery factor = 0.11 based on 10% interest, 25 years lifetime)	ID 26.9
maintenance, 2% of construction costs	ID 4.9
total 10%–15% of the construction costs	ID 31.8

The above costs are of the same order as reported for composite systems for salinity control of irrigated land in a number of countries in the Middle East and South Asia (investment costs all about US $ 1 000 per ha during the 1980–2000 period).

Notes

1 cif = cost insurance and freight vs fob = free on board
2 Local price, below world market price; ID = Iraqi Dinar (1 ID = 3.50 US $, 1980)

Part II

Investigations

Chapter 5

Climate, land, soil and environment

5.1 Climate

Most drainage problems result from rainfall exceeding the evapotranspiration during short or long periods. Climatic analyses, therefore, are able to contribute a great deal to a better understanding and diagnosis of these problems. More specifically such analyses contribute to:

*   an assessment of the scope for solving the problem as well as identification of the most appropriate drainage method
*   the formulation of the design discharges.

Useful types of climatic analyses in this respect are outlined below.

5.1.1 Climate; soil moisture balance calculations

As an example, a soil moisture balance calculation for Croatia (temperate zone) is presented in Table 5.1 and Figure 5.1.

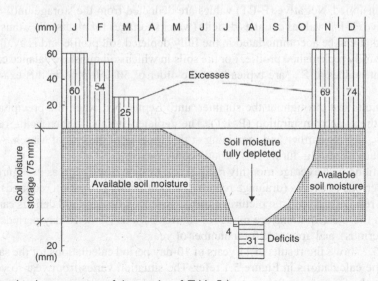

Figure 5.1 Graphical presentation of the results of Table 5.1

Table 5.1 Simple soil moisture balance calculations for Croatia

	P Precipitation mm	ET Evapo- transpiration mm	P-ET mm	S Storage mm	Excesses mm	Deficits mm
End of January			$S = S_{max} = 75$			
February	54	–	+54	75	54	
March	49	24	+25	75	25	
April	69	70	–1	74		
May	90	102	–12	62		
June	101	116	–15	47		
July	85	136	–51	0		4
August	81	112	–31	0		31
September	80	65	+15	15		
October	67	30	+37	52		
November	102	10	+92	75	69	
December	79	5	+74	75	74	
January	60	–	+60	75	60	

In this case the starting point for the calculations is the situation at the end of January when under the prevailing climatic condition, it can be assumed that the watertable is at its highest permitted level while the overlying soil profile is at field capacity (FC). The soil moisture storage (S) thus is maximal ($S = S_{max}$).[1] The soil moisture storage remains at S_{max} as long as P > ET. As no more water can be stored in the soil, all positive (P-ET) values count as excesses, to be drained. From April onwards P<ET and the stored soil moisture starts to become consumed. Negative (P-ET) values are deducted from the storage until S = 0. Further negative (P-ET) values count as deficits (water shortages for plants). S_{max} thus represents the water that can be accommodated in the fully depleted soil profile and is available to the crop in the fully replenished profile. For the soils in which soil moisture balance calculations are relevant, values of S_{max} are typically of the order of 50-150 mm (in this example taken as 75mm).

The deficits last throughout the summer, until September, when the precipitation again overtakes the evapotranspiration (P>ET). The depleted storage is then being replenished, reaching S_{max} in November and remaining at this level until the end of March. All positive (P-ET) values occurring during this period count as excess.

Calculations with average monthly rainfall and evapotranspiration as in Figure 5.1 show the main periods of excess (drainage required) and of deficit (irrigation required).

A more refined and realistic picture of the occurrence of excess and deficits can be developed when these calculations are made for shorter periods (5- to 10-day periods instead of monthly periods), and are made for a number of years.

Figure 5.2 shows the results of 11 years of 10-day period calculations for the same station to which the calculations in Figure 5.1 refer. The situation varies from year to year, giving insight into the frequency of occurrence of excess and deficits.

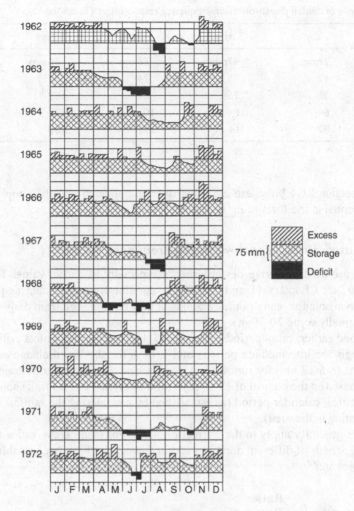

Figure 5.2 Soil moisture balance calculations for 10-day periods (Zagreb, Croatia)

Projected against calendars of cropping and farm operations, critical periods can be identified, growing conditions for crops and workable days for farm machinery can be assessed and drainage and irrigation requirements defined. Information as in Figure 5.2 is especially useful in establishing how adapting the farming of the land, e.g., changing the crop calendar to avoid a coincidence of planting or harvest with a wet period, can alleviate drainage problems. Although primarily intended to analyse drainage problems in a qualitative sense, soil moisture balance calculations also provide useful quantitative information for drainage design.

The procedure used in the example is straightforward. All the rainfall is assumed to infiltrate and evapotranspiration is assumed to be independent of the moisture status within the soil. The calculation can be done by hand or be set up in a spreadsheet. Other more sophisticated methods of analysis exist, using e.g., the agro-hydrological computer models

Table 5.2 Example of rainfall depth-duration-frequency relationships (Tanzania)

Frequency	1 day	2 days	3 days	5 days	10 days
1 × 1 yr.	27mm	47mm	56mm	76mm	90mm
1 × 2 yr.	42	65	74	97	111
1 × 3 yr.	52	75	87	109	118
1 × 5 yr.	64	84	102	127	143
1 × 10 yr.	82	114	132	160	218

presented in section 21.4 but these are often over complicated and the simpler hands-on approach presented in the fore-going is recommended.

5.1.2 Climate: rainfall depth-duration-frequency studies

These studies are used to derive design criteria, especially to derive values for q (drainage coefficient, see Chapters 11 and 12). Depth-duration-frequency data as presented in Table 5.2 are available for many rainfall stations or may be compiled from daily rainfall data (covering minimally some 20-30yr.).

As mentioned earlier, rainy periods of some 3-5 days are normally most critical for subsurface drainage, the intermediate periods of 1-2 days for shallow drainage and the short duration storms (< 6-12 hrs) for runoff from sloping land. For agricultural drainage, design is commonly based on the control of 1 × 2-10yr. rainfall events. The rainfall should of course apply to the critical calendar period (as established by considering the rainfall data in relation to the farming in the area).

High values generally apply to the warmer climates. For many areas, ratios between the rainfall during periods of different durations and/or frequencies have been established. Some typical examples are:

	Ratio:
6 hrs/24 hrs rainfall	= 0.5-0.7
12 hrs/24 hrs rainfall	= 0.6-0.8
5yrs/1yr frequency	= 1.5-2.0
10yrs/1yr frequency	= 1.7-2.5

Such ratios are useful for interpolation and extrapolation purposes, especially in situations where only rough estimates of order of magnitude are required.

5.2 Topography

Good topographical maps, showing the lie of the land, are indispensable in drainage planning and design. For feasibility studies, maps with a scale of 1:10 000/25 000 showing 0.5-1.0 m interval contour lines, will generally suffice for the planning of the main system; sample areas may be mapped in more detail, to be used to assess aspects of field drainage (including cost estimates).

For final planning and design, more detailed maps are required with map scales usually 1:5 000 to 1:10 000 and contour line intervals of 0.25-0.50 m. Contour lines at 0.50 m interval will generally suffice for sloping land. For regular, flat land an interval of not more than 0.25 m is normally required. Map scales should match the contour line interval so that the contour lines (recorded on the map) are not more than 5-10 cm and not less than 0.5-1.0 cm apart (this implies the use of a smaller scale map for sloping land than for flat land).

Detailed topographic maps are especially needed for the design of surface drainage systems for flat land. Small differences in elevation are important and contour lines should be based on a sufficient number of points to provide a good picture of the meso/micro-topography. For the planning and design of runoff control systems for sloping land, the maps should show such features as slope pattern, length and degree of slope, uniformity, etc. For the design of subsurface drainage systems somewhat lower standards for the mapping of the in-field topographic situation apply. The mapping should, however, be sufficient to establish alignments and grades.

The topographic maps should also show the main elements of any existing drainage systems and all relevant infra-structural features such as roads, power lines settlements, etc. To assess whether existing drains can be used, longitudinal profiles (scale 1:5 000/10 000) with cross-sections (every 100-200 m, scale 1:100) will be needed as well as the characteristic dimensions and levels of all structures in these systems (bridges, culverts, etc.). To assess outlet conditions it may be necessary to extend the topographic mapping to well outside the project area.

Map preparation from aerial photography and state of the art remote sensing images, incl. Digital Elevation Maps (DEM) from LiDAR are generally sufficiently detailed for feasibility level study and planning. Concurrent and complementary terrestrial surveys are necessary for ground-truthing and to meet the requirements for detailed design.

It is essential that all maps use the same common reference elevation, for instance, MSL (Mean Sea Level) or NAP (Normaal Amsterdams Peil) in the Netherlands. If a drainage project is covered by multiple maps and construction drawings one should not use a different reference level for each map. For instance, a separate reference level was used for each of the drainage units shown in Figure 21.1. Each polygon in the figure represents an individual drainage unit. Because of the map based individual reference elevation, the drainage base of polygon 41 was higher than the surrounding drainage units. When the surrounding pumped drainage units were operating, watertables in unit 41 were always below the drainage base of unit 41 and hence the system never needed to operate.

5.3   Soil and land conditions

Information on the soil and land conditions in the area is used for many purposes in drainage planning and design, e.g.,

- to diagnose the drainage problem
- to suggest/evaluate possible solutions
- to formulate design criteria

The main soil and land characteristics to be considered are:

a)   Soil/land surface

   Signs of wetness such as indicator plants (sedges, rushes, etc.), poor decomposition of organic matter, water on the land after rain, water in the tracks, springs/springlines, seepage water, day-lighting of rocks, iron ochre deposits, condition of crops, smell, etc.

   Slope features such as overall relief, length, degree and regularity of slope, etc. (partly also to be derived from topographic maps), to assess overland flow conditions.

Retention/detention features such as micro-relief, surface roughness, type of cover, presence of terraces or bunds, etc.

b)   Topsoil (0-20/30 cm)

Erodibility, e.g., expressed by the erodibility index of the USDA universal soil loss equation, or by some other integrated erodibility parameter, or to be deduced from related characteristics such as soil texture and structure, rainfall intensity, length and degree of slope, etc.

Infiltrability, e.g., expressed by measured infiltration rates/curves or to be deduced from soil texture, structure, etc. (see section 6.6), liability of the soil to capping, swelling and cracking characteristics, etc.

FC-PL relation, to assess workability as influenced by soil moisture conditions (see Figure 5.5).

Thickness and hydraulic conductivity of the topsoil, to assess conditions for lateral, topsoil drainage (usually interflow).

c)   Main soil profile/rootzone (0-75/150 cm)

General stratification and main properties of the different layers to assess:

* root development potential
* available moisture, including extent and limitations of capillary rise
* limitations to earth movement
* most appropriate drainage base depth, including required measures for the installation of pipe drains (method, envelope, pipe material, etc.), mole drains (soil stability, method, depth, etc.) and ditches (side slopes, bank protection, etc.)
* subsidence to be expected; state of ripening
* degree of clogging of pipes by iron ochre to be expected.

Hydraulic conductivity (K) and/or KD value of the different layers (see Chapter 7), including any existing anisotropy, mainly to assess:

* conditions for vertical downward percolation, with emphasis on the identification of impeding layers; depth of occurrence, thickness, nature, genesis, etc., of these layers; these observations enable the feasibility of sub-soiling or other measures to improve vertical flow to be assessed and can guide the planning and execution of such measures
* vertical flow conditions in the soil may also partly be deduced from the hydromorphic characteristics in the profile (gley features, mottles, concretions, colours, etc.) and from the rooting pattern
* conditions for horizontal and radial flow towards pipe drains and ditches (to be assessed in conjunction with the substratum characteristics).

Soil-water-air relations of the different layers:

Study of pF-curves (section 6.3) to assess the harmful effects of excess water in the profile and the beneficial effects of its control (from pore size distribution, field capacity, available moisture, drainable pore space, etc.).

d)   Substratum (75/150cm-5/20m)

KD values of the substrata and the depth of an underlying impermeable base, to assess drainage flow conditions below the drainage base.

e) Soil salinity/groundwater salinity

Visual signs (see section 14.3.2); EC and ESP data of the soil at different depths; EC and SAR values of the groundwater; salt composition; seasonal variations; critical capillary height and critical watertable depth (to be deduced from profile characteristics). See also salinity determination and mapping in sections 14.4 and 14.5.

f) Watertables

Real and perched watertables are described in Section 6.1.

Standard soil surveys usually do not provide all of this information and additional investigations are required to make a satisfactory diagnosis and to establish and specify the required measures. These investigations may be integrated with a normal soil survey or be carried out separately. The first procedure is usually followed in drainage projects where no soil mapping has yet been done, while the second procedure refers to plans for improved drainage made for an area for which soil maps are available. The results of these drainage investigations may be incorporated in the soil maps, but some aspects may also be conveniently compiled as 'single value' maps:

- hydraulic conductivity maps showing the flow conditions above and below the drainage base, the latter, e.g., in the form of the composite KD value (directly suitable for the calculation of the required drain spacing)
- mapping of the depth and thickness of impeding layers in the profile (basis for the planning of a sub-soiling programme)
- mapping of the depth of soil ripening (basis for the calculation of the subsidence to be expected).

Many other characteristics may be 'single value' mapped such as watertable depth (including seasonal fluctuations), soil and groundwater salinity, infiltrability, slope features, etc.

5.4 Soil parameters and properties

This section details a number of soil parameters and properties which are of specific importance to aspects of drainage design.

5.4.1 Texture

Soil texture refers to the size distribution of the constituent soil particles. Three size classes are distinguished: clay (< 2 μm), silt (2-50 μm) and sand (2 000-50 μm). Sand may be subdivided (coarse, medium, fine); particles > 2 000 μm = 2 mm are not soil but rock particles (gravel). Soil texture, especially the relative presence of clay, largely determines the rating of important soil properties such as: structure, consistency, workability, permeability, infiltration, erodibility, soil moisture holding characteristics and fertility. The particle size distribution curve (PSD curve) provides the details needed in many formulae that relate particle size to a particular soil property. On the basis of the size distribution, soils may be classified into eleven to fourteen textural classes, depending on which classification is used: USDA, USC, AASHTO or that of SRWSC.[2] The soil texture triangle shown in Figure 5.3 used the basic USDA classifications.

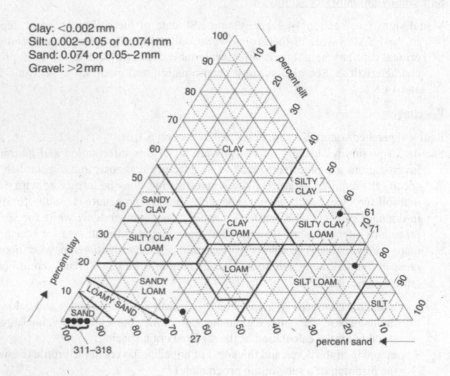

Clay: <0.002 mm
Silt: 0.002–0.05 or 0.074 mm
Sand: 0.074 or 0.05–2 mm
Gravel: >2 mm

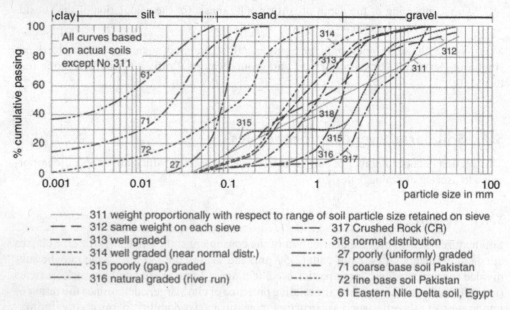

311 weight proportionally with respect to range of soil particle size retained on sieve
312 same weight on each sieve   317 Crushed Rock (CR)
313 well graded   318 normal distribution
314 well graded (near normal distr.)   27 poorly (uniformly) graded
315 poorly (gap) graded   71 coarse base soil Pakistan
316 natural graded (river run)   72 fine base soil Pakistan
                                       61 Eastern Nile Delta soil, Egypt

Figure 5.3 Selected base soils and gravels in texture triangle and the semi-logarithmic gradation plot (afterVlotman et al. 2000)

To determine soil texture and particle size distribution standard hydrometer and sieve analysis is performed (Gee and Bauder 1986, USBR[3] 1974). Hydrometer analysis, or sedimentation analysis, is used for particle sizes smaller than 0.053 mm (standard sieve no 400 of the US standard sieve set). Soils are not always uniformly graded. Some soils are missing certain ranges of particle size due to their particular geological formation and these are known as *gap graded soils*. Gap graded soils can pose problems when they are used as drain envelope material. The successful application of a granular material as a filtering material relies in part on how well the material is graded. Using a standard seven-sieve analysis it is not always possible to identify a gap graded soil and for selection of a drain envelope granular material Vlotman et al. (2000) recommend that the full set of 21 standard sieves is used at least for the initial selection of the material. For quality control in the field the standard set of 7-8 sieves can be used.

Soil bandwidth

From a practical point of view, it is desirable to have the soils in the region represented by a band on the particle size distribution plot. From field surveys there are often hundreds of soil sample sieve analysis results available. It is not practical to display all these graphically and hence a statistical methodology using quartiles is used to select representative bandwidths. The 25% and 75% quartiles are used as is shown in Figure 5.4.

Soil samples were collected from different depths, based on a grid of 150 × 150 m, in anticipation of drains being installed at depths ranging between 1.2 and 1.9 m (Figure 5.4). The gradings were analysed statistically. Soils at depths of 100–125 cm were heavy textured, with a ratio between the size of sieve allowing 75% percent of the material to pass (the 75% quartile) and the size allowing 25% to pass (the 25% quartile) between 1.7 and 3.4. The ratio is known as the "bandwidth ratio". Soils deeper down were lighter textured with bandwidth ratios of around 1.5. Figure 5.4 shows little difference between bandwidths at the various depths. For this reason, final bandwidths are based on the 25 and 75% quartiles of all soil samples taken at different depths. The fine boundary of the soils at a depth 100–125 cm at $d_{90}, d_{85}, \ldots d_{15}$ (the 25% quartile values) and the coarse boundary of the soils of the 175–200 cm range (the 75% quartile values) were used to establish a fine and coarse boundary for the final bandwidth (the grey band).

Broad bandwidths with coarse/fine boundary ratios >10 will not result in satisfactory drain envelope design (see section 8.5, Table 8.4).

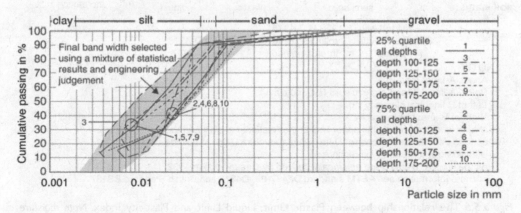

Figure 5.4 Representative soil particle size bandwidth

5.4.2   Plasticity index

Soil consistency is an expression for the plasticity of the soil and as such its resistance to mechanical deformation and rupture. The state of plasticity of a soil is mostly determined by its clay and its moisture content and may be expressed by determining the *Atterberg consistency limits*. For drainage, the most important of these limits is the Plastic Limit (PL), Figure 5.5. The PL may be determined by a simple hand kneading/rolling test (also called the thread method).

Fine textured soils are easily moulded without breaking up when their moisture content is above PL, and the soil is in the so-called *plastic state of consistency*. In this state the soil easily smears, puddles or compacts when driven or walked on by cattle or when the soil is tilled, resulting in a poor soil structure for plant growth. For most fine textured soils the Field Capacity (FC) is greater than the Plastic Limit (PL) which means that after drainage, further evaporative drying is required to reach the workable stage (Figure 5.6). This takes longer the further PL is on the dry side of FC. The FC-PL relationship of a soil may thus be used to assess the scope for drainage to improve soil workability.[4]

Soil management, however, also has an influence. Smearing and compaction of the soil resulting e.g., from working the soil under too wet conditions, destroys macropores and as

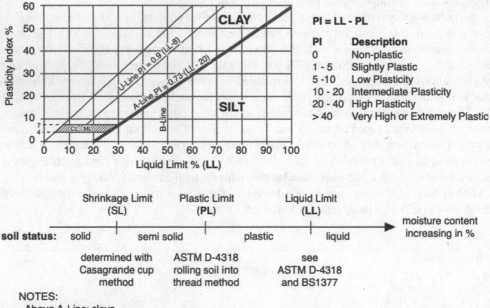

NOTES:
- Above A-Line: clays
- Below A-line silts and organic soils (silt and clays)
- Left of B-line: low plasticity
- Right of B-line: high plasticity
- U-Line is approximately the upper limit of the relationship of PI and the LL for any soil used to derive this relationship. Data falling above the U-line should be considered as likely in error and should be rechecked
- All the lines A, B and U are *empirically* derived by Casagrande in 1932.
For more information see ASTM standard D2487-93, D4318 and British Standard BS1377.

Figure 5.5 The relationship between Plastic Limit, Liquid Limit and Plasticity Index. *Note moisture content from dry to wet*

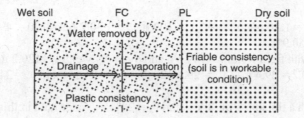

Figure 5.6 The relationship between PL and FC and their effect on soil workability. *Note moisture content from wet to dry*

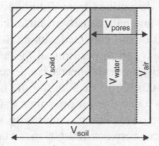

Figure 5.7 Bulk density and soil moisture content

such tends to increase FC. Addition of organic matter to the soil tends to increase the PL value of the soil. The first activity thus has a negative effect on soil workability; the second a positive effect.

5.4.3 Bulk density and soil moisture content

On the basis of Figure 5.7 (showing the solid, pores, water and air phases of the soil), the following soil constants and parameters may be defined: V = volume in cm^3, W = weight in g, ρ = density in g/cm^3, BD = bulk density in g/cm^3, θ = soil moisture content by volume or by weight in %.

$porosity = V_{pores}/V_{soil}$ Eq. 5.1

where
$V_{soil} = V_{solid} + V_{pores}$
$V_{pores} = V_{water} + V_{air}$
$W_{dry\ soil} = V_{solid} \times \rho_{solid}$ (particle density = 2.65 g/cm^3)
$W_{water} = V_{water} \times \rho_{water}$ (density of water = 1.0 g/cm^3)

Bulk Density BD = $W_{dry\ soil}/V_{soil}$ Eq. 5.2
Water content by weight $\theta_w = (W_{water}/W_{dry\ soil}) \times 100\%$ Eq. 5.3
Water content by volume $\theta_v = (V_{water}/V_{soil}) \times 100\%$ Eq. 5.4

note that $\theta_v = \theta_w \times BD$

The above relationships may be used to perform the following calculations:

1) Soil with BD = 1.30 g/cm^3 → V_{solid} = BD/ρ_{solid} = 1.30/2.65 × 100 = 49.1 % and V_{pores} = 100 – 49.1 = 50.9%

2) Soil sample with field weight = 149.6 g and oven dry weight = 120.2 g → water content is 149.6 – 120.2 = 29.4 g → θ_w = 29.4/120.2 × 100 = 24.5% and θ_v = 24.5 × BD = 24.5 × 1.30 = 31.9%

3) Soil layer with a thickness of 30 cm and θ_v = 24% → water held in this layer, expressed in equivalent depth: 300 × 0.24 = 72 mm.

5.4.4 Sample quantity and density

Soil texture and plasticity may be determined from disturbed soil samples, but the determination of hydraulic conductivity and bulk density requires undisturbed soil samples (see section 7.1). Generally, the amount of soil collected with an 8-10 cm diameter auger over a 30 cm length will provide enough soil to perform all the necessary analyses required. The sampling density depends on the particular needs, but generally a density of one measurement per 10-25 ha (grids of 300x300 m or 500x500 m) is commonly used (see also Table 7.4). Gallichand et al. (1992) found that for preliminary surveys a grid of 900x900 m would provide adequate information on hydraulic conductivity, while optimum results were obtained with grids that had distances between 400-600 m. Detailed soil analysis for drainage design in Pakistan was performed by using a 150x150 m grid.

5.4.5 Data requirement for drain envelope design

The following data need to be collected:

*   sieve analysis results with typical particle sizes at certain percentage cumulative passing selected sieves
*   Plasticity Index (PI) values, as an indicator of soil stability and to determine the Hydraulic Failure Gradient (HFG, see section 8.4.1) without the need for direct laboratory determination
*   the soil texture analysis using the hydrometer method (sand, silt and clay percentages)
*   chemical analysis to determine the susceptibility of soils to dispersion (deflocculate): EC_e, EC_w, SAR, etc. (see section 14.2.3)
*   iron ochre, calcium carbonate, sulphur and manganese content in the soil
*   the saturated hydraulic conductivity (K_s) at drain depth. If the hole from which the soil samples have been taken for sieve analysis etc. remains open until the next day, the same hole can be used for the auger hole method (see section 7.2.1) for the determination of the Saturated Hydraulic Conductivity. Instability of the auger hole is a first indicator of potential construction problems and the need for an envelope
*   the depth to the impermeable layer. This information, which is part of the standard pre-drainage investigation, is needed to determine the hydraulic gradients to be expected at the drain and to compare these with the HFG. The accuracy of the depth to the impermeable layer is less critical then the determination of the hydraulic conductivity.

For certain assessments (e.g., the critical gradient), the porosity and the bulk density of the soil may be required. In this case, undisturbed samples need to be taken. Data may be readily available from pre-project investigations and can be used, or if laboratory tests have been performed with typical soils, indicator values may be obtained from these.

5.5 Watertable and groundwater

Watertable surveys provide valuable information on the subsurface drainage conditions in the area. Watertable levels reflect the prevailing balance between the different groundwater recharge/discharge components. As the balance changes, so does the watertable level. When the watertable is permanently or seasonally too close to the soil surface, control by subsurface drainage systems may be required.

5.5.1 Watertable observation wells

Watertable levels (phreatic levels) may be measured in boreholes reaching deep into the groundwater (see Figure 6.2). In an area-wide survey, these observation wells may be placed in a grid or another regular pattern, although the (conceived) groundwater flow pattern in the area should also be taken into account. Groundwater flow will generally be down the slope, towards depressions and (natural) drains. A few judiciously located observation wells may provide as representative a picture of the phreatic surface as that obtained using a dense grid system.

Water levels in (natural or excavated) wells and open drains are also closely related to the *phreatic surface*. During periods of discharge, these levels will tend to be lower than the watertables in the adjoining land. Only when there is static equilibrium (several days after rain) are these levels more or less the same. Well and canal levels may also be higher than the watertable in the adjoining land, e.g., when the well or canal walls have become sealed by sediment in the water.

In watertable surveys, the main interest is primarily in the long-term, seasonal variations (short term fluctuations after rain or irrigation may be studied separately, usually in connection with the functioning of a drainage system). The seasonal trends in watertable levels are often closely related to the rainfall or irrigation regime in the area as well as the water levels in open drains (as was observed in the Fourth Drainage Project in Pakistan, Vlotman et al. 1993 and 1994) and may be identified adequately by observations at a few key sites on key occasions (selected on the basis of a few previously undertaken detailed studies). For most drainage surveys, simple observation wells consisting of a 2.5-5 cm diameter perforated pipe installed in a 8-10 cm bore hole (with the annular[5] space filled with sand) will generally suffice. For long-term observations, more sturdy construction is warranted.

The prevailing seasonal watertable regime in alluvial soils can often also be deduced fairly accurately from the hydromorphic characteristics of the soil profile. The soil below the average low watertable (= average summer/dry season watertable level) typically has a uniform light grey to dark blue colour associated with permanent non-aerated soil. The soil above the average high watertable (= average rainy/winter season level) is typically uniformly brown to grey-brown. In between a *mottled gley-zone* occurs, resulting from the intermittent oxidation and reduction of iron and other elements. The upper and lower gley levels can usually be identified in a soil profile (after some initial guidance on the basis of available watertable records) and these recorded in soil surveys provide valuable information on the current watertable regime in the soil. Hydromorphic characteristics in the soil only change slowly and it should be confirmed that the observed *gley features* reflect the present groundwater regime (fossil gley can occur in alluvial soils well below the present low watertable levels, dating from a previous sedimentation and water regime period).

5.5.2 Piezometric studies

Watertable wells provide no information on potential differences within the groundwater body. Where such differences are expected to occur, piezometers should be used instead of watertable wells. Piezometers measure the pressure in a water body at the point where its filter is placed (the last 5-10 cm of the piezometer tube, see Box 5.1 and Figure 6.2).

A typical use of piezometers is shown in Figure 5.8a. A battery of piezometers, 1-2 m apart, is installed at different depths.[6] The readings in the piezometers (1) and (2) show that the pressure distribution with depth in the highly permeable layer III is hydrostatic, indicating that there is no flow in a vertical direction in this layer (or the flow is too small to generate any significant head loss in this highly permeable soil). Piezometers (2) and (3), however, show that there is an upward flow from layer III to the rootzone (generally termed *upward seepage* flow or in this specific case also *leakage* from an underlying semi-confined aquifer). The rate of this flow may be determined using Eq. 12.1: $q = K_{II}(h/D)$. This leakage adds to the regular drainage load due to rainfall or irrigation.

Piezometers can also be used to diagnose problems of impeded infiltration and impeded percolation (Figure 5.8b). By installing a piezometer arrangement as shown in Figure 5.8a (a nested set of piezometers), it can be established whether the occurrence of water on the soil surface, or of excess wetness at certain depths in the soil is due to high watertables, impeding layers or infiltration problems at the soil surface.

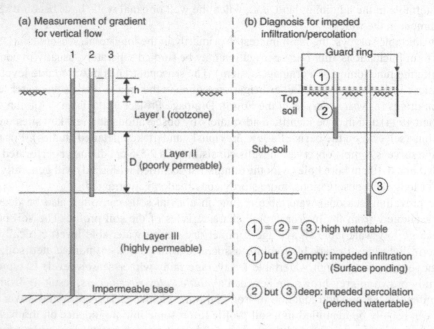

Figure 5.8 Typical use of piezometers

Box 5.1   Simple piezometer suitable for drainage studies

Made from common plastics electrical conduit with an inside diameter of about 2 cm, closed at the end by melting or by means of a stopper.

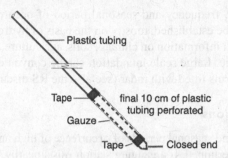

The final 10 cm of the lower end is perforated with some ten holes of 2mm diameter. This part is covered with durable gauze (nylon stocking will do), taped to the tube. The piezometer is installed in an 8 cm diameter augerhole. The initial backfilling is with sand (surrounding the whole of the perforated lower end), followed by bentonite powder that will swell on becoming wet to form a sealing ring. Further backfilling may then be done with normal soil.

5.5.3   Groundwater sampling

For most routine groundwater quality assessment and monitoring work, simple retrieval techniques (bailing or pumping from an augerhole or well) will generally be quite acceptable. Suction techniques can only be applied down to depths of about 9 m. Stagnant water should generally be refreshed at least twice and inflow equilibrium be re-established before sampling. While generally a larger sample will be retrieved, most analyses require < 0.5 l of water. Sample containers should not be contaminated and be rinsed with the water to be sampled before a sample is extracted.

The applied retrieval procedures should take account of the fact that the chemical composition of the groundwater may vary with depth. For example, when the groundwater has recently been recharged by rainfall or by irrigation water losses, the salinity in the upper fringe may, temporarily, be somewhat less than in the deeper layers. In the upper 5-10 m of the phreatic groundwater, which is of most relevance to agricultural drainage, these variations are however mostly quite small and gradual. Samples taken from a shallow augerhole should generally be fairly representative for conditions around the watertable while samples from a deeper hole would reflect the average composition over the installed depth. Stratified sampling (taking samples at specific depths) in an augerhole or well is almost always flawed as during the sampling, mixing cannot be prevented. When detailed insight into the variation with depth is required, sample retrieval is best done using piezometers installed at different depths (see Figure 5.8a; a nested set of piezometers).

5.6   Hydrology and geohydrology

In these two areas of expertise the following aspects should be investigated, both for diagnosis and for planning purposes:

5.6.1   Runoff and flooding

The origin of floodwater, frequency and seasonal period of occurrence, affected area and resultant damage should be established, mostly on the basis of hydrological and topographical surveys. Together with information on climate, soils, agriculture, etc., they form the basis for a flood control scheme. Large scale inundation can be conveniently mapped by remote sensing (RS) from platforms fitted with radar (see also the RS discussion in section 3.2).

5.6.2   Outlet conditions

The duration, frequency and seasonal periods of occurrence of high and low water levels at the outlet point should be determined. If a gauging station exists nearby, available data should be compiled into stage-frequency-duration relationships for selected calendar periods. Where such a station does not exist, data from a more remote station with a similar character may be used. The stage heights recorded at such a station may be transposed to the outlet point either by:

- hydraulic calculations, using the Manning/Strickler formula for canal flow (with measured or estimated roughness values)
- carrying out a short measurement programme, correlating the corresponding stage heights at the two sites.

Where data from gauging stations are unavailable, a new station may be established at the selected outlet point. A short measuring period of 1-2 years may yield sufficient useful information while extreme conditions not covered by this period may be assessed by correlative analysis. For instance, one could correlate the stage height and/or discharges recorded in the 1-2 years measuring period to the rainfall data of this period, the latter forming part of a long series (10-20 years). Alternatively, one could estimate discharges based on rainfall-runoff relationships and then determine stage heights by either extrapolating the short period rating curve or by hydraulic calculations. The procedures described refer to outlets into a river, but rather similar procedures apply to outlets into a lake or the sea (see section 12.3).

5.6.3   Geohydrological conditions

Satisfactory diagnosis and solution of subsurface drainage problems will often require information on the substrata beyond the depth investigated in normal soil survey work. The latter investigations do not as a rule extend beyond some 2-5 m depth whereas groundwater flow to drains may extend down to depths of 1/6-1/4 L below the drainage base where L is the drain spacing (see also section 7.5). Where widely spaced drains are used, but also in many other subsurface drainage situations, the normal soil investigations should be supplemented by geohydrological investigations, collecting information on:

- the overall groundwater flow pattern in the area (NB a network of watertable observation wells could yield useful information as to seepage and recharge areas.)

- the hydraulic characteristics of the substrata (e.g., to some 10 m below the drainage base for L = 50 m when no shallower impermeable layer occurs)
- piezometric levels and gradients at different depths and sites
- patterns and rates of natural drainage and seepage flow
- groundwater salinity at different depths.

Deep geohydrological investigations (beyond 5-10 m) require special equipment and experience and as such are often contracted to specialised firms.

5.7   Agriculture and irrigation

The agricultural investigations should:

a) establish the land use alternatives for the area, taking into account the present drainage conditions and the technical/economic feasibility of improved drainage (see section 1.9)
b) help to formulate the drainage requirements on the basis of these land use alternatives and the prevailing soil and climatic conditions
c) help to assess the anticipated benefits from drainage (implicit in the objectives a) and b) above but needed explicitly to carry out economic evaluations as described in Chapter 2).

Investigations to be carried out would normally include:

- survey of present land use and farm practices, incl. type of irrigation system
- survey of damage done by excess water/salinity
- various crop and farm management studies
- interviews with farmers and other relevant stakeholders.

As pointed out in section 1.9, the relationships between drainage and farming are complicated and even experimental results are of limited value since the actual conditions may differ in almost any way from the experimental conditions. An in-depth understanding of the local agricultural conditions in combination with a sound analysis of the prevailing soil and climatic conditions, offers the best basis and hope of arriving at the correct drainage requirements and a sound drainage plan. The collection of the views and experiences of the farmers, residents, officials and other local knowledge is almost always indispensable to gain this insight.

For drainage of irrigated land additional studies should be undertaken to acquire a sound understanding of the existing/planned water distribution systems, operational rules and farm irrigation practices. Water losses of different types and at various system levels as well as options for reducing the losses should be assessed (see also Chapter 16).

5.8   Pilot areas and field testing

Since the design requirements and the impact of drainage systems can often not be fully assessed in advance, it is recommended for countries and areas with limited drainage experience, to establish one or more pilot areas before embarking on a large-scale drainage investment program. This applies especially to cases where a new subsurface drainage technology like pipe drainage, is to be introduced (most of the discussion refers to this case).

Experience from other countries/areas, expert advice, modelling, etc. all can help to guide new developments but cannot entirely replace onsite observations of the functioning of the new systems and observed yield and other responses. Pilot areas can in particular help to provide guidance on:

- appropriate technology, design and materials
- expected yield and other relevant responses
- best implementation, construction and O&M practices
- construction costs, operational costs and monetary benefits.

However, in order the test the limits of under and over design deliberate failure of some of the experiments may need to be built-in with appropriate compensation for the farmers involved.

At the pilot area, typically a selected number of alternative technologies, designs, drainage materials and construction and management practices are compared and tested. This may e.g., involve pipe drainage vs. mole drainage, different drain depths and spacing, need for and type of envelope material, manual vs. machine installation (including trench vs. trenchless), controlled vs. uncontrolled drainage. These comparisons and tests can help to assess the validity of the drain spacing equations and other drainage theories. Pilot areas can, however, also help to create awareness of the need for and the benefits of improved drainage, to build up national or regional research capacity and to train the technical staff involved in the construction and O&M of drainage projects. Pilot areas can also be used to hold field days for extension personnel and farmers.

5.8.1   Types of pilot areas

Different types of pilot areas may be distinguished. A first distinction is between *pioneer* type pilot areas (to provide guidance at the initial stage of drainage development in the area) and *second stage* type pilot areas (to work out the details after first options have been assessed and first choices have been made). Another distinction is between pilot areas conducted under *controlled conditions* or under *farmers conditions* whereby in the first case the land use and all other relevant operational decisions are controlled by the pilot area management while in the latter case these decisions are left to the farmers (in line with the terminology of the conventional agricultural research, controlled condition type pilot areas may be more appropriately classified as *drainage experiments*). Pilot areas managed under controlled conditions may provide more straightforward scientific results, but these results may not always be representative of the real world.

Operational pilot areas are typically for the second stage when the findings of the initial stage pilot projects need to be made operational for large scale application. While with the initial stage pilot projects, the focus would be on establishing the critical physical project parameters (technology, design criteria), the focus in the second stage would be on the implementation and on the future O&M of the installed drainage works. The subsurface drainage development in the State of Haryana in NW India constitutes a good example of the above described approach and relationships. It began with three small drainage experiments (» 10-25 ha each) conducted by the CSSRI (Central Soil Salinity Research Institute) to establish the principal design parameters (drainage materials, drainage coefficient, drain depth and spacing, confirmation of applied theories, installation techniques, etc.). This phase was followed by an 1 000 ha operational pilot project in which the mechanics and logistics of mechanical installation of pipe drainage systems, the requirements of the desired participatory development approach and the real world socio-economics of pipe drainage were established. The State has been working on establishing the institutional framework for the development, implementation, financing and management of large-scale pipe drainage investment projects.

The Drainage Research Institute in El-Kanater, Egypt, has several second stage pilot areas in the Nile Delta (i.e., Haress, Mashtul and Mit Kenana). These areas were established around 1994 and are still operational and have received support over the years from the Netherlands, Japan and most recently from Australia. Experiments with controlled drainage (Figure 20.3) and V-plough drain installation (Figure 8.20) have been conducted.

5.8.2 Analysis of results of pilot areas

The standard type of experimental layout and statistical analysis as e.g., used for agricultural experiments (Latin squares and other factorial layouts to evaluate the impact of fertiliser, to evaluate yield performance of new species, etc.) are generally not applicable to drainage experiments. In most pilot areas, the number of alternatives (treatments) and repetitions that can be feasibly accommodated in almost all cases is far below the statistical requirements (a statistically sound experiment with three spacing treatments (L = 30, 60 and 90 m) and two installation depths (W = 1.50 and 2.0 m) would require a pilot area of some 300-400 ha). Results of controlled experiments offer the best hope for statistical analysis but even here the number of variables and inter-actions is considerable. Most pilot areas do not provide and are not meant to provide statistically significant results. The findings are indicative only, to be interpreted and extrapolated on the basis of sound judgement.

5.8.3 Visual drainage need assessment

An interesting design process has been implemented by Tuohy et al. 2016a. The method is based on a Visual Drainage (soil) Assessment method (VDA) whereby an approximation of the permeability of specific soil horizons is made using seven indicators (water seepage, pan layers, texture, porosity, consistency, stone content and root development) to provide a basis for the design of a site-specific drainage system for grassland/pasture drainage. The incentive was the ability to design a suitable system for each of the stakeholders at the lowest possible costs. Soil pits were dug at each of the six sites and a scoring system was developed for the seven indicators (Figure 5.9). Water seepage and the presence of pan layers were weighted by 10 in the scoring system, texture by 4 and the other parameters by 1. Porosity and consistency could score 0, 1 or 2, while the other indicators scored 0 or 1. The resulting total VDA score was < 5 for poorly permeable soils, 5-10 for moderately permeable horizons and > 10 for highly permeable soils. The VDA-based design was compared with an ideal design (utilising soil physical measurements to elucidate soil hydraulic parameters) and a commonly used standard design (0.8 m deep drains at a 15 m spacing) by model estimate of watertable control and rainfall recharge/drain discharge capacity. Data were analysed using ANOVA with VDA permeabilty classification as a fixed effect. The flow chart in Figure 5.9 was used to select the appropriate site-specific drainage system for grassland of dairy farms in Ireland. The comparison of drainage design methods showed that the VDA based design performed as well as the ideal system and considerably better than the standard system.

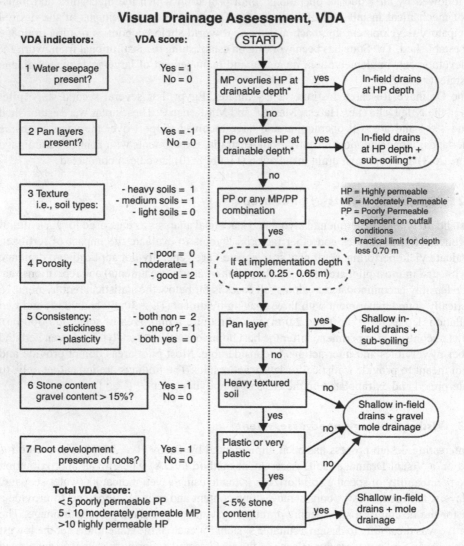

Figure 5.9 Flow chart to determine type of drainage system to be considered with VDA permeability classification (after Tuohy et al. 2016a)

5.8.4 Statistical analysis

To compare the results of various indicators of different experimental sites and to compare the results of pilot areas various statistical methods are available:

Coefficient of Variation (CV): this is a simple calculation of the relative magnitude of the average value of a data-series with the aims to indicate how representative the average is to represent a data set (see section 25.3.1).

Nash-Sutcliffe Efficiency (NSE) coefficient: the NSE is a measure to assess how well a computer program simulates reality. The NSE assesses the predictive power of a hydrological model and its value can vary from minus infinity to 1; with 1 indicating a perfect fit. It is calculated from:

$$NSE = 1 - \frac{\sum_1^n (O_i - P_i)^2}{\sum_1^n (O_i - \bar{O}_i)^2}$$ Eq. 5.5

The percent bias or error (PE): The PE value can vary from minus infinity to positive infinity. A negative value indicates under-prediction and a positive value overprediction. It may be calculated from:

$$PE = \frac{\sum_1^n P_i - \sum_1^n O_i}{\sum_1^n O_i} * 100$$ Eq. 5.6

The correlation coefficient (R^2): The value of R^2 can vary from zero to 1, with 1 indicating a perfect linear relationship between the observed and simulated values. It is calculated as follows:

$$R^2 = \sqrt{\frac{\sum_1^n (O_i P_i - n\bar{O}\bar{P})^2}{(\sum_1^n (O_i^2 - n\bar{O}^2))(\sum_1^n (P_i^2 - n\bar{P}^2))}}$$ Eq. 5.7

O_i is the daily observed value, P_i is the daily simulated value, $\bar{O}_i$ is the average of observed values, $\bar{P}$ is the average of simulated values and n is the number of observed values (Saadat et al. 2018).

The goodness of fit statistics can be used to evaluate drain flow estimation from watertable height above a drain or to compare the relationship between drain flows from different experimental plots.

5.9 Environmental impact

Avoidance and amelioration of adverse environmental impacts of drainage measures starts at the project planning and preparation stage. At this stage, the possible critical changes in hydrological regime and land use should be identified and the necessary information collected for assessing the environmental impact of these changes. The assessed changes would generally refer to:

Waterlogging and Land Use: improved drainage may be expected to lower watertables, reduce surface ponding and eliminate various other forms of waterlogging. These changes may also be expected to lead to considerable changes in land use. Ecological conditions will change, and valuable flora and fauna may disappear. Existing wetland areas, even when not fully reclaimed, are likely to be affected. These changes may not be restricted to the drained area but also affect some of the surrounding area.

Disposal: drainage disposal may change the hydrological regime and water quality of the receiving streams or other water bodies. Changes in high and low flows may influence ecological and morphological regimes/conditions in the river/lake beds and adjoining flood plains. Drainage effluent carrying nutrients and toxic elements may greatly affect the aquatic life in the receiving water bodies, which in turn may affect the dependent habitat in the area. The changes may extend into adjoining riparian lands and to downstream areas (salt intrusion and coastal erosion).

Positive impacts: improved land drainage may help to combat various forms of land degradation by the control of erosion and sedimentation, by the reclamation/prevention of irrigation induced land salinisation. Improved drainage may also help to combat waterborne diseases, to improve rural sanitation and to protect rural infrastructure. Drainage may also, by design or default, create new ecological regimes and enhance biodiversity. Drainage can lead to improve social well-being of stakeholders.

Almost all of the aforementioned investigations (soil conditions, groundwater, hydrology, agriculture, etc.) will provide useful information for the environmental assessment. They may however have to be supplemented by specific ecological and biological investigations. In a wider interpretation of environmental assessment, impact on human well-being, impact on women and minority groups, impact on poverty, settlement, cost recovery, etc. would also be considered, adding a distinct socio-economic impact and sustainability dimension to the assessment.

5.9.1   Environmental impact assessment

For large-scale projects in ecologically sensitive areas, it may be obligatory that an *Environmental Impact Assessment* (EIA) be conducted in which the possible environmental impacts are assessed on the basis of a prescribed methodology and standards. An EIA would investigate how environmental damage can be avoided/mitigated and how drainage measures could possible help to enhance local/regional environmental values. Properly used, EIA's may evolve from a narrow obligatory requirement for project approval into a wider tool for the design of a more valuable project for society. For further guidance on EIA objectives and procedures, reference is made to FAO 1995a.

5.9.2   Miscellaneous investigations

Various infra-structural features may influence the drainage plan by imposing specific drainage requirements. Outside interests may also be affected by the improved drainage. Administrative and legal aspects need to be studied and rights of way established. Information on local construction conditions need to be collected (available machinery, materials and skills, labour situation, accessibility and workability conditions, unit costs, etc.).

Notes

1 In dry climates calculations may start at the end of the dry season when it may be safely assumed that the soil moisture storage is fully depleted (S = 0).
2 USDA – United States Department of Agriculture (www.usda.gov).
   USCS – Unified Soil Classification System (USBR 1985, ASTM D 2487-93).
   ASTM – American Society for Testing and Materials, West Conshohocken, PA, USA.
   AASHTO – American Association of State Highway and Transportation Officials, USA.
   SRWSC – State Rivers and Water Supply Commission, Victoria, Australia.

3 USBR – United States Bureau of Reclamation (www.usbr.gov).
4 *Workability = tillability + trafficability;* although criteria for soil tillage and vehicular movement over the soil differ, present knowledge does not justify such a refinement in criteria.
5 Pertaining to, or having the form of, a ring; ring-shaped.
6 Where there is also horizontal flow, the line of piezometers should be aligned at right angles to the direction of this flow to avoid horizontal head losses over the 1-2 m distance between the piezometers being falsely interpreted as vertical head losses. Also referred to as nested piezometers. They may be installed as close as practicable from installation point of view.

Chapter 6

Water in the soil

The design of an agricultural drainage system requires a good understanding of the occurrence, nature and movement of water in the soil as well as of the drainage-related hydrological processes. These subjects are described in this Chapter.

6.1 Forms and nature of occurrence of water in the soil

Figure 6.1 illustrates the different forms of occurrence of water in the soil.

Groundwater

Groundwater fills the pores of the permeable strata in the earth's upper crust *(aquifers)*. The groundwater in these strata may be under normal static or dynamic pressure or it may be subjected to an over-pressure. The latter may occur when an aquifer is overlain by a poorly permeable layer. Such aquifers are termed *confined aquifers*. Aquifers of different types may occur at varying depths, over or underlying each other. The free groundwater found in or directly below the soil layers is called *phreatic* groundwater and its surface is termed *phreatic level* or *watertable*.

Most land is underlain at some depth by phreatic groundwater. When it occurs deep below the soil surface (say at 25-50 m) it is of little direct concern to agricultural drainage. This condition exists across much of the land in arid/semi-arid climates and also in fairly humid climates when permeable, freely draining strata underlie the land, located high above the local subsurface drainage base. Groundwater at shallow depths (watertables <5-10 m below the soil surface) has a much greater relevance to agricultural drainage. This situation is to be expected where the groundwater recharge is high (high rainfall or high irrigation losses, seepage, etc.) or the subsurface drainage is slow (poorly permeable substrata, poorly developed drainage system, high drainage base, etc.).

Watertable (phreatic level)

The watertable marks the division between the groundwater and the moisture zones in the soil. Its location is found by sinking a borehole into the groundwater body. Water from the surrounding soil will flow into the hole and fill it to the watertable level (Figure 6.2). For regular watertable depth measurements, the borehole may be fitted with a perforated pipe in order to give the hole a degree of permanence.

When a poorly permeable layer impedes the deep percolation, a so-called *perched watertable* may develop (Figure 6.2b). Its occurrence is temporary as the impeded water will continue to seep through to deeper layers or drain laterally. It may exist long enough to cause a

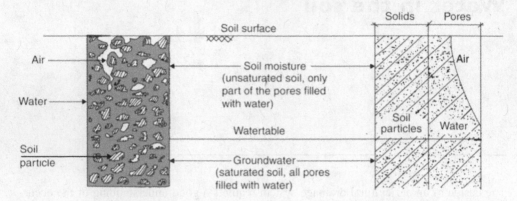

Figure 6.1 Forms of occurrence of soil water

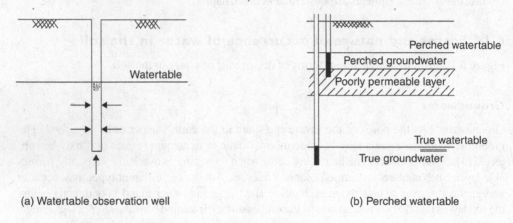

(a) Watertable observation well   (b) Perched watertable

Figure 6.2 Watertables in the soil

serious problem of excess water in the soil. Perched watertables may be detected by drilling a borehole into (but not through) the impeding layer.

Soil moisture

In the unsaturated soil above the watertable the pores are partly occupied by water and partly by air. The soil water in this zone is commonly referred to as *soil moisture*. The amount of soil moisture varies greatly with depth and in time. The soil moisture profile as depicted in Figure 6.1 would be typical of a soil with a fairly high watertable (within 3 m from the soil surface) shortly after a prolonged period of rainfall. In a narrow zone above the watertable, pores fill by capillary rise from the groundwater. In the lower part of this capillary zone all pores are filled with water, making the soil in this so-called *capillary fringe* as saturated as in the groundwater below the watertable. Nevertheless, the water in the capillary fringe is under negative pressure while the pressure in the groundwater is positive (section 6.2). Above the capillary zone, pores fill with water mostly by retaining part of the percolating rain (or irrigation) water.

When the groundwater is very deep (> 10-20 m below the soil surface) the soil may be moist in the upper layers (retained rain or irrigation water) and in a zone immediately above the watertable (capillary zone) while, in between, the soil may be much drier. The soil moisture content in the upper soil layer (down to 0.5-1.0 m) is particularly variable, mainly due to variations in daily weather conditions (especially rainfall variations). Deeper down, variations occur over a longer term in parallel with seasonal weather variations.

6.2   Pressures in the soil water

Pressures are forces per unit area (symbol P). In the SI system *(Système International)* the unit of force is the Newton (= force needed to give 1 kg mass an acceleration of 1 m/s²); the unit of pressure is the Pascal = Newton per m². In a (soil) water system, pressures are often more conveniently expressed as an equivalent water gauge reading (hydraulic head or just head):

$$P = H \rho_w g \qquad\qquad\qquad\qquad \text{Eq. 6.1}$$

where:
P = pressure in Pascal, Pa
H = pressure expressed as meter head, m
ρ_w = density of water (~ 10^3 kg/m³ at normal soil temperatures)
g = gravitational acceleration (~ 9.81 m/s²)

For large pressures, the bar and the atmosphere are useful units: 1 bar = 10^5 Pascal and 1 atm = 1.0132 bar or roughly 1 bar = 1 atm = 10 m head. Negative pressures in the soil moisture are also expressed by their pF-value where pF = log |P|, with P in cm head.

Pressures below the watertable

At the watertable level, the soil water is always at atmospheric pressure (P = P_{atm}). The forces prevailing in the groundwater below the watertable are the normal forces encountered in a standing or moving body of water. Flow velocities in the groundwater are always too low to generate any significant kinetic forces. Normal positive hydrostatic pressures prevail (P > 0 with P_{atm} = 0 as reference) which may conveniently be measured with a piezometer (Figure 6.3).

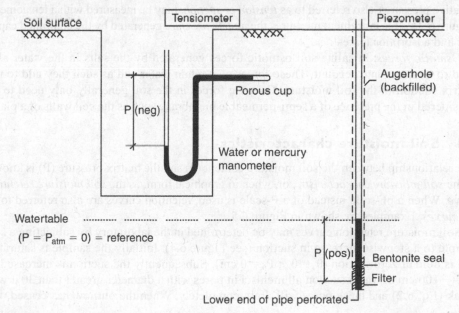

Figure 6.3 Pressure measurements in the soil water

Pressures above the watertable

In the unsaturated soil above the watertable, two types of forces prevail: capillary forces and adsorption forces.

Capillary forces: these are essentially surface tension forces, in this case activated by the adhesion between water and the soil and by the fineness of the pores. In unsaturated soils, capillary forces manifest themselves by the formation of curved air-water interfaces *(menisci)*. The water pressure beneath the meniscus is negative, becoming increasingly so with increasing curvature of the meniscus as shown in:

$$P_{cap} = \frac{-2\tau}{R \, \rho_w \, g}$$
Eq. 6.2

where:
P_{cap} = water pressure beneath meniscus (m head)
τ = surface tension (0.073 N/m for water at 15°C)
R = radius of curvature of the meniscus (m)
ρ_w = density of the water ($\sim 10^3$ kg/m³)
g = gravitational acceleration ($\sim$ 9.81 m/s²)

In a pore, the curvature of a meniscus attains a maximum value when its radius equals the radius of the pore, in which case $P_{cap} = -4 \, \tau/D\rho_w g$ where D = diameter of the pore. Such a meniscus can hold a water column of length equal to P_{cap}.

Adsorption forces: these include van der Waal's and electrostatic forces exerted on the water by the charged colloidal surfaces of the soil particles. These forces exercise increasing influence with increasing colloidal surface area within the soil (small in sand but large in clay).

Both the capillary and the adsorption forces bind the soil moisture to the soil particles making up the soil skeleton *(soil matrix),* thereby retaining it above the watertable against the gravitational downward pull. Pressures in the soil moisture are negative ($P < P_{atm}$). These negative pressures, also referred to as *tension* or *suction,* may be measured with a tensiometer (Figure 6.3). This instrument measures the *matrix pressure* generated by the combined capillary and adsorption forces[1].

Osmotic forces: In saline soil osmotic forces generated by the salts in the water also need to be taken into account. These forces also retain water and as such they add to the matrix suction of the soil moisture. Osmotic forces in the soil generally only need to be considered in the presence of a semi-permeable membrane such as the cell walls of a plant.

6.3   Soil moisture characteristics

The relationship between the soil moisture content (θ) and the matrix pressure (P) is known as the *soil moisture characteristic,* or, when in graphical form, as the *soil moisture retention curve.* When a pF-scale instead of a P-scale is used, retention curves are also referred to as *pF-curves.* Examples are shown in Figure 6.5.

Soil moisture retention curves may be determined in the laboratory by subjecting a soil sample to a stepwise increase in suction (see Figure 6.4). Initially the sample is saturated and is held at zero suction ($\theta_0 = \theta_{sat}$; $P_0 = 0$ cm). Subsequently the suction is increased to $P_1 = -100$ cm at which suction all menisci in pores with a diameter greater than 30 μ will break (Eq. 6.2) and these pores will drain completely. When the outflow has ceased, the

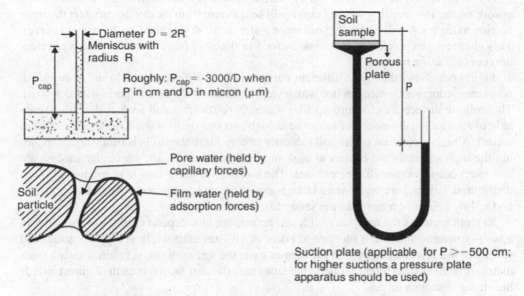

Figure 6.4 Soil moisture retention mechanism and measuring apparatus

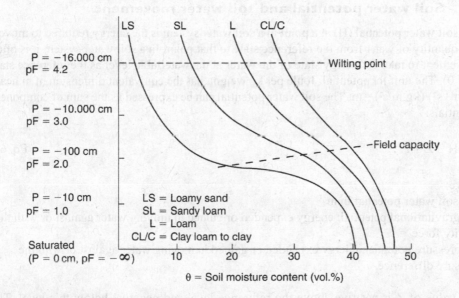

Figure 6.5 Soil moisture retention curves

moisture content of the sample is determined (θ_t). This value represents the moisture held in the still finer pores and as film water. As the suction increases, increasingly finer pores will drain until eventually even some of the film water (held mostly at strong suctions) is removed. The corresponding suction and moisture content values are plotted enabling the soil moisture retention curve to be drawn.

A P-θ curve may also be obtained by starting with a dry soil sample and letting the soil absorb water. The resulting wetting curve will be different from the drying curve (at the same suction value a soil will normally hold more water on the drying than on the wetting curve). This phenomenon is referred to as *hysteresis*. For drainage purposes the drying type retention curve is more relevant.

Figure 6.5 shows that quite different curves apply to different soils. In the loamy sand, pores are predominantly medium fine which yield to suctions > 70–80 cm (pores with D > 40 μ). The colloid surface area, adsorbing film water, is relatively small so that the soil is only able to retain a small amount of moisture at high suctions (most water held at pF > 3 is film water). A high proportion of the soil moisture in clay loam/clay soils is film water, explaining the high soil moisture content at high suction in these soils. The curves for sandy loam and loam occupy intermediate positions. The moisture in these soils is proportionally well distributed, held as pore water and as film water, the loam having somewhat more film water and tightly held pore water than the sandy loam.

Retention curves not only vary with soil texture but also depend on the soil structure. In a well-structured soil, a fair number of pores of all sizes exist while in a poorly structured soil there are few macropores. In the former case, the soil moisture is retained over a wide distribution of suction forces while in the latter case the distribution is centred more towards the stronger suction forces.

6.4   Soil water potential and soil water movement

The soil water potential (H) at a point in a soil water system is the energy required to move a unit quantity of water from the reference state to that point. In a soil water system it is often convenient to take the energy state of the water at the watertable level as the reference state (H = 0). The unit for potential, Joule per kg weight, has the equivalent dimension of m head: $(kg.m^2/s^2)/(kg.m/s^2) = m$. The soil water potential can be expressed as the sum of component potentials:

$$H = Z + P$$
Eq. 6.3

where:
H = soil water potential in m
Z = gravitational potential: energy expended or gained in moving water against or with the gravity force
P = pressure potential: energy expended or gained in moving water against or with the pressure difference.

The value of Z is positive above the reference level and negative below this level. The value of P is positive when the water at the point under consideration has a positive pressure as is normally the case in the groundwater below the watertable; P is negative when the water has a negative pressure as is normally the case in the soil moisture above the watertable. A soil-water system will always strive to attain a minimum energy level. As long as $H_A > H_B$ water will move from point A to point B as this will lower the energy level of the system (Figure 6.6). The water movement will continue until $H_A = H_B$ (equilibrium).

Darcy's Law

The rate of water movement through the soil from point A to point B obeys the Law of Darcy (developed in 1856 by the French engineer Henri Darcy, Figure 6.6):

$$v = -K \frac{H_B - H_A}{L} = Ki$$ Eq. 6.4

$$Q = vA = KiA$$ Eq. 6.5

$$i = \frac{|\Delta H|}{L} = \text{hydraulic gradient}$$ Eq. 6.6

where:
v = flow velocity (the so-called filter velocity) through a porous medium where water is only able to move through the pores; when the porosity of the medium is for example 40% the real average flow velocity in the pores is 2.5 × the filter velocity, cm/s or m/d
H = head, cm or m
L = distance between point A and B, cm or m
i = gradient, dimensionless
Q = discharge, cm³/s or m³/d
K = hydraulic conductivity of the soil (popularly, also referred to as permeability), cm/s or m/d
A = cross-sectional flow area, cm² or m²

Darcy's Law states that the rate of water movement through a soil is proportional to the gradient of the soil water potential (which is the driving force behind the movement). The hydraulic conductivity (Chapter 7) is the constant of proportionality in this relationship. The negative sign accounts for the fact that the potential decreases in the flow direction. Darcy's Law also applies to the movement of water through the unsaturated soil (unsaturated flow), whereby the K-value depends on the soil moisture content (K_θ). Darcy's Law for unsaturated flow combined with the conservation of mass equation is referred to as the Richard's equation.

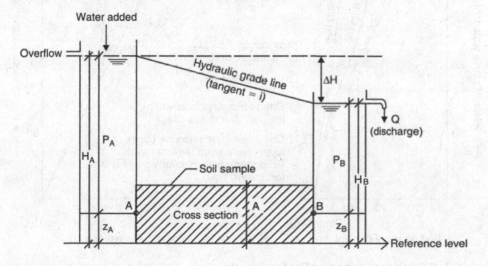

Figure 6.6 Illustration of Darcy's Law

For horizontal flow through a layer in the soil, the cross-section (A) may be replaced by the thickness of the layer D (i.e., the cross-sectional area of a unit width of the layer, equal to D m²). Darcy's Law then becomes: Q = KiD where Q is the discharge per m width (Q/m). Together, K and D determine the opportunity for horizontal flow through a soil layer and as such are often combined into one parameter, the *KD-value (transmissivity)* of the soil.

6.5   Unsaturated zone; soil moisture constants

Soil moisture conditions vary greatly in time and space. The soil moisture profile shows the variation in soil moisture content with depth in the soil. Indirectly it expresses the prevailing pressures in the soil water at different depths since pressure and moisture content are related through the soil moisture characteristic. When the pressure changes, soil moisture content will adjust, and vice versa.

Three typical pressure profiles and corresponding moisture profiles are depicted in Figure 6.7. In the no flow situation, the pressure potential at each point is equivalent (but oppositely valued) to the height above the watertable (H = Z + P = 0). The moisture profile in this case has the same shape as the soil moisture characteristic.

The other two profiles hold respectively for steady state downward flow to the watertable (percolation) and steady state upward flow from the watertable (capillary rise). The pressure profile for downward flow shows that the pressure potentials at all heights are higher (less negative) than in the no-flow situation and the moisture contents are correspondingly higher. In the case of upward flow, pressure potentials are lower, and the profile is drier than in the no-flow situation.

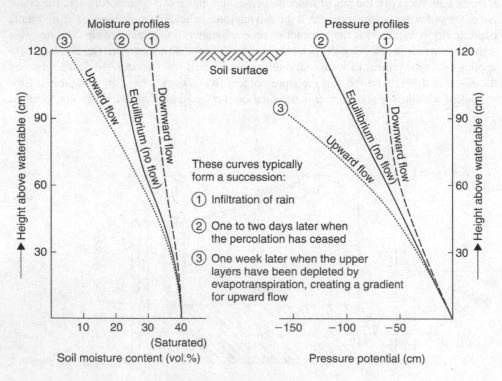

Figure 6.7 Moisture and pressure profiles for three situations

For the downward flow case, the soil moisture content well above the watertable becomes uniform and the gradient converges to one (no pressure gradient; water moves by gravity only). The moisture content in this zone settles at such a value that its corresponding K-value equals the infiltration rate (according to Darcy's Law: $v = Ki$, so for $i = 1 \rightarrow v = K$). The steady state percolation reaches its maximum attainable value when the soil profile has become saturated and the percolation rate equals K_{sat}. When the rainfall exceeds this rate surface ponding will occur.

Under the prevailing field soil moisture regimes, steady state flow conditions are the exception and non-steady state conditions are the rule. The corresponding moisture profiles may vary almost endlessly. A number of special soil moisture situations are recognised, the corresponding indicative soil moisture parameters being known as the *soil moisture constants*.

Field capacity (FC), wilting point (WP) and available moisture (AM)

The soil is said to be *at field capacity* when after a good watering of the soil the deep percolation flow has virtually ceased while no significant evapo(transpi)rative soil moisture depletion has yet occurred. When the watertable is at a shallow depth (< 1-2 m below the soil surface), the field capacity situation approaches the no-flow soil moisture profile (equilibrium profile in Figure 6.7). When the watertable is deep, such a no-flow profile never becomes established in the field since considerable evapo(transpi)rative depletion has already occurred by the time the deep percolation process has ended. In such cases field capacity refers to the situation where the rapidly draining macro-pores in the upper soil layers *(rootzone)* have been drained. In a readily draining soil profile this situation is reached within one day after rainfall or irrigation, leaving the upper soil layers at a moisture content roughly corresponding with $P = -200/300$ cm (pF = 2.2–2.5). In a slowly draining profile, field capacity is reached somewhat later and/or the corresponding suctions are somewhat lower ($P = -60/100$ cm; pF = 1.8–2.0).

Actually, the field capacity situation is only a momentary situation in the profile drainage process; this process continues after field capacity has been reached, although at a much slower rate (very low hydraulic conductivity, see Figure 7.1). Its real significance is practical rather than physical: in most soils, it constitutes a readily identifiable starting point for soil moisture depletion calculations (see available moisture below).

A soil is said to have reached *wilting point* when plants are no longer able to extract enough water from that soil to survive. At wilting point the soil has lost all of its pore water and part of its film water, the remainder being held close to the colloid surface under suctions lower than the plant roots are able to overcome. For most crops this occurs when the suction has risen to -16 atm $= -16\ 000$ cm (pF = 4.2). The soil water held between FC and WP is called the *available moisture*. Actually, crop growth is retarded well before the soil moisture content has reached wilting point.

Drainable pore space (μ)

When in a situation with a no-flow moisture profile existing above a shallow watertable, the watertable is lowered, some pore space will drain and a new equilibrium profile will develop (Figure 6.8).

A similar shifting of the equilibrium profile occurs when the watertable rises and pore space is filled. The pore space which drains/fills with a fall/rise of the watertable is called the drainable pore space (μ).

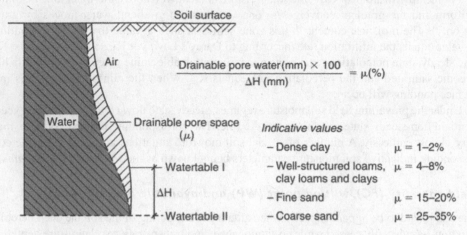

Figure 6.8 Illustration of the drainable pore space

Typical μ-values for field soils vary from 2 to 10 %[2]. When the value is e.g., 4 % the watertable falls or rises by 10 cm for each 4 mm of water extracted from or percolated to the groundwater. These responses are small when μ-values are large, and greater when μ-values are small. The drainable pore space can be measured in the field by observing the drain discharge and the watertable recession over a period when the evaporation is insignificant. Alternatively, the μ-value may be estimated as the percentage of air in the soil at field capacity as read from the moisture characteristic.

6.6   Infiltration and percolation

The maximum rate at which water can infiltrate into a dry soil decreases normally from an initially high value to a much lower value as the infiltration process continues, becoming nearly constant some 1 to 3 hours after the start. The value of the nearly constant final infiltration rate reflects the pore-geometrical situation in the topsoil which varies with the soil texture but is also very much influenced by the current state of the soil structure. Under normal conditions the following orders of magnitude apply (initial soil moisture content of topsoil approaching WP, so these values would apply at about the time for the next irrigation):

	After about 3 hrs infiltration	
	Total infiltrated	Final infiltration rate
Coarse textured soils	150–300 mm	15–20 mm/hr
Medium textured soils	30–100 mm	5–10 mm/hr
Fine textured soils (not cracked)	30–70 mm	1–5 mm/hr

A cracked clay soil may absorb almost instantaneously 50 to 100 mm of water but once the cracks have been filled and closed because of the swelling of the soil, infiltration virtually ceases. In compacted or otherwise densely structured soils, very little water is able to infiltrate, and rainfall easily results in ponding. Soils may also disperse under the impact of rainfall and clog the surface pores (*surface sealing* or *capping* of the soil); this happens particularly in many silty and fine sandy loam soils in the semi-arid zones at the beginning of the rainy season when they are exposed to intense storms with little vegetative protection. As a result, much of the rainfall runs off the land.

Darcy's Law also applies to infiltration of water into soil. Water enters the soil from the surface under the combined gravitational and pressure gradients in the soil water. The latter is very strong during the initial infiltration of water into a dry soil when in the free water on the soil surface the pressure is atmospheric (P = 0) while in the soil just below the surface it is strongly negative (P = −5 000 to −10 000 cm). The resistance for the entry of water in the soil, however, is also high as the hydraulic conductivity of the dry soil is very low. Most of the available head is dissipated in the advancing wetting front where the resistance is greatest and as a result this front becomes very sharp (Figure 6.9). In a moist soil, the wetting front is much more diffuse.

As the infiltration process continues, the flow paths become longer and consequently the pressure gradient becomes smaller. This partly explains the decline of the infiltration rate in time. By the time the infiltration rate has become constant, a pronounced transition zone has established with nearly uniform moisture content close to saturation. Pressure differences in the zone are small and the water movement is dominated by the gravity force. The final infiltration rate thus becomes approximately equal to the K_{sat} of the soil (see also section 7.3.1).

Percolation refers to the downward movement of excess water in the soil profile to underlying layers. In the process the upper layers drain to field capacity. The post-infiltration percolation in a dry soil (redistribution, see Figure 6.9) is a special case. Moisture profiles during percolation to a shallow watertable are described in section 6.5.

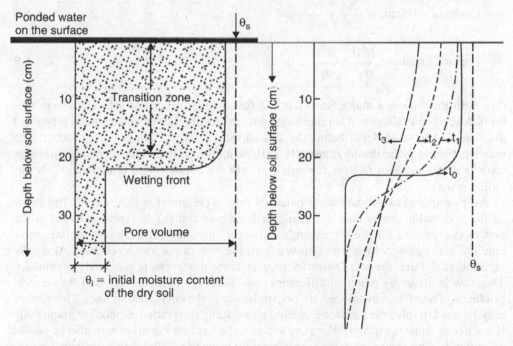

Figure 6.9 Soil moisture during the infiltration of water into a dry soil and the post-infiltration redistribution of the infiltrated water

In field soils the downward flow may be impeded due to the occurrence of poorly permeable dense layers while stratification in general may impede downward movement of excess water. This may lead to perched ground water and similar cases of near-saturated moisture conditions above the watertable (Figure 6.2). Where this occurs, horizontal gradients for water movement may exceed the vertical gradients, resulting in lateral water movement above the true watertable, referred to as *interflow*. Conditions under which interflow occurs are often ill-defined, this type of flow rarely lending itself to the sort of physical-mathematical analysis used for other types of soil water movement. Interflow, however, plays an important role in the functioning of shallow drainage systems described in Chapter 9.

6.7   Groundwater flow; Laplace Equation

Groundwater flow refers to saturated flow below the watertable. The rate and direction of flow are controlled by differences in soil water potential (combined gravitational and pressure potential). Consequently, groundwater flow may be described by the classic equations of steady state potential flow such as the *Laplace Equation*. This equation combines the two basic laws of groundwater flow, viz Darcy's Law and the Law of Conservation of Mass (also known as the *Continuity Equation*). For two-dimensional flow these equations are:

Darcy's Law: $v_x = -K\dfrac{\partial H}{\partial x}; v_y = -K\dfrac{\partial H}{\partial y}$
Eq. 6.7

Continuity Equation: $\dfrac{\partial v_x}{\partial x} + \dfrac{\partial v_y}{\partial y} = 0$
Eq. 6.8

Laplace Equation: $\dfrac{\partial^2 H}{\partial x^2} + \dfrac{\partial^2 H}{\partial y^2} = 0$
Eq. 6.9

Groundwater flow is a major feature of the functioning of the subsurface drainage systems described in Chapter 8. In these systems, drains (ditches or underground pipelines) are installed to some depth below the soil surface, reaching into the groundwater. The water potential in the drains is zero (H = 0), so all the groundwater above this drainage base is under gradient to flow towards and into the drains (these drains constitute so-called sinks).

An example of two-dimensional potential flow is presented in Figure 6.10. The shape of the watertable shows how the available head/potential (h) is expended (most at the end to compensate for the increasingly narrower cross sections). The streamlines indicate the paths along which water moves from the high to the low-level drain. Along the equipotential lines, the total potential (= gravitational + pressure potential) is constant. The flow is driven by potential differences and is always in the direction of the steepest gradients. Therefore, streamlines are perpendicular to the equipotential lines. This feature may be used to solve the Laplace Equation numerically (relaxation method) or graphically (flow net or squares method). For simple cases, the Laplace Equation may also be solved analytically. One such case is the steady state drainage to parallel drains described in section 11.1 and 11.2.

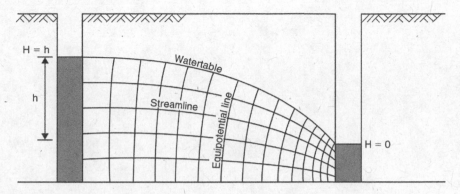

Figure 6.10 Typical potential flow pattern

Notes

1 Positive pressures in the soil water below the watertable as measured with a piezometer, are also referred to as *piezometric pressures*; similarly, negative pressures in the soil moisture above the watertable may be termed *tensiometric pressures*.

2 The drainable pore space and the hydraulic conductivity both depend primarily on the pore geometry of the soil and their values are generally correlated. For rough estimates use may be made of the formula: $\mu = \sqrt{K}$, with μ in % and K in cm/day.

Chapter 7

Hydraulic conductivity

The hydraulic conductivity value (K-value) of the soil is a very important characteristic in relation to drainage (especially in relation to subsurface drainage). The natural drainage of the soil and the scope for and costs of improved drainage all depend greatly upon it. The determination of the K-value of the soil is thus an important aspect of almost any drainage investigation.

The hydraulic conductivity of the soil depends mainly on the geometry and distribution of the water-filled pores[1]. Values are low when the water has to follow a tortuous path through fine pores. This will generally be the case when the soil is unsaturated since under these conditions the water will mainly be present in the finer pores and as film water, forming an irregular hydraulic continuity with many bottlenecks.

Differences in the hydraulic conductivity between soils under saturated conditions reflect differences in the geometry of the total pore space of the soil. As such K_{sat} values are often well correlated with soil texture and structure (Table 7.1).

The presence of bio-pores (root channels, worm holes and other small conduits left by biological processes in the soil) also greatly influences the hydraulic conductivity. This also applies to the presence of cracks and fissures. These typically form when the soil dries out but close again when re-wetted and can therefore generally not be relied on for drainage purposes. Some cracks and fissures, however, have become permanent structural features of the soil. Bio-pores, cracks and other major conduits of water may create distinct flow paths in the soil, by-passing the mass of the soil. This phenomenon is known as *preferential flow*.

The hydraulic conductivity of a field soil may vary considerably across an area as well as in depth due to variations in soil texture and soil structure. In a layered soil, K-values will generally differ between layers. Some soils also show a marked difference between hydraulic conductivity in the horizontal and vertical directions *(anisotropic soil)*.

Figure 7.1 presents two examples of the decrease in K-value with decreasing soil moisture content. The general shape of the K-θ curves reflects the proportional distribution of the soil water held in macropores, in micro pores and as film water. This distribution also underlies the P-θ curve (soil moisture characteristic) and the two curves can generally be related. The light textured soil has a higher proportion of macropores through which water is able to flow rather easily, resulting in a higher K-saturated. At low moisture content the finer textured soil has a higher K-value because it has a larger proportion of interconnected water filled micro pores and water films than the lighter soil.

The hydraulic conductivity generally drops to a very low value when most of the macropores have been drained and the soil moisture content has decreased to the extent that the water in the soil no longer forms an effective hydraulic continuity. This situation occurs roughly at field capacity (section 6.5). Although under these conditions water movement over long periods (some months) may still be considerable, over short periods (few days) it is insignificant. Measurement of saturated hydraulic conductivity in the field is described in section 7.2.

Table 7.1 Indicative hydraulic conductivity values of different soils

Texture	K_{sat} (in order of magnitude) in m/d
Gravel, crushed rock	1 500–3 500
Gravel, natural river run material (section 8.5.1)	100–1 500
Sand and gravel mixtures	5–100
Gravely coarse sand	10–50
Very coarse texture (cS)	5–10
Coarse texture (fS, LS)	2–6
Medium sand	1–5
Sandy loam/fine sand	1–3
Loam/clay loam/clay, well structured	0.5–2.0
Medium coarse texture (fLS, SL)	0.5–2
Medium texture (fSL, SiL, L)	0.25–0.8
Very fine sandy loam	0.2–0.5
Medium fine texture (SCL, CL, SiCL)	0.10–0.40
Fine texture(SC, SiC, C)	<0.05–0.10
Clay loam/clay, poorly structured	0.02–0.2
Dense clay, not cracked and no bio-pores	< 0.002

See Figure 5.3 for texture (NB c = course, f = fine, C = Clay, CL = Clay loam, S = Sand, SC = Silty Clay, Si = Silt, SL = Sandy Loam, L = Loam, LS = Loamy Sand.

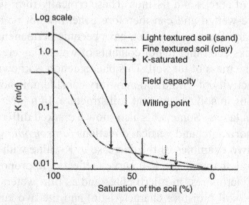

Figure 7.1 Hydraulic conductivity of soils as a function of the soil moisture content

7.1   Laboratory measurement

With these methods the K-value is measured in the laboratory on a soil sample taken in the field. The quality of the results depends very much upon the quality of the sample, which should be representative of the site under investigation. Owing to soil variability, a large number of locations may have to be sampled and at each location 2-3 replicate samples

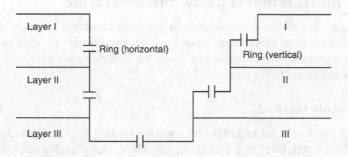

Figure 7.2 Core sampling in soil pit

may be required. It is also vitally important to minimise soil disturbance since the K-values depend to a large extent upon the soil's pore geometry.

Undisturbed samples (also referred to as *core samples*) may be taken by pressing an open cylinder ring into the soil (Figure 7.2). Rings in common use have a depth and inside diameter of about 5 cm (volume exactly 100 cm³, a convenient volume for bulk density determination). Samples may be taken vertically or horizontally, depending on whether K is to be measured in the vertical (K_v) or in the horizontal (K_h) direction.

The principle of all measurement methods is that water is arranged to flow through the sample while the rate of flow and the corresponding head loss are recorded. During the measurement the head may be kept constant (*constant head method*), resulting in a steady flow through the sample, or the head may decrease (*falling head method*), resulting in decreasing rates of flow during the measurement. The hydraulic conductivity may then be determined from the flow and head record, using Darcy's Law (see for example the arrangement in Figure 6.6). Of course, the sample must be fully saturated during the measurement, especially making sure that no entrapped air remains.

Evaluation

a) The main drawback of laboratory measurements is that the K-value only relates to a small part of the flow domain (sample of only 100 cm³). Many samples could be taken but this becomes laborious. Equipment has been developed to take large, relatively undisturbed samples of the entire rootzone although for routine drainage investigations this type of equipment is unlikely to be available.

b) Inconsistent, highly variable values are obtained in clay soils in which K depends almost wholly upon the development of cracks. A sample taken in between the cracks may have virtually zero conductivity whereas an adjacent sample containing a crack might have a very high value. Clays are also difficult to saturate and are liable to swell, when wetted, rather more in the unconfined situation in the laboratory than in the field situation.

c) In general determination of the K-value in the field is preferable to laboratory measurements, the latter only being used to supplement field methods to:

   • determine the K above the watertable; field measurements for this purpose are almost equally as unsatisfactory as laboratory methods;
   • separate K_h and K_v;
   • help to predict how salt affects the hydraulic conductivity and other physical properties, especially whether deflocculation will occur (see also section 7.3.1).

7.2   Field measurements below the watertable

A distinction is made between determination of hydraulic conductivity below the watertable and above the watertable. Reliable methods have been developed for measurement below the watertable whereas methods employed above the watertable are generally less satisfactory. The latter are described in section 7.3.

7.2.1   Augerhole method

Hooghoudt (1936) developed an empirical equation relating hydraulic conductivity to the rate of rise of the water level in a bailed out augerhole. Hooghoudt's analysis was subsequently refined by Kirkham and van Bavel (1948) and by Ernst (1962).

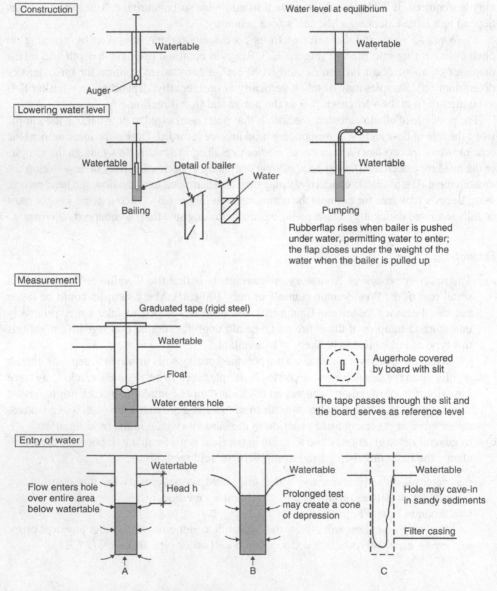

Figure 7.3 Procedures for the augerhole method

Figure 7.3 illustrates the basic steps involved in carrying out the augerhole method. An augerhole is made to a depth well below the watertable and deep into the layer to be measured.

The water level should then be allowed to rise in the hole until it establishes itself at an equilibrium with the watertable level in the surrounding soil. In highly permeable soils this will take 10 to 15 minutes whereas in poorly permeable soils it may take days. Bailing or pumping water out of the hole then lowers the water level in the hole. Flow immediately commences from the soil into the augerhole driven by the hydraulic gradient created by the difference in level between the prevailing watertable and the water level in the hole. A detailed record of the (rising) water level in the hole is then maintained over an appropriate time period.

The equations used are based upon the assumptions that the watertable level outside the hole remains constant and horizontal throughout the test and that flow enters the augerhole over its entire area below the watertable (situation A in Figure 7.3). This situation exists initially after lowering of the water level, although a cone of depression (B) will eventually develop invalidating the assumptions. No significant errors are made provided the measurements are completed before 25% of the water removed from the hole, has been replaced. This condition will be complied with provided the test is completed before $h \leq \frac{3}{4}h_0$ (see Figure 7.4).

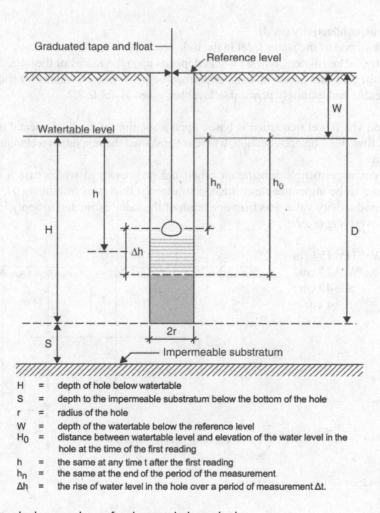

H	=	depth of hole below watertable
S	=	depth to the impermeable substratum below the bottom of the hole
r	=	radius of the hole
W	=	depth of the watertable below the reference level
H_0	=	distance between watertable level and elevation of the water level in the hole at the time of the first reading
h	=	the same at any time t after the first reading
h_n	=	the same at the end of the period of the measurement
Δh	=	the rise of water level in the hole over a period of measurement Δt.

Figure 7.4 Standard nomenclature for the augerhole method

The value of h_0 must also be chosen in relation to soil stability and hydraulic conductivity. In highly permeable soils, take $h_0 \sim 40$ cm for a larger value may result in caving-in of the hole. Caving-in, always a problem in sandy soils, should be prevented by use of a filter casing (situation C in Figure 7.3). For soils of low K value take h_0 up to 80 cm to create sufficient head for a good rate of inflow.

In clay soils smearing of the wall of the hole may occur hindering the inflow. This can sometimes be overcome by prior installation in dry soil (during the summer or before irrigation). The adverse effects of smearing may also be partly overcome by bailing out the hole several times to flush out clogged pores prior to the actual test.

Calculations

Kirkham and van Bavel (1948) and Ernst (1962) separately analysed the flow to auger holes and developed the following equation:

$$K = C\frac{\Delta h}{\Delta t}$$

Eq. 7.1

where:
K = hydraulic conductivity (m/d)
$\Delta h/\Delta t$ = rate of rise of the water level in the hole (cm/sec)
C = geometry factor (dimensionless) which depends upon the radius of the hole, the depth of water in the hole and the depth to an impermeable substratum. C values for soil above an impermeable and infinitely permeable layer are given in Table 7.2.

Kirkham and van Bavel's equation is based upon a solution of the fundamental differential equation of flow, the Laplace equation, whereas Ernst used the relaxation technique to arrive at a solution.

A clear-cut impermeable substratum might not be present in which case a layer may be considered to be impermeable if it has a substantial thickness (minimum 1-2 m) and a hydraulic conductivity value less than one tenth of the value estimated to apply in the more permeable overlying layer.

Example: W + H = 154 cm
W = 74 cm
S = 40 cm
r = 4 cm

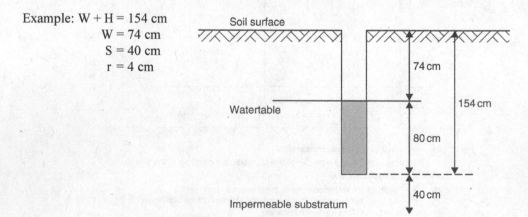

Table 7.2 Values of C for use with the augerhole method (Boast and Kirkham 1971)

H/r	h̄/H*	S/H for impermeable substratum									S/H for infinitely permeable substratum			
		0	0.05	0.1	0.2	0.5	1	2	5	∞	5	2	1	0.5
1	1	447	423	404	375	323	286	264	255	254	252	241	213	166
	0.75	469	450	434	408	360	324	303	292	291	289	278	248	198
	0.5	555	537	522	497	449	411	386	380	379	377	359	324	264
2	1	186	176	167	154	134	123	119	116	115	115	113	106	91
	0.75	196	187	180	168	149	138	133	131	131	130	128	121	106
	0.5	234	225	218	207	198	175	169	167	167	166	164	156	139
5	1	51.9	48.6	46.2	42.8	38.7	36.9	36.1		35.8		33.5	34.6	32.4
	0.75	54.8	52.0	49.9	46.8	42.8	41.0	40.2		40.0		39.6	38.6	36.3
	0.5	66.1	63.4	61.3	58.1	53.9	51.9	51.0		50.7		50.3	49.2	46.6
10	1	18.1	16.9	16.1	15.1	14.1	13.6	13.4		13.4		13.3	13.1	12.6
	0.75	19.1	18.1	17.4	16.5	15.5	15.0	14.8		14.8		14.7	14.5	14.0
	0.5	23.3	22.3	21.5	20.6	19.5	19.0	18.8		18.7		18.6	18.4	17.8
20	1	5.91	5.53	5.30	5.06	4.81	4.70	4.66		4.64		4.62	4.58	4.46
	0.75	6.27	5.94	5.73	5.50	5.25	5.15	5.10		5.08		5.07	5.02	4.89
	0.5	7.67	7.34	7.12	6.88	6.60	6.48	6.43		6.41		6.39	6.34	6.19
50	1	1.25	1.18	1.14	1.11	1.07	1.05			1.04			1.03	1.02
	0.75	1.33	1.27	1.23	1.20	1.16	1.14			1.13			1.12	1.11
	0.5	1.64	1.57	1.54	1.50	1.46	1.44			1.43			1.42	1.39
100	1	0.37	0.35	0.34	0.34	0.33	0.32			0.32			0.32	0.31
	0.75	0.40	0.38	0.37	0.36	0.35	0.35			0.35			0.34	0.34
	0.5	0.49	0.47	0.46	0.45	0.44	0.44			0.44			0.43	0.43

* h̄ is the value of the mean head during the measuring period.

Table 7.3 Recorded values

Time (sec)	Water level W + h (cm)	Head h (cm)	Change in head Dh (cm)	Time interval Dt (sec)
0	116.8	42.8	–	–
20	115.6	41.6	1.2	20
40	114.4	40.4	1.2	20
60	113.3	39.3	1.1	20
80	112.2	38.2	1.1	20
100	111.2	37.2	1.0	20

Calculations for the complete time period:

$$\Delta h = 42.8 - 37.2 = 5.6 \text{ cm}; \Delta t = 100 - 0 = 100 \text{ s}$$

$$\bar{h} = \frac{42.8 + 37.2}{2} = 40 \text{ cm (mean head)}$$

H = 154 – 74 = 80; H/r = 80/4 = 20, S/H = 40/80 = 0.5, $\bar{h}$/H = 40/80 = 0.5, so C = 6.60
(Table 7.2)

$$K = C\frac{\Delta h}{\Delta t} = 6.60\frac{5.6}{100} = 0.37 \text{ m/d}$$

Use of the augerhole method in layered soils

The augerhole method can also be used to estimate the K-value of two layers separately (when both layers occur below the watertable). In this case two auger holes are made, a shallow one wholly within the upper layer and a deeper hole reaching into the underlying layer. A measurement is first carried out in the deeper hole (No 2 in Figure 7.5) yielding a hydraulic conductivity K_{ab}. A second measurement is then made in the shallower hole (no 1) yielding $K = K_a$.

The unknown K_b may then be calculated using the known values of K_{ab}, K_a, H_a and H_b in Eq. 7.2 (the derivation of this equation is given in section 7.4):

$$K_b = \frac{K_{ab}(H_a + H_b) - K_a H_a}{H_b}$$

Eq. 7.2

Of course, when $K_a \gg K_b$, values of K_b thus obtained should not be expected to be very accurate.

Evaluation

a) Inflow into the augerhole (Figure 7.6) will be mostly by horizontal flow passing through the vertical walls and the calculated K-value will as a result be mainly an estimate of the K_h of the soil.

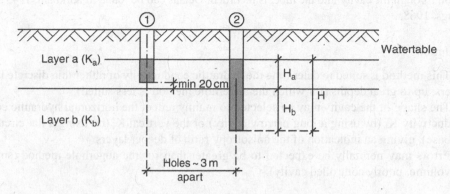

Figure 7.5 Arrangement of augerhole method in two-layer profiles

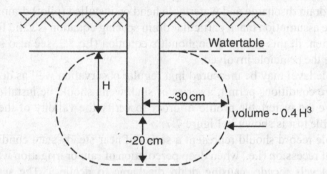

Figure 7.6 Volume of soil normally included in the augerhole method

b)   A much larger volume of soil is included in the measurement compared for example to the 100 cm³ of a normal sample for laboratory measurement. This averages out much of the micro-variability in soil structure. The remaining variability should be addressed by the selection of sites and the number of measurements. Considerable variability will, however, always remain due to:

- measurement errors (~ 10-20% error)
- soil variability (~ 10-50% error in silty/sandy soils, much greater in clays)   in total approximately 10-30% variation between averages of 5 replicates

7.2.2   Piezometer method

The piezometer method is similar in many respects to the augerhole method. A tube (ID = 3 to 8 cm) is installed to some depth below the watertable. A cavity is formed at the lower end of the tube, which may simply consist of the flat circular soil base under the tube, or it may be cylindrical in shape and extend below the tube. Measurements can start once the water level within the tube has reached an equilibrium with the prevailing watertable level. The water level in the tube is then lowered by bailing or pumping and its subsequent rate of rise, as the water flows from the soil through the cavity into the tube, is recorded. Details can be found in Kirkham, 1945 and Youngs, 1968.

Evaluation

a)   This method is suited to determine the hydraulic conductivity of rather thin discrete layers, up to great depths (for which the augerhole method is less suited)
b)   The shape of the cavity may be selected to mainly reflect the horizontal hydraulic conductivity K_h (by using a long narrow cavity) or the vertical K_v (by using a flat circular base), giving an indication of the anisotropy ratio of deeper layers
c)   Errors may normally be expected to be greater than for the augerhole method (small volume, poorly controlled cavity).

7.2.3   Drain outflow method

With this method, the hydraulic conductivity at a site is determined by analysing a simultaneous record of drain discharge and watertable head on installed (pilot) drains. The analysis is based upon the assumption that a particular drain spacing equation is valid for the situation under investigation. In this section Hooghoudt's equation (Eq.7.3, see also section 11.3) is used to illustrate the principle involved.

The watertable level may be measured in a regular observation well as described in section 5.5.1. Where conditions permit, a series of such wells should be installed between the drains to observe the watertable profile in order to verify the validity of the analysis. The layout of a suitable test is shown in Figure 7.7.

The watertable record should represent a series of near steady state conditions obtained during periods of recession (i.e., when deep percolation of rain or irrigation water has ceased and watertable levels recede, causing drain discharge to decline). The series preferably should be continuous but may also be compiled from single measurements (collected over say a few months). Records obtained during periods of recharge to the watertable are often erratic and should not be used. It is also important to ensure that the watertable profile is approximately uniform along the whole drain length so that the measured discharge can be assumed to have resulted from a uniform inflow.

The discharge may be measured most simply by means of a bucket and stopwatch. A continuous record may also be obtained by using a drain flow meter (V-notch weir, freely discharging orifice or purpose designed instrument), which can later be linked with a continuous record of the watertable elevation. The degree of sophistication of the measurements depends upon many factors although manual observation on a regular basis and using simple equipment has much to recommend it.

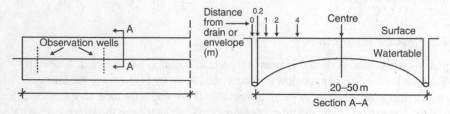

Figure 7.7 Test layout and watertable observation wells

Analysis

The data given apply to a test with drains at 2 m depth and a spacing of 50 m. Investigations have indicated that the soil is a uniform silty clay loam, underlain by a thick layer of impervious clay some 4-5 m from the surface.

Figure 7.8 illustrates typical features of the drainage response. Rainfall in the first five-day period creates a discharge hydrograph, which rises to a distinct peak value. Analysis of the recession period after the soil has been completely wetted and significant recharge to the watertable has ceased (1-3 days after rainfall) yields the relationship between drain discharge (q) and watertable head (h) as illustrated in Figure 7.9.

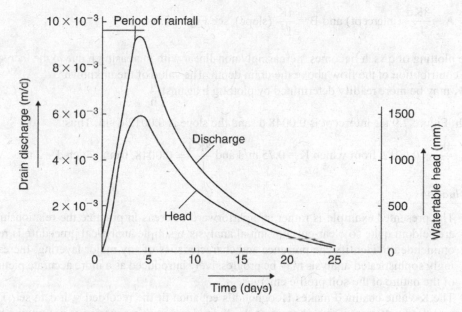

Figure 7.8 Observed watertable head and discharge in a drain outflow test

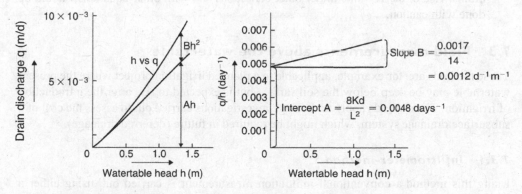

Figure 7.9 Relationship between watertable head and drain discharge

In this case, it is assumed that the relationship between q and h obeys Hooghoudt's equation for drain spacing (see section 11.3):

$$q = \frac{8Kdh}{L^2} + \frac{4Kh^2}{L^2}$$

Eq. 7.3

The first part of this equation describes the flow to the drain below drain depth and the second part the flow to the drain above drain depth. The equation may be reformulated as:

$$q = Ah + Bh^2 \text{ or } q/h = A + Bh$$

where:

$$A = \frac{8Kd}{L^2} \text{ (intercept) and } B = \frac{4K}{L^2} \text{ (slope), see Figure 7.9.}$$

The plotting of q vs. h becomes increasingly non-linear with increasing h due to the increasing contribution of the flow above the drain depth. The value of the unknown K may be more readily determined by plotting h against $\frac{q}{h}$

In Figure 7.9 the intercept is 0.0048 d^{-1} and the slope is 0.0012 d^{-1}/m. Thus:

$$\frac{4K}{L^2} = 0.0012, \text{ from which K} = 0.75 \text{ m/d and } \frac{8Kd}{L^2} = 0.0048, \text{ from which d} = 2 \text{ m}$$

Evaluation

a)   The presented example is rather straightforward whereas in practice the relationships are seldom quite so clear-cut. For initial analysis, a simple analytical procedure is recommended to identify the presence and characteristics of any major layering. Increasingly sophisticated analysis may be progressively introduced as a more accurate picture of the nature of the soil profile emerges.

b)   The K-value obtained makes Hooghoudt's equation fit the recorded q, h data set. The real flow pattern remains obscure as the K in Hooghoudt's equation is not unique but rather a combination of K_h and K_v, mainly the former.

c)   The estimated K value refers to the entire body of the soil involved in the flow to the drains. Use of the K value under other conditions and with other equations, should be done with caution.

7.3   Field measurements above the watertable

These methods are, for example, applicable to a planned irrigation project where the present watertable may be deep below the soil surface but is expected to rise after the introduction of irrigation. The methods apply to drainage investigations carried out to assess the cost of a subsurface drainage system, which might be required in future (deferred drainage).

7.3.1   Infiltrometer-method

Using this method a conventional infiltration measurement is carried out, using either a single (Figure 7.10) or a double ring infiltrometer. Measurements can be made at the surface

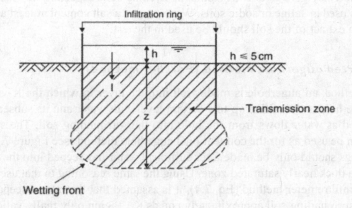

Figure 7.10 Single ring infiltration measurement

or at different depths below the soil surface (on steps in a profile pit see Figure 7.2). The infiltration rate (I) of water into the soil is governed by Darcy's Law:

$$I = K_\theta \frac{h+z-p}{z}$$ Eq. 7.4

where:
I = infiltration rate (m/d)
K_θ = hydraulic conductivity of the soil at moisture content θ (m/d)
h = water depth on the soil surface (m)
z = depth to the wetting front (m)
P = soil water pressure at the wetting front, inside the transmission, zone (m)

The moisture content in the transmission zone becomes virtually saturated so that θ→$θ_{sat}$ and P→0 (this applies to most medium/heavy textured soils, but not for coarse textured soils). When θ→$θ_{sat}$, also K_θ→K_{sat}. After prolonged infiltration, z becomes relatively large compared to (h-P) so that the hydraulic gradient approaches unity:

$$I_{final} = K_\theta \frac{h+z-P}{z} \approx K_{sat}$$

Therefore: $\left(\frac{h+z-p}{z} \to \frac{z}{z} \to 1 \right)$

Evaluation

a) This method measures K_v (vertical hydraulic conductivity)
b) The method is simple but is not very accurate due to:

   • I_{final} only approximates K_{sat} (see above)
   •   soil variability (small volume of soil involved; at least 3 replicates should be made to arrive at a reasonably reliable value)
   •   disturbance of soil when driving the infiltrometer ring into the soil

c)   The method may give erroneous results in swelling soils especially if water of low salt content is used in saline or sodic soils. Water having a salt content at least as high as the saturation extract of the soil should be used in the test.

7.3.2   Inverted augerhole method (Porchet method)

Under this method, an augerhole is made well into the layer in which the K-value is to be measured. The hole is then filled up to a certain level with water and its subsequent rate of fall is recorded as water flows from the hole into the surrounding soil. The same type of equipment can be used as for the conventional augerhole method (see Figure 7.3).

The readings should only be made after sufficient water has seeped into the surrounding soil to create a thick nearly saturated zone. Using the same reasoning to that used in connection with the infiltrometer method (Eq. 7.4), it is assumed that the rate of seepage from the hole into the surrounding soil approximately equals K_{sat} (again only really valid in medium to heavy textured soils).

Using the notation detailed in Figure 7.11 one may derive for time t: $Q = K(2\pi r h + \pi r^2) = 2K\pi r(h + r/2)$

- outflow from hole:

- from fall in water level in hole: $Q = -\pi r^2 \dfrac{dh}{dt}$

Equating these two expressions for Q yields: $-\dfrac{2K}{r} dt = \dfrac{dh}{h + \dfrac{r}{2}}$

and integration between the limits: $t = 0, h = h_0; t = t, h = h_t$

gives: $\dfrac{2K}{r} t = \ln\left(h_0 + \dfrac{r}{2}\right) - \ln\left(h_t + \dfrac{r}{2}\right) \rightarrow K = \dfrac{1.15 r\left[\log\left(h_0 + \dfrac{r}{2}\right) - \log\left(h_t + \dfrac{r}{2}\right)\right]}{t}$

Recorded values of t and h_t may be plotted on log normal paper where they should form a straight line. The value of K may then be determined from the slope of this line (comparable to the determination of the reaction factor α in Figure 11.11).

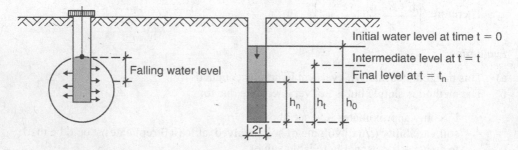

Figure 7.11 The inverted augerhole method

Evaluation

a) This method is essentially inhibited by the same drawbacks as the infiltrometer method (section 7.3.1). In fact, since the flow from the hole is more horizontal (through the walls) than vertical (through the bottom of the hole), there is less validity to the claim of unit gradient than for the infiltrometer method (this claim only applies to vertical flow)

b) The nature of the measured K-value is also rather obscure but is closer to the K_h than to the K_v value of the measured soil layer.

7.4 Composed K-values

Soils are often composed of layers of different hydraulic conductivity while individual layers are frequently highly anisotropic, especially below the root zone. Composed K-values have been derived for some typical situations (Saadat et al. 2018).

Horizontal flow through layered soil (Figure 7.12)

Darcy's Law: $Q_1 = K_1 D_1 i$
$Q_2 = K_2 D_2 i$
$Q = Q_1 + Q_2 = K_1 D_1 i + K_2 D_2 i$

Composed K for the two layers $(\bar{K})$: $Q = \bar{K}(D_1 + D_2)i$

Therefore

$$\bar{K} = \frac{K_1 D_1 + K_2 D_2}{D_1 + D_2} \text{ and } \bar{K}\bar{D} = K_1 D_1 + K_2 D_2 \qquad \text{Eq. 7.5}$$

Vertical flow through layered soil (Figure 7.13)

Darcy's Law: $v = K_1 \dfrac{h_1}{D_1}, h_1 = \dfrac{vD_1}{K_1}$ and $v = K_2 \dfrac{h_2}{D_2}, h_2 = \dfrac{vD_2}{K_2}$

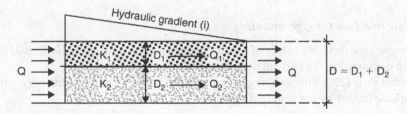

Figure 7.12 Horizontal flow through layered soil

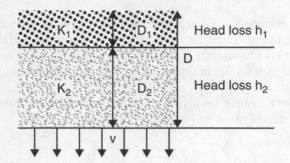

Figure 7.13 Vertical flow through layered soil

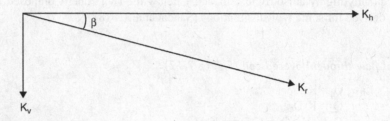

Figure 7.14 Composed K-value for radial flow through anisotropic soil

Therefore:

$$h = h_1 + h_2 = v\left(\frac{D_1}{K_1} + \frac{D_2}{K_2}\right)$$

Since: $v = \bar{K}\dfrac{h}{D_1 + D_2} \rightarrow h = v\dfrac{D_1 + D_2}{\bar{K}}$

The composed K-value is:

$$v = \bar{K}\frac{D_1 + D_2}{\dfrac{D_1}{K_1} + \dfrac{D_2}{K_2}} = \frac{K_1 K_2 D}{D_1 K_2 + D_2 K_1}$$ Eq. 7.6

Radial flow and flow through anisotropic soil

In anisotropic soils different values of hydraulic conductivity exist in the horizontal direction (K_h) when compared to the vertical direction (K_v).

For radial flow (e.g., flow converging on a drain) the hydraulic conductivity value may best be approximated by the geometric mean:

$$K_r = (K_h K_v)^{1/2}$$ Eq. 7.7

Strictly this value of hydraulic conductivity will only apply to flow with the particular orientation b to the horizontal where $\tan\beta = \dfrac{K_v}{K_h}$. Transformations of scale may also be carried out which permit anisotropic soil to be treated as being isotropic (section 11.3.3).

7.5 Surveys and data processing

The K-value should preferably be determined as part of a soil survey or against the background of an available soil map. By detailed studies in sample areas, correlations may be established between measured K-values and readily observable soil characteristics. In general, coarse textured soils are more permeable than fine textured soils and useful correlations between K-values and soil texture may be established for the area under investigation (Table 7.1).

Soils with identical texture may have quite different K-values due to differences in structure. This applies especially to fine textured soils (some heavy clay soils for example with well-developed structures have much higher K-values than those indicated in the table). Hydromorphic features such as colour mottling, etc., are also often helpful is assessing K. All these soil characteristics can of course be observed most clearly in a pit, although fair estimates of K may also be derived from examination of the soil obtained with an Edelman auger (an 8 cm diameter, 1.2 m long open spiral auger). Pertinent soil features (texture, structure, hydromorphy, impeding layers, etc.) should be noted on the field form.

These deduced K-values are used to supplement the network of measured K-values. The ultimate aim should, however, be to establish how the main soil characteristics influence the hydraulic conductivity of each soil type. Although not all soil characteristics used in soil mapping are relevant to drainage, many of them are, and so boundaries between drainability classes will often coincide with soil boundaries. In new projects where both the soil and the drainage survey still have to be done, the soil mapping criteria should be planned jointly by the soil surveyors and the drainage engineer (and possibly other specialists) to make the resulting maps as relevant to drainage as possible (see also section 5.3).

Density and depth

Initially, a random distribution pattern, for example a grid pattern and using the site densities recommended in Table 7.4 may be adopted (see also section 5.4.4). Once correlations with soil conditions have been established, a stratified selection based on the known variation in soils within the area, should be adopted. Even where delineation of drainability classes

Table 7.4 Densities of K measurements (FAO 1980)

Soil condition	Areas of land per test site	
	Drain spacing about 30 m	Drain spacing about 75 m
Heterogeneous	<5 ha	10 to 15 ha
↓	↓	↓
Homogeneous	10 to 25 ha	40 to 75 ha

Table 7.5 Depth of site Investigations for drainage

Soil conditions	Depth of investigation in situations where, there is no impermeable layer at shallower depth		
	on all sites	on 20% of sites	on 10% of sites
	Below drain depth in relation to spacing L		
Deep, reasonably permeable soil	1/10 L	1/6 L	
Stratified substrata	1/20 L	1/10 L	1/10 L
- $K_v < K_h$	1/20 L		
- $K_v \ll K_h$			
Poorly permeable substrata	1/20 L		1/10 L

can be largely based on soil boundaries, a low-density area wide measurement programme should be completed to check on the validity of the established correlation.

The required depth of investigation depends on the drain spacing and on the soil conditions. When the soil is homogeneous and permeable to great depth, the main drainage flow extends to depths of 1/6L to 1/4L beneath drain depth. In such soils the investigations should ideally extend down to e.g., 5 to 7.5 m below drain depth when L = 30 m. In most soils, however, the flow below drain depth is limited by poorly permeable layers and/or by the marked anisotropy ($K_v \ll K_h$) of the substrata (common in most alluvial soils). The general guidelines of Table 7.5 are suggested which also take some account of the cost of deep investigations.

It should be emphasised that there is little point in focussing on the shallow soil layers in a situation where most of the drainage flow passes beneath the level of the drains. In these circumstances it is sensible to reduce the number of tests whilst increasing the proportion of tests carried out at greater depth. Guidelines should also remain flexible, permitting increase/ decrease of the density and/or the depth in the course of the survey when the results indicate this to be warranted. Guidelines may for example often be relaxed once reliable correlations with soil conditions have been established.

Two-layer measurements as described in section 7.2.1 (Figure 7.5), are most relevant for cases when a highly permeable upper layer is underlain by a poorer permeable subsoil. In the opposite case (low K over high K), the spacing will depend mostly on the high K layer and a single measurement in a deep hole will generally suffice.

Data processing

The preceding text has emphasised the practical difficulty of evaluating hydraulic conductivity on a field scale. The number of measurements is generally small with limited or no statistical significance. The values of hydraulic conductivity are, however, the foundation for rational drainage design and it is therefore absolutely essential that they should be critically appraised. For a first appraisal, the values should be arranged in order of magnitude per soil unit and the variation be analysed, in the first instance to remove or adjust extreme values (i.e., a very high value of 5 m/d in a clay known to have poor drainage, should be ignored).

Maps should be prepared at a suitable scale (1:5 000/10 000 for final design work) showing the K-values differentiated for K above and K below drain depth and the depths to the impermeable substratum. Observed values may be grouped in classes (say three to four

K-classes, two to three D-classes) and be given a colour code on the map for easy overview. Areas having similar drainability properties (similar K and depth to impermeable substratum) may then be delineated and drain spacing calculated. Spacing calculations may be done per individual site but may also be done per land unit on the basis of an actual or synthesised representative soil profile. The resulting spacing may later be rationalised into a limited range of spacing e.g., 15, 20, 25, 30, 40 m etc. Generally, tertiary layout units of 50-100 ha, e.g., all the land on one side of a collector, should have fairly uniform spacing (two classes at most).

Note

1 Temperature also has an influence on K. With rising temperature, water becomes less viscous and the K-value increases. In the deeper soil layers through which most drainage flow takes place, temperatures are rather uniform and steady, and its influence on K may generally be neglected.

Part III

Systems and technology

Chapter 8

Subsurface drainage systems

Traditionally these systems consist of a network of deeply installed field drains (1–2 m below the land surface) establishing a deep drainage base in the soil, well below the main rootzone. In the drain the soil water potential is effectively zero so that all groundwater above the drainage base has a higher potential and is under gradient to flow towards and into these drains. Low watertable depths may be maintained by selecting a combination of suitable drain spacing and drain depth. Provided the downward movement of excess water is not impeded, the overlying soil profile will drain to field capacity.

Either the field drains may be perforated pipes or deep ditches. Pipe drains are also frequently referred to as buried or covered drains, in contrast to open ditch drains. Low watertables may also be maintained by a network of pumped wells. This method of subsurface drainage is referred to as *vertical drainage* in contrast to the *horizontal drainage* provided by pipe drains or ditches. The use of vertical drainage is mostly limited to irrigated land (section 16.4), while the horizontal method is by far the most common method in rainfed land.

Subsurface drainage is applicable in soils where:

a) the rootzone is underlain by strata of reasonable hydraulic conductivity K and/or thickness D (the KD-value of these strata should be reasonably high);
b) the excess water on or in the soil is able to infiltrate and percolate through the rootzone to the underlying watertable at reasonable rates.

Figure 8.1 illustrates the flow pattern of the excess water to the field drains for the case of a parallel pipe drainage system. When the pipe drains are underlain by deep permeable soil, as is assumed in respect of Figure 8.1, the flow of groundwater will extend well below the drains. The height 'h' of the watertable above the drainage base midway between two drains constitutes the head driving the groundwater towards the drains. When the KD-value of the soil at and below drainage base depth is high, the excess water may be discharged with a low head and so low watertables can be maintained even with rather wide drain spacing (> 50–100 m). When the KD-value is low, either more head must be provided (e.g., by laying the drain deeper) or a smaller drain spacing must be used to generate the same discharge. In soils with poor infiltration or percolation characteristics and/or with poorly permeable substrata, excess water control in the rootzone by means of subsurface drainage is both technically and economically impossible, and other drainage measures have to be used (Chapter 9).

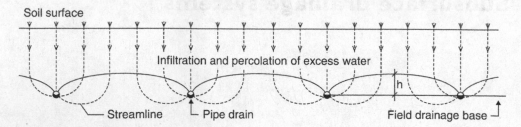

Figure 8.1 Typical flow pattern to parallel pipe drains

8.1 Pipe drain systems

Most subsurface drainage for modern farming in temperate climates is achieved by means of pipe drain systems. These systems are also extensively used for salinity control in irrigated areas. Very little pipe drainage has yet been installed in the (semi) humid tropics. The reasons for this are partly economic (the present low output level does not warrant a high investment in drainage as required for pipe drainage) and partly technical (open ditches are often preferred as these drains can also serve to evacuate excess surface water).

A considerable amount of research has been done on the functioning of the pipe drain, the materials used and the installation methods, all of which has added to the sophistication and dependability of the pipe drainage system.

Layout patterns

The alignment of field drains and the collector drains into which they discharge are mutually dependent. Their alignment, however, is in the first instance determined by the topography of the system and the land to be drained (Figure 8.2a). Drains are most effective when they are passing through the lowest areas in the land since these are the sinks to which water gravitates naturally. Such alignments may, however, not be optimal for reuse of drainage water (Chapter 20).

For example, a field in which waterlogging is confined mainly to a number of depressions is drained most efficiently by aligning the drains so that they pass through and connect these depressions. The irregular pattern of field drains which results is termed a *natural system* (Figure 8.2a). As a drain running through a depression (*depression drain*) is liable to collect some surface runoff and interflow as well as groundwater flow, an open ditch is often used in preference to a buried pipe.

As the size of the depressions decrease and their number increases, the natural layout increasingly loses its advantage compared to a regular network of field drains uniformly covering the field. Common examples are the *parallel grid system* and the *herringbone system* towards the collector (Figure 8.2b).

In the herringbone system the collector drains are aligned down the main slope and the field drains (*laterals*) are aligned across the slope but at a slight angle to the contours, so that the pipes slope downwards towards the collector drain but remain at a constant depth below the surface. In a parallel grid system, the laterals are given slope by increasing the installation depth along the drain or along the land to be drained. As typical slopes are 5–10 cm per 100 m length and the maximum drain length is seldom in excess of 250 m, the difference in drain depth below the soil surface between the ends is only 10–20 cm. The depth uniformity, therefore, is hardly a consideration in favour of the herringbone layout. This also applies to

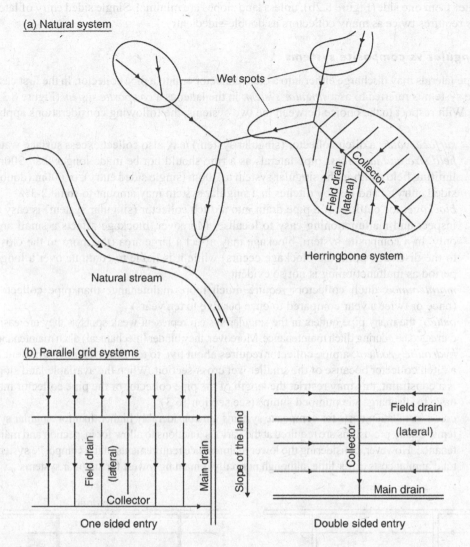

Figure 8.2 Field drainage layout patterns

the alleged advantage of an oblique entry by laterals into the collector drain in the herringbone system. The choice of one system in favour of another depends mainly on the field situation. In flat plains where large rectangular fields are the norm, the field drainage systems are commonly of the grid type. Herringbone systems have an application in more irregular situations where only part of a field requires drainage. The situation depicted in Figure 8.2a is typical, with the collector passing through a depression whilst laterals cover the lower slopes.

In situations where there is a distinct groundwater flow in a certain direction, field drains are best aligned as *interceptors* across the direction of flow. In sloping fields this means that the laterals should be aligned across the slope herringbone-wise or parallel to the contours with the collector running down the slope (section 17.1).

Alignment with the collector across and laterals parallel to the contours is also the most suitable choice on most flat land since this enables collectors to be entered by laterals from both sides. When collectors run parallel to the contours, laterals are normally only able to enter from one side (Figure 8.2b), unless land slopes are minimal. Single sided entry of laterals requires twice as many collectors as double-sided entry.

Singular vs composite systems

Pipe laterals may discharge either into a ditch collector or into a pipe collector. In the first case, the system is referred to as a *singular system;* in the latter as a *composite system* (Figure 8.3).

With respect to the choice between the two systems, the following considerations apply:

a) *surface water:* a ditch collector (singular system) may also collect excess surface water
b) *field size and land loss:* pipe laterals as a rule should not be made longer than 300m, limiting field width in the singular system to 300m (single sided entry), or 600m (double sided entry). Land loss by ditches in a singular system may amount to some 2-3%
c) *blockage*: the outflow of a pipe drain into a ditch collector (singular system) is easy to inspect and malfunctioning easy to localise. Moreover, blockage affects a small area only. In a composite system, blockage may affect a large area (the more so the closer to the discharge point the blockage occurs) while it is liable to continue over a longer period as malfunctioning is not so evident
d) *maintenance*: ditch collectors require much more maintenance than pipe collectors (once or twice a year compared to once per five to ten years)
e) *outlets*: the many pipe outlets in the singular system represent weak spots as they are easily damaged i.e., during ditch maintenance. Moreover, they hinder (mechanical) ditch maintenance
f) *hydraulic gradient*: a pipe collector requires about five to ten times as much gradient as a ditch collector because of the smaller wet cross-section. When the available land slope is a constraint, this may restrict the length of the pipe collector or the pipe collector may have to discharge via pumped sumps (see section 16.3)
g) *costs*: installation costs for composite systems are considerably higher than for singular systems. Costly provisions are required at the various junctions to allow for inspection and maintenance. However, considering the lower maintenance requirements of the composite system, total annual costs differ little, although normally remaining lowest for singular systems.

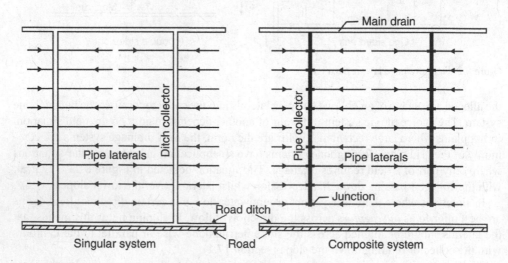

Figure 8.3 Singular and composite pipe drain systems

Considering all factors, singular systems are as a rule most suitable for flat plains under humid climatic conditions. Composite systems often have significant advantages in drainage for salinity control in irrigated areas due to the non-interference of the underground drainage system with the irrigation canal system (section 16.1). The lower maintenance requirements of composite systems are also an important consideration. Not only does the ditch maintenance become costly with rising labour costs but in practice it is frequently not carried out up to the required standards, to the detriment of the functioning of the whole drainage system.

The above layout considerations may not all apply to controlled drainage systems. Alignment through the depressions may limit the possible reuse of drainage water. To give farmers more control over sections of the drainage system, more collectors or sub-collectors are desirable (see section 20.3). To achieve farmer-controlled drainage systems that are sustainable, higher installation costs are acceptable up to 15–25% of the traditional composite system.

8.2   Deep ditch systems

These systems consist of ditches laid out in various patterns. To function as effective groundwater drains, ditches must be rather deep so that they reach well below the depth at which the watertable is to be maintained (water level ~ 1.0–1.5 m below soil surface). The ditches normally have a trapezoidal section with side slopes varying from 1:1 to 1:2, depending on the soil conditions.
  A number of factors enter into the choice between pipe drains and ditches as the most suitable field drain for an area. The influence of each of these factors depends very much on the prevailing local conditions, although the following main considerations apply:

* surface water: ditches can also collect excess surface water, a function which pipe drains are only incidentally able to perform
* land loss: can add up to 10% for ditches (when narrowly spaced and constructed with gentle side slopes) as compared to no loss in the case of pipe drainage
* hindrance to field operations: ditches restrict machine manoeuvrability resulting in higher farm costs compared to pipe drainage
* maintenance: ditches require frequent maintenance (a few times a year as compared to once every few years for pipe drains). In practice maintenance is either not done or it frequently fails to meet required standards, resulting in poor performance of many ditch drain systems
* installation: pipe drains can be installed by machine under good grade, guaranteeing good discharge performance. Similar machines are not available for ditch construction, and ditches often have adverse bed grades resulting in poor discharge and functioning
* soil conditions: may favour or prevent the use of ditches or pipe drains (caving in of ditches; clogging of pipes by iron compounds; etc.).

Generally, pipe drains result in better drainage than ditches due to better installation and mostly trouble-free functioning, while the fact that pipe drains do not hinder (mechanical) farm operations is often of overriding importance. However, the use of ditches may well be preferable when:

* the land in question requires both surface and subsurface drainage (applies to large areas of land in the humid tropics)
* ditches do not severely hinder farm operations as for example would be the case where ditches are widely spaced, where the land is used for perennial crops like sugarcane, bananas, etc., or where farm operations are mostly done manually

- standard of drainage required is not high (applies to many grassland areas)
- cost advantages overshadow all disadvantages.

Many peat soils are best drained by ditches, as pipes are liable to become clogged by iron compounds, become misaligned due to uneven subsidence and chemical deterioration. Newly reclaimed, non-ripened marsh and sea or lake bottom soils are also best drained by ditches; after some years when the soil has ripened sufficiently, the ditches may be replaced by pipe drains (see also Chapter 18).

Most factors set out in section 8.1 with respect to alignment and layout of field drains are equally applicable to ditches. As noted, ditches are often well suited to a natural system of field drainage but are also commonly used in parallel systems. The field ditch system presented in section 9.2 as a shallow drainage system for heavy soils, may also function as a subsurface drainage system by making the ditches extra deep (75–100 cm). This system would for example be applicable where intense rainfall occurs on poorly infiltrating soils underlain by a fairly permeable subsoil/substratum.

Widely spaced (100–200 m), deep (2.0–3.0 m) ditches have also been used in irrigated areas to provide subsurface drainage for salinity control (widely used in Egypt and Iraq until the 1950/60's). These days, however, most subsurface drainage of irrigated land is by 'horizontal' pipe drains or by vertical drainage: tubewells (Chapter 16).

8.3   Drainpipes

The drainpipe has evolved considerably since its first use some 300 years ago. Originally, these consisted of no more than a trench filled at the bottom with stones or brushwood (and then further backfilled with soil). Later, regular underground conduits were made, first in the form of wooden box spouts and since the turn of the 18th century, mostly in the form of clay-ware pipes (also called *tiles*). In places without an established tile industry, the drainpipe was usually made of concrete. Plastic pipe was introduced around 1960 and has since conquered almost all of the drainpipe market in many countries. Figure 8.4 and Figure 8.5 depict some specific features of commonly (or widely) used pipes.

Clay tile pipe: standard sizes vary between countries although typical pipe sections are 30 cm long and have internal diameters (ID) equal to 5, 6.5, 8, 10 up to 20 cm. Special pipes with collars are available for use in soils in which consolidation is likely to occur, e.g., peat or non-ripened marine soil, although these are 25–30% more expensive than regular pipes. The clay tile is highly resistant to deterioration in aggressive soil conditions. Pipe sections are abutted against each other and water enters through the joints that exist due to the result of the imperfect fit between the ends of the pipe sections, see Figure 8.6a.

Concrete pipe: mostly medium to large size with diameters of 15–20 cm or more and section lengths of 30 cm for small diameter pipe and up to 50 cm for large diameter pipe; water entry occurs through the joints. Pipes made with ordinary (Portland) cement are liable to deteriorate in acidic or salt affected soils (especially attacked by sulphates) and in these circumstances special resistant cements should be used.

Plastic pipe: made from PE (polyethylene) or PVC (polyvinyl chloride) of which the latter is the most common (higher bearing capacity, lower costs). It is very durable but subject to deterioration by long exposure to the ultra-violet radiation of sunlight. PVC becomes brittle with freezing temperatures and can then easily fracture. PE has a specific density of less than one and floats which may cause problems with construction of drains below the watertable.

(a) Clay/concrete Tiles (water collection), laterals

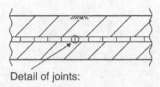

Detail of joints:

Junctions in a pipe drain system

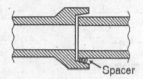

Cross-piece   Blind junctions   T-piece

(1) With spacing lug

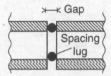

(2) Bell and spigot with spacer

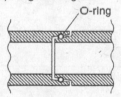

(b) concrete pipes for conveyance, collectors

(1) Flush against each other

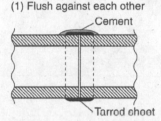

(2) Tongue and groove

Figure 8.4 Clay and concrete drainpipes

Most plastic drainpipe is supplied as corrugated pipe with either spiral or parallel corrugations. Pipes with spiral corrugations (Figure 8.5) have the advantage that with partial blockage of perforations or the drain envelope, water can flow more easily via the spiral corrugations to non-blocked sections. Corrugated pipes have a higher hydraulic roughness and require a 20% increase in diameter to carry the same flow as smooth pipes of the same diameter. The cost per metre length of smooth and corrugated pipes having similar diameter is about equal, since more material goes into a smooth pipe compared to a corrugated pipe of the same strength.

Standard corrugated drainpipes usually have outside diameters (OD) of 50, 65, 80, 100, 125, 160 and 200 mm, whereas ID~0.9xOD. At small diameters, the pipe is flexible and is delivered in coils of 30–50 kg, containing up to 200 m of pipe. The price of plastic pipe increases sharply with increasing diameter (a 10 cm diameter pipe is for example roughly four times as expensive as a 5 cm pipe). Large diameter pipe (>20 cm) as required for collector drains, is available in plastic but is not always competitive with concrete pipe. Plastic collectors are generally preferred under difficult installation conditions (see section 8.7).

The pipes are perforated for water entry as is shown in Figure 8.6 (0.6–2.0 mm wide slots, usually in the grooves of the corrugations). The perforations may be arranged in any pattern that provides an even distribution around the whole circumference. With parallel corrugations, one should make sure perforations are in every groove (not every other groove as this doubles the entry resistance). The open area should be minimally 800 mm^2 per m pipe length (see Box 8.1).

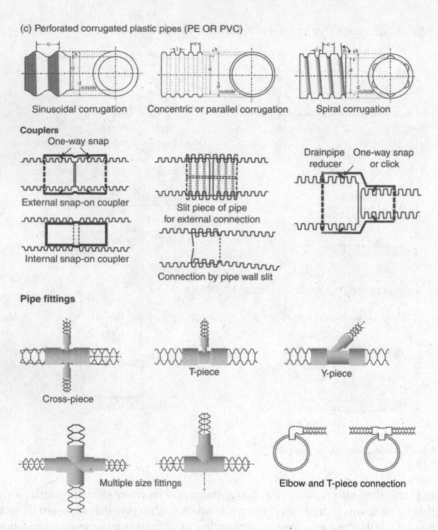

Figure 8.5 Plastic drainpipes (PE and PVC) and fittings

Entry losses

Head may be lost as the water flows towards and through the rather limited open areas into the pipe. This head loss constitutes the entry loss and it follows that, for the same inflow, pipes with a small entry area will have a higher head loss than pipes with a large entry area. Corrugated plastic pipes generally have sufficiently large open entry areas. When a drain has an envelope around it, entrance losses may be reduced significantly as the open area of drain envelopes is significantly greater than that of the pipe (compare values in Box 8.1 with those of Figure 8.7). High entry resistance is mostly due to clogging/blocking of the openings of the pipe and/or the envelope material or may be caused by low permeability adjacent to the pipe due to smearing of clayey soil in the trench wall. Smearing was also a problem with the vertical plough (Spoor 1995, see also section 8.7.2).

Box 8.1   Typical and standard pipe drain opening sizes

General remarks: minimum opening area for subsurface drains 800 mm²/m (ISO)
Typically, openings between clay tiles of 250–300 mm long are assumed to be 1.6 mm. With a 130 mm diameter tile, this results in $2\pi \times 65 \times 1.6 \times 3.3 = 2\,156$ mm²/m.

Plastic perforated pipes typically have openings 0.5–2 mm width and 5 mm long in the valleys of the corrugations. Wavelength of corrugations is typically 5–15 mm for diameters up to 100 mm, 15–30 mm for up to 200 mm, and as much as 50–75 mm for pipe diameters of 300–750 mm (Cavelaars et al. 1994).

Typically opening area would be 1–2% of the pipe surface area.

USA   Tiles: recommended crack widths 3–6 mm in stable soils; unstable soils as close as possible. In practice crack width of 0.8–3 mm is usually observed. To assure crack widths USBR (1978, 1993) recommend using spacer lugs of 3 mm as standard.

Corrugated plastic pipe (USBR 1993): a minimum of 2 120 mm²/m (1 sq. inch per foot). To meet this criterion manufacturers, provide pipe up to 250 mm diameter with 5 mm round holes and pipes >250 mm with 10 mm round holes!

Europe   Arranged in any pattern, not less than 4 perforations in the valley of each corrugation, with at least 2 perforations per 100 mm of each single row. Nominal perforation width 1–2.3 mm by increment of 0.1 mm, average perforation width shall not deviate more than 0.2 mm, single perforation more than 0.4 mm. Entrance area >1 200 mm²/m. Nominal diameters 50–30 mm. (CEN/TC 155 N 1259 E 1995).

Standards   Common standards for pipe qualities (clay, concrete, plastic and steel) are given in the ASAE Engineering Practice guidelines, the American Society of Testing Materials, ASTM, the Netherlands Engineering Norms (NEN, in Dutch), the Deutsche Industry Norms (DIN, in German), the British Standards (BS) and in the European and International Norms, commonly referred to as CEN and ISO norms. Typical terms used in the titles of the norms are concrete or clay drain tile, corrugated polyethylene tubing and fittings, and corrugated unplasticised PVC pipes.

Entry characteristics of different pipe/envelope combinations may be compared using a standardised entry resistance W_e which is related to the head loss at entry as:

$$h_e = QW_e \qquad\qquad\qquad \text{Eq. 8.1}$$

where:
h_e = head loss at entry, m
Q = the inflow rate to the pipe per m length of pipe m³/m/d
W_e = the entry resistance in d/m

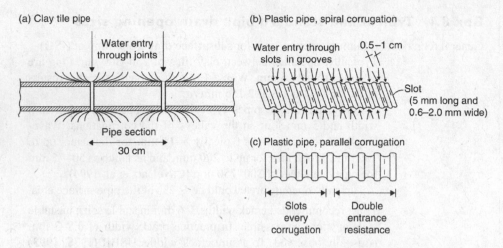

Figure 8.6 Entry flow pattern for clay/concrete pipe and for corrugated plastic pipe

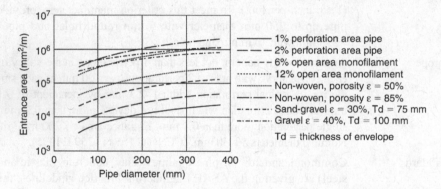

Figure 8.7 Typical open area of water entry per unit length for different envelope materials (Vlotman et al. 2000)

The entry resistance of the pipe-envelope combination should be as close as possible to zero (the ideal drain); see further section 11.1. A critical factor in comparing entry losses is the distance at which the closest observation well is placed relative to the pipe (Figure 8.7).

Selection criteria

As far as their practical drainage performance is concerned, there is very little to choose between the different types of drainpipe, provided they have been installed correctly. Selection, therefore, is mostly based on cost comparison and on local availability. In addition, the following considerations may be relevant:

a) where pipe drains are not locally available, local manufacture of concrete pipes is the most straightforward and easiest to organise. It requires little skill and can be economically done even on a very small scale. Local manufacture of clay tile drains requires a considerable investment and skill and is economically feasible only for large quantities. Plastics occupy an intermediate position; local manufacturing (i.e., with mobile plants), or imported (plastic) material is feasible for reasonable quantities

b)   where all types are available the use of corrugated plastic pipe often has distinct advantages. These are due to its light weight and delivery in long lengths, making transport costs (to the site as well as on site) relatively low and on-site handling relatively easy (can be carried by hand, as may-be required under wet conditions)

c)   plastic pipe has the general advantage that performance is less affected by poor installation. Plastic pipe is particularly suitable for machine installation

d)   costs of small diameter pipe (<10 cm) are usually of the same order of magnitude for tile, concrete and plastics. For large diameter pipe, concrete is normally the cheapest and plastic the most expensive.

8.4   Envelopes

An envelope is porous material placed around a perforated pipe drain to perform one or more of the following functions:

- *Filter function*: to provide mechanical support or restraint of the soil, at the drain interface with the soil, to prevent or limit the movement of soil particles into the drainpipe where they may settle and eventually clog the pipe. Initially some fines and colloidal material may pass through the envelope into the drain. When, after construction, the soil-envelope combination has stabilised, a limited flow of clay and other suspended particles, which are expected to remain in suspension in the drained water and leave the drain, is acceptable. The filter function may be temporary i.e., long enough to allow the disturbed soil to stabilise (organic envelopes have been used successfully for this purpose in the Netherlands)

- *Hydraulic function*: to provide a porous medium of relatively high permeability around the pipe to reduce entrance resistance at or near the drain openings

- *Mechanical function*: to provide passive mechanical support to the pipe in order to prevent excess deflection and damage to the pipe due to soil load

- *Bedding function*: to provide a stable base to support the pipe in order to prevent vertical displacement due to soil load during and after construction.

The latter two functions can only be achieved with gravel and sand envelopes.

8.4.1   Envelope need

The decision about the need for a drain envelope in a particular soil can be based on local experience or on empirical relationships between measurable soil properties. Soils in temperate humid areas, unless they are sand, generally have a strong soil structural strength and drains can be installed in such soils without envelopes. Soils with a high clay and/ or organic matter content also have higher structural strength. Simple correlation of soil structural strength with organic matter content or clay content have not been conclusive in determining whether drain envelopes will be needed for a particular soil, but this information, coupled with local experience, can give dependable predictions. For soil conditions found in the Netherlands, van Zeijts (1992) developed relationships between clay and silt contents of soils and the need for a drain envelope, as well as the appropriateness of envelope types (organic, synthetic, thin or voluminous) for certain soil types (Table 8.1).

Soils of the (semi) arid tropics are generally less stable, so clay content alone is not a good indicator of soil strength and stability. A parameter called *Hydraulic Failure Gradient*, HFG, has been developed to determine the resistance of soils to flowing water. The expected inflow to the drains and the area of openings in the drains can be used to calculate the exit gradient (related to the velocity of water in the soil) of water entering the drain openings for the case where there is no envelope. If the exit gradient (i_x) exceeds the hydraulic failure gradient of the soil, an envelope is needed.

The Plasticity Index (PI) and the saturated hydraulic conductivity (K_{sat}) of the soil are used in an empirical equation to determine the hydraulic failure gradient (Samani and Willardson 1986 in Vlotman et al. 2000):

$$HFG = e^{0.337 - 0.132K + 1.07\ln(PI)} \qquad \text{Eq. 8.2}$$

where:
HFG = the Hydraulic Failure Gradient
K   = the saturated hydraulic conductivity in m/d
PI   = the Plasticity Index (see section 5.4.2)
e   = the base of natural logarithm (2.7183)

The exit gradient is determined from the Hooghoudt equation (Eq. 12.7) solving for the maximum expected discharge (Q_{max}) as a function of the hydraulic conductivity of the soil (K_s), the maximum midpoint watertable height above drain level (h = W, see Figure 11.5), and the drain spacing. Based on Darcy's Law the discharge per unit length of the drain becomes:

$$q_{max} = \frac{Q_{max}}{L} \qquad \text{Eq. 8.3}$$

and the exit gradient can be determined from:

$$i_x = \frac{q_{max}}{K_{sat} A_{pu}} \qquad \text{Eq. 8.4}$$

where:
Q_{max} = the maximum possible discharge under free flow conditions in the drainpipe, m³/d
q_{max} = the maximum possible discharge per unit length, m²/d
K_{sat} = the saturated hydraulic conductivity at drain depth i.e., immediately adjacent to the pipe-envelope interface, m/d
A_{pu} = the actual area of inflow into the drainpipe in m²/m. Only the bottom half of the drainpipe is assumed to contribute. So $A_{pu} = \frac{1}{2} A_p$ which is the actual wetted area of inflow as a function of drain radius. See Figure 8.8 and Box 8.1 for actual values and suggested porosity value of synthetic materials. To calculate the exit gradient at the soil-envelope interface replace A_{pu} with the appropriate value of the selected envelope material.

Table 8.1 Recommendations on the use of the drain envelopes in the Netherlands based on soil type (Vlotman et al. 2000)

Soil				Envelopes[1]				
					Material			
					Gravel	Voluminous[4]		Thin[5]
						Organic	Synthetic	
						100 mm	4–7 mm	3–10 mm	< 1 mm
Type based on percentage clay and silt particles[2]	Geological formation	Remarks	Characteristics related to envelopes[3]	Function	thickness:			
					O90[7]:	N/A	650 μm	650–1750 μm	200–400 μm
> 25% clay	Alluvial; marine/fluvial	Ripe[8]	Stable; high K	–	No envelope necessary			
		Unripe[8]	Stable; low K	Hydraulic (temp.)	+	+	+	–
> 25% clay[6]	Alluvial; marine/fluvial	Ripe	Unstable; high K	Filter	+	–	+	+
		Unripe	Unstable; low K	Filter and hydraulic	+	–	+	–
< 25% clay < 10% silt	Marine	d_{50} < 120μm	Unstable; high K	Filter	+	–	+	+
< 25% clay < 10% silt	Aeolian	d_{50} > 120μm	Initially unstable; high K	Filter (temp.)	+	+	+	+
< 25% clay > 10% silt	Aeolian, fluvial or (fluvio) glacial		Initially unstable; low K	Filter (temp.) and hydraulic	+	+	+	–

1 + = suitable; – = not suitable; temp. = temporary
2 textures in soil profile above drain level, clay particles are < 2 μm and silt particles are 2–50 μm
3 high hydraulic conductivity: K ≥ 0.25 m/day, low K ≤ 0.05 m/day
4 voluminous envelopes in the Netherlands are primarily pre-wrapped loose coconut or synthetic fibres
5 only suitable if there is no risk of biochemical clogging (ochre/iron primarily)
6 at drain level > 25% but lighter layers with < 25% clay in soil profile above drain level
7 the O90 is the size of the envelope material opening at which 90% of the openings are smaller
8 see section 18.1

Darcy's Law is only valid for laminar flow conditions. When the flow is turbulent or in transition between laminar and turbulent, head losses and exit gradients will be higher for the same discharge.

The stability of the soil may also be expressed by means of the coefficient of uniformity (C_u) of the soil gradation curve:

$$C_u = \frac{d_{60}}{d_{10}}$$
Eq. 8.5

where:
C_u = the coefficient of uniformity
d_{xx} = the particle size at 60 and 10 percent passing (see Box 8.2 and Figure 8.10)

The coefficient of uniformity and the Plasticity Index are indicators of the tendency to siltation and with caution may be taken as an indicator of the need for an envelope (Table 8.2).

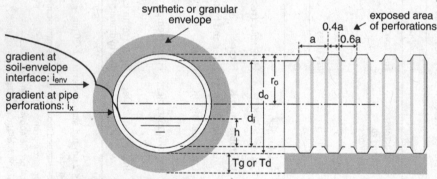

A_p = area of perforations per unit length
A_{pu} = ½A_p, or actual area as function of wetted perimeter (u)
A_{pe} = open area of envelope material
Td = envelope thickness granular material
Tg = thickness of synthetic material at ground pressure at drain depth

Figure 8.8 Exit gradients at pipe-envelope-soil interface

Table 8.2 Coefficient of Uniformity or Plasticity Index as indicator of soil siltation tendency (FAO 1976, Vlotman et al. 2000)

Silting tendency	C_u	or	PI (section 5.4.2)
no tendency	≥ 15		>12
limited tendency	5–15		6–12
high tendency	≤ 5		< 6

Placing a properly designed envelope around the drain to protect the drain openings will also reduce exit gradients. Increasing the drain diameter and increasing the area of perforations are other alternatives for decreasing exit gradients. The soils that do not require a filtering envelope are:

- Heavy clay soils (heavy clay soils can be defined as having a clay percentage > 60% and hydraulic conductivities < 0.1 m/d)
- Clay soils with the percentage clay exceeding 25–30% in humid climates
- Soils with a Plasticity Index (PI) greater than 12
- Soils with a Coefficient of Uniformity (C_u) > 15 and
- Coarse soils with 90% of the particle sizes larger than the maximum drainpipe perforation width.

For a soil to be well-graded the coefficient of uniformity must be greater than 4 for gravel and greater than 6 for sand (USBR 1974). The coefficient of curvature is a measure of asymmetry of the soil or gravel sample and is calculated from:

$$C_c = \frac{d_{30}^2}{d_{10} * d_{60}}$$

Eq. 8.6

C_c = 1 for a normally distributed sample. The coefficient of curvature must be between 1 and 3 for both gravel and sand to be used as envelope material, while also conforming to the C_u criteria above.

8.4.2   Material selection

Based on the material used the following types of envelopes can be distinguished (for detailed specifications see sections 8.5.1 through 8.5.3):

- *Granular envelopes.* Gravel or sand/gravel combinations
- *Organic envelopes.* A wide range of organic materials have been used in the past, including peat, top soil, sod, building paper, hay, straw, corn cobs, cloth, leather, wood chips, rice husks, etc. More recently, primarily wire coir[1] or coconut fibre, have been used. Coconut fibres in combination with synthetic fibres are also used. In Europe, Pre-wrapped Loose Material (PLM) is commonly used
- *Fabric envelopes.* A wide range of (synthetic) fabric material have been used as drain envelopes (besides limited use of natural fabrics; jute, cotton). These are the above-mentioned PLM envelopes with synthetic fibres only, thin knitted materials (also known as socks, or available as sheets), and a variety of non-woven materials, mostly thin to thick needle punched.

To select the type of envelope material a set of general conditions should be checked:

- availability of materials, and hence the likely cost
- expected function: hydraulic, filter, mechanical, bedding
- loading on the pipe and envelope
- handling characteristics during transport and transportation

- danger of biochemical fouling (iron ochre)
- ripening process of the soil
- organic matter and pH of the soil
- calcium carbonate content (of soil and granular envelope) and pH of the water
- climatic conditions, and finally
- required thickness as explained in the next section.

Box 8.2   Soil terminology

Soil texture

Refers to the size distribution of the constituting soil particles. Three size classes are distinguished: clay (< 2 μm), silt (2–50 μm) and sand (2000–50 μm). Sand may be subdivided (coarse, medium, fine); particles > 2000 μm = 2 mm are not soil but rock particles (gravel). On the basis of the size distribution, soils may be classified into textural classes (textural triangle, Figure 5.3).

Soil particle size

To determine the particle size distribution sieve analysis is performed (wet or dry sieving, see section 18.4). The material retained on the various sieves is calculated as the cumulative percent passing and shown graphically in a semi-logarithmic plot (e.g., Figure 5.3, Figure 8.10). Particle size is given as the d_{10}, d_{15}, d_{60}, d_{85}, d_{90}, etc. The value represented by d_{xx} is the diameter of the particle for which xx% on dry weight basis has a smaller diameter.

Soil bandwidth

Based on field investigations many soil particle size distribution curves will be available resulting in a wide band of soils on the semi-logarithmic particle size distribution plot. To select a representative bandwidth for drain envelope design, the 25 and 75 percent quartile values can be selected as representing the upper and lower boundary (Figure 5.4 and Figure 8.10).

Soil structure

Refers to the combination or aggregation of soil particles into aggregates or clusters (soil peds) that are separated from each other by weak forces. For particle size determination aggregates need to be broken up.

Soil consistency

This expresses the plasticity of the soil and as such its resistance to mechanical deformation and rupture. The state of plasticity of a soil is mostly determined by its clay and its moisture content and may be expressed by determining the *Atterberg consistency limits*. For drainage, the Plastic Limit (PL) and the Plasticity Index (PI) are relevant (see section 5.4.2).

8.4.3   Envelope thickness

The thickness of a drain envelope is determined by the need to reduce the exit gradients to acceptable levels as mentioned in the foregoing section, and by practical matters related to construction. For instance, it is not possible to reliably construct a granular envelope surrounding a drainpipe to a uniform thickness of 3 cm. Hence, when granular material is used the thickness described in specifications is typically 7–10 cm. For 100 mm (4 inch) corrugated plastic lateral pipe drain widely used for salinity control drainage of irrigated land, the gravel requirements (including wastage) typical runs in the order of some 0.05–0.10 m³ per m drain length.

The larger the circumference of the pipe and envelope material, the smaller the gradients that develop at the soil-pipe, the soil-envelope, and the envelope-pipe interfaces.

If it has been decided that the exit gradient at the soil-pipe interface (i_x) exceeds the Hydraulic Failure Gradient (HFG), implying that an envelope is needed, then a thin envelope will reduce the exit gradient at the soil-envelope interface considerably simply because of the much larger open area (Figure 8.7, and Eq. 8.4). If the exit gradient at the soil-envelope interface (i_{env}) is less than the HFG, use a thin envelope (thickness less than 5 mm), but if i_{env} > HFG determine the pipe plus envelope radius (r_{env}) such that $i_{env} \leq$ HFG. If the resulting thickness[2] is between 1 and 5 mm use a voluminous synthetic envelope, if the required thickness is >5 mm then synthetic materials may become too expensive and granular materials should be considered instead.

It should be noted that the thickness of synthetic materials as reported by manufacturers is determined at a pressure of 2 kPa (ASTM D1599–88) to an accuracy of 0.02 mm, but that under actual field conditions the material will be even more compressed. The pressure of 2 kPa is approximately equivalent to 0.1 m of soil pressure. Compressibility factors (i.e., the ratio of thickness at 2kPa and thickness at 1–3 m soil depth) depend on the type of material used but the most common ones for drains at 1 to 3 m depth are: nonwoven needle punched materials 0.7–0.58; for needle punched heat bonded and woven monofilament materials 0.94–0.9; and for woven slit film material 0.88–0.81 (Figure 8.9). The thickness of commonly used geotextiles is between 0.25 and 7.5 mm (or 10–300 mils, where 1 mil = 0.001 inch = 0.0254 mm, see also Box 8.3). Besides the foregoing theoretical determination of the required envelope thickness various organisations prescribe minimum thickness for different materials (Table 8.3).

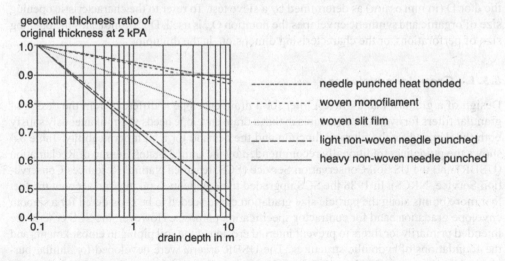

Figure 8.9 Compressibility of geotextiles as a function of load. Vlotman et al. 2000

Table 8.3 Required thickness of drain envelopes

Description	Min. thickness in mm	Remarks
Vegetative material	150	ASAE EP260.4 (ASAE 1984) recommends use only for rigid pipes and not plastic which depends on lateral support provided by granular envelope
Pre-wrapped Loose Material (PLM):		CEN/TC 155 WI 1261, 1994
synthetic fibrous	3	Thickness determined according ISO 9863
synthetic granular	8	or as in Annex B of this standard. Deviation
organic fibrous	4	no more than 25% from declared thickness
organic granular	8	by manufacturer.
Coconut fibre:		NEN-7047, 1981 (Dutch Standard)
Type 750 g (mass > 750 g/m^2)	6	but not greater than 10 × actual mass/750
Type 1000 g (mass > 1000 g/m^2)	8.5	but not greater than 13 × actual mass/1000
Gravel envelope:	75	SCS 1971, SCS 1988 and 1991 do not describe a minimum thickness.
	>75	Depends on construction methodology and type of material i.e., crushed rock, or river run material and gradation required. Also, trenchers have a fixed trencher box for different pipe diameters hence width may vary from 75–200 mm.

8.5   Envelope design guidelines and criteria

The standard notation used herein to refer to particle size of the base soil material and the envelope material will be d_{xx} and D_{xx} respectively. The number following each letter (xx) is the percentage of the sample, by weight (cumulative percentage passing), that is finer than the d or D (in mm or μm) as determined by a sieve test. To refer to the characteristic opening size of organic and synthetic envelopes the notation O_{xx} is used. $D_{opening}$ indicates the opening size of perforations, or the characterising dimension, in the drainpipe.

8.5.1   Granular envelopes

Design of a granular (sand-gravel) filter for a drain envelope is different from the design of granular filters for hydraulic structures in that a drain envelope needs to simultaneously satisfy both the demand for the filtering function and the demand for a high permeability. Table 8.4 shows the guidelines traditionally recommended by the United States Bureau of Reclamation (USBR) and the US Soil Conservation Service (SCS, renamed Natural Resources Conservation Service, NRCS). In 1988 the SCS upgraded their guidelines, taking into account the fact that more points along the particle size gradation curve needed to be considered for a smooth envelope gradation band for contractor specification purposes. However, the SCS criteria are intended primarily for filters to prevent internal erosion of the soil piping in embankments and the foundations of hydraulic structures. The USBR criteria were developed for similar purposes. Vlotman et al. (2000) took these criteria as well as insights obtained with agricultural drainage conditions in Pakistan and Egypt into account, and came up with revised guidelines specifically for granular drain envelope material (Table 8.4d).

It is assumed that from a pre-drainage soil investigation a base soil bandwidth for soils in need of a drain envelope has been determined (for details see section 5.4.1). The finer boundary of the base soil bandwidth will be used for the filter/retention criteria and the coarser boundary to satisfy the permeability or hydraulic criteria.

Table 8.4 Existing and recommended guidelines for granular envelope design

(a) USBR GRAVEL DRAIN ENVELOPE DESIGN (USBR 1978, 1993)

Non-filter envelope	$D_{100} < 38$ mm	(1.5" US standard sieve series).
	$D_5 > 0.3$ mm	(sieve N°50 US standard sieve series).
For well graded material [1]	$C_u > 4$	for gravel.
	$C_u > 6$	for sands.
	$1 < C_c < 3$	for both gravel and sand.
new in 1993:	$K_{env} \geq 10\ K_s$	when $K_{env} > 150$ m/d material difficult to place without segregation.

gradation relationship between base material and diameters of graded envelope material:

Base soil limits for d_{60} (mm)	Lower limits (mm) percentage passing						Upper limits (mm) percentage passing					
	100	60	30	10	5	0	100	60	30	10	5	0
0.02–0.05	9.52	2.00	0.81	0.33	0.30	0.074	38.10	10.00	8.70	2.50	–	0.59
0.05–0.1	9.52	3.00	1.07	0.38	0.30	0.074	38.10	12.00	10.0	3.00	–	0.59
0.1–0.25	9.52	4.00	1.30	0.40	0.30	0.074	38.10	15.00	13.10	3.80	–	0.59
0.25–1.0	9.52	5.00	1.45	0.42	0.30	0.074	38.10	20.00	17.30	5.00	–	0.59

(b) SCS DRAIN ENVELOPE CRITERIA (SCS 1971)

SCS criteria for envelope (SCS 1971, revised in 1988, see below)

Graded envelope	$D_{50}/d_{50} = 12–58$	minimal thickness 3" (75 mm).
	$D_{10} \geq 0.25$ mm	0.25 mm = US standard sieve N°60
	$D_{15}/d_{15} = 12–40$	
Uniform envelope	$D_{15}/d_{85} < 5$	
	$D_{85} \geq 0.5 \times$ Diameter of the perforations.	

(c) SCS REVISED DRAIN ENVELOPE CRITERIA (SCS 1988 and also in SCS 1991)

SCS criteria for filter gradation (SCS 1988)

	$D_{15} < 7\ d_{85}$	but D_{15} need not be smaller than 0.6 mm (2).
	$D_{15} > 4\ d_{15}$	
	$D_5 > 0.074$ mm	% passing US standard sieve N° 200 less than 5%.

SCS criteria for envelope (surround) (SCS 1988)

	$D_{100} < 38.1$ mm	whole sample should pass sieve of 1.5" (38.1 mm).
	$D_{30} > 0.25$ mm	% passing US standard sieve N° 60 less than 30%.
	$D_5 > 0.074$ mm	% passing US standard sieve N° 200 less than 5%.

(d) RECOMMENDED DRAIN ENVELOPE CRITERIA (Vlotman et al. 2004)

Control points coarse boundary envelope material bandwidth (c and f in subscript refer to coarse and fine base soil bandwidths):

1. Filter, retention criterion	$D_{15c} < 7\ d_{85f}$	SCS (1988)
2. Gradation curve guide	$D_{60c} = 5\ D_{15c}$	
3. Segregation criterion	$D_{100} < 9.5$ mm	based on Pakistan findings (Shafiq-ur-Rehman 1995).

(Continued)

Table 8.4 (Continued)

Control points fine boundary envelope material bandwidth:	
4a. Hydraulic criterion	$D_{15f} > 4d_{15c}$
4b. Gradation curve guide (bandwidth)	$D_{15f} > D_{15c} / 5$ Based on $C_u \leq 6$ and bandwidth ratio ≤ 5.
5. Hydraulic criterion	$D_5 > 0.074$ mm
6. Gradation curve guide (bandwidth)	$D_{60f} > D_{60c} / 5$ bandwidth ratio ≤ 5 below 60% passing on PSD curve.
7. Retention criterion (bridging)	$D_{85} > D_{opening}$
8. Construction criterion	All openings should be covered by at least 76 mm (3") of filter material. The envelope should not contain deleterious material.
9. Crushed material criterion [3]	No particles to be disproportionably larger in one direction by a factor of 2 compared to the shortest dimension. Twenty-one sieve analysis to check for missing particle ranges. $K_{env} < 300$ m/d.

NB For detailed references see Vlotman et al. 2000. SCS is now Natural Resources Conservation Service of USDA (www.nrcs.usda.gov).

(1) C_u is Coefficient of Uniformity Eq. 8.5, C_c is Coefficient of Curvature Eq. 8.6

(2) Original text read: 'but not smaller than 0.6 mm'

(3) For more description see below.

The criteria will result in control points on the Particle Size Distribution (PSD) curve through which the coarse and fine boundaries of a granular envelope bandwidth can be drawn (subscripts c and f mean coarse and fine boundary respectively in Table 8.4). Gradation curve guides and bandwidth guides determine the recommended shape of the curves. They are not criteria rather they are guidelines.

Example granular envelope design

Figure 8.10 shows two examples of how to determine the envelope gradation band for specifications from the bandwidth of soils that require drains to be constructed with an envelope. The first example is based on soils of the Fourth Drainage Project area near Faisalabad in Pakistan. The second example uses what is typically reported in the literature as the range of UK problem soils from a stability point of view.

From the graph the d_{xx} values of the base soil (see Table 8.4d) can be found. With the guidelines the seven control points D_{xx} for the fine and coarse envelope material boundaries can be determined. Note that for some of the guidelines the coarse base soil boundary determines the fine envelope material boundary and vice versa.

In some cases, there will be conflict with respect to the relative position of control points 4a, 4b and 5. This is caused by the conflicting objectives of high permeability and high filtering capability. For instance in both cases it was decided that guideline 4b should be more dominant, implying that in the Pakistan case a higher permeability was accepted in favour of a more practical envelope bandwidth and in the UK case a lower permeability than would follow from the hydraulic criterion (control point 4a).

In particularly for the UK case, using control point 4a would have resulted in specifications that would have been hard to satisfy by the contractor. It remains the judgement of the design engineer to decide the final shape of the envelope gradation band at the 15% passing.

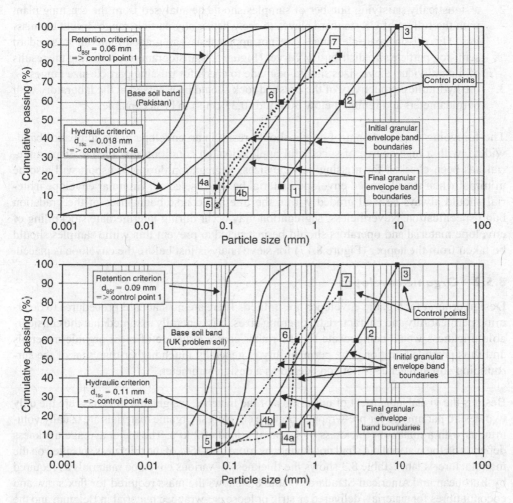

Figure 8.10 Sample designs of granular drain envelopes; Pakistan top figure, UK bottom figure

One may want to use some of the USBR and SCS criteria in Table 8.4 to see if this would result in a more acceptable gradation band at the 15% passing but bearing in mind that the main objective of those criteria is the filter function.

Granular material is either naturally available alongside the upper reaches of rivers (River-Run material) or from other sources and is the most widely used, generally with good results. The use of crushed rock for granular envelope material has been quite acceptable in most cases, although it failed to function as a filter with subsurface drains in Pakistan. This was because of missing ranges of particle sizes and hence the envelope material was not well graded, and internal bridging amongst different particle sizes did not occur. Therefore, crushed materials are acceptable provided the following provisions are adhered to:

1. there should be no particles that are disproportionately larger in one direction by a factor >2 compared to the shortest dimension (long time requirement in specs. Coming from USBR and Corp of Engineers criteria and found to have merit; this prevents segregation during transport and large pore spaces)

2. a statistically satisfying number of samples should be analysed from the crushing plant with the full set of US standard sieves to see whether any particle ranges might be missing. The missing particle ranges are not apparent as gap-graded material in standard semi-log particle size distribution curves (Figure 5.4); histograms representing the results of the individual sieve sizes should be made to check for missing particle size ranges
3. the hydraulic conductivity of the crushed rock should be assessed in the laboratory with permeameters and remain below 300 m/d (0.35 cm/s) to be acceptable.

The foregoing steps and control points should result in an envelope gradation band (bandwidth), with a high degree of successful application, on condition that the envelope material has been checked for missing particle sizes. Segregation during transport will also be minimal when a well graded envelope material has been selected. Granular envelope material should always be well graded within the fine and coarse boundaries of the gradation band specification. Nevertheless, segregation may occur during intermediate stockpiling of envelope material and operators should be instructed to prevent this while samples should be taken from the hopper (Figure 8.17) for sieve analysis just before the envelope is placed.

8.5.2 Organic envelopes

Design criteria for organic envelopes are limited. There are no standard procedures to determine permeability and characteristic opening sizes. It is generally assumed that the considerable thickness would compensate for the rather coarse structure of most organic materials and thus provide filtration. The permeability is generally much higher than that of the surrounding soil and hence is not considered as a design parameter.

Besides the visual judgement of uniformity of the material, weight and thickness are the only two design parameters for pre-wrapped organic envelopes. Organic materials are usually voluminous with a minimum thickness of 4 mm at a load of 2 kPa. The mean average thickness depends on the material but shall not deviate by more than 25% of the thickness specified on the manufacturer's label. Table 8.3 shows the thickness of various envelope materials as required by European and American Standards. Table 8.5 shows the mass required for flax straw and coconut fibres for materials delivered as strip or loose pre-wrapped material in Belgium and the Netherlands. Both weight and thickness ensure a proper functioning of the organic envelope.

8.5.3 Synthetic envelopes

There are more than twenty-two sets of criteria for geotextile filter design (Koerner 1994, Vlotman et al. 2000). The design criteria most appropriate for drain envelope design are shown in Table 8.6. There are four classes of criteria for the design of synthetic

Table 8.5 Guidelines for the required mass of organic material around drainpipes

	Flax straw		Coconut Fibres	
	strip	pre-wrapped	strip	pre-wrapped
Nominal mass	2 000 g/m²	1 500 g/m²	1 000 g/m²	750 g/m²
Minimal mass	1 800 g/m²	1 350 g/m²	900 g/m²	675 g/m²

Table 8.6 Selected existing criteria for geotextile filter design

Geotextile	Soil	Flow type	Criteria	Remarks
(a) Christopher and Holtz (1992) in Wilson-Fahmy et al. 1996				
not specified	$d_{50} < 74$ μm (sieve No 200)	steady	$O_{95}/d_{85} \leq B$	Federal Highway Admin. (FHwA)
	$2 \leq C_u \geq 8$ [1]		$B = 1$	Origin: USA
	$2 < C_u \leq 4$		$B = 0.5$ [2]	
	$4 < C_u < 8$		$B = 8/C_u$	
woven	$d_{50} \geq 74$ μm (sieve No 200)	steady	$O_{95}/d_{85} \leq 1$	
non-woven	$d_{50} \geq 74$ μm (sieve No 200)	steady	$O_{95}/d_{85} \leq 1.8$	
both	$d_{50} \geq 74$ μm (sieve No 200)	steady	$O_{95} \leq 300$ μm	(US standard sieve No. 50)
all materials	all soil types	dynamic	$O_{95}/d_{85} \leq 0.5$	incl. pulsating and cyclic [3]
(b) Dierickx (1992a, 1993a, 1994, 1996)				
Origin: Belgium, Egypt, Pakistan				
all materials	all soil types	steady	$O_{90}/d_{90} > 1$	permeability criterion
thin geotextiles ($T_g \leq 1$ mm)		–	$O_{90}/d_{90} < 2.5$	retention criterion
voluminous ($T_g \geq 5$ mm)		–	$O_{90}/d_{90} < 5$	retention criterion
			$O_{90} > 200$ μm	hydraulic criterion and anti-clogging

(c) RECOMMENDED DESIGN CRITERIA (Vlotman et al. 2000)			
Geotextile	Flow type	Criteria	Remarks
thin geotextiles[5] ($T_g \leq 1$ mm)	–	$O_{90}/d_{90} < 2.5$	retention criterion
voluminous ($T_g \geq 5$ mm)	–	$O_{90}/d_{90} < 5$	retention criterion
$1 \leq T_g \leq 5$ mm	–	interpolate between $O_{90}/d_{90} = 2.5$ and 5	
	–	$O_{90} > 200$ μm	hydraulic & anti-clogging criterion
	dynamic & steady	$K_e \geq a K_s$	hydraulic criterion where $a = 0.1$ no safety, $a = 1$ for non-critical conditions[4], and $a = 10$ for reverse flow conditions
	–	$O_{90}/d_{90} > 1$	anti-clogging criterion
		$O_{90} > 100$ 100–200 μm anti clogging criterion	

for mechanical strength and other criteria see Box 13 in Vlotman et al. (2002).

NB For detailed references see Vlotman et al. 2000.

(1) There is some confusion in the literature whether it should be $1 < C_u < 2$ (Williams and Luettich 1990) or $2 \geq C_u \geq 8$ (Wilson-Fahmy et al. 1996)

(2) Williams and Luettich 1990 report $B = 0.5 C_u$

(3) Christopher and Holtz (1989) presented slightly different criteria for dynamic soils which must have been superseded by the 1992 reference: if soil can move beneath geotextile $O_{95}/d_{15} \leq 1$ or $O_{50}/d_{85} \leq 0.5$

(4) A non-critical condition is where flow is steady and in one direction only (no reverse flow)

(5) The true thickness is actual uncompressed thickness as measured or the thickness specified by manufacturers at 2 kPa, times the compressibility factor (see section 8.4.3).

envelopes: (1) the retention criteria, (2) the hydraulic criteria, (3) the requirement to avoid long-term clogging, and (4) mechanical and strength criteria. See Box 8.3 for explanation of some of the standard terms, e.g., fabric, geosynthetics, etc.

As with granular envelopes the permeability and soil retention criteria are often in conflict with each other and it is up to the judgement of the designer to allow the most important ones to prevail. The criteria are applicable for all soils (sand and clay). Although the term criterion is used, all are guidelines based on practical experience and often with limited validity outside the conditions under which, or for which, they were developed.

Retention criteria. Essentially three types exist: (1) O_{90}/d_{90} ratios; (2) O_{95}/d_{85} ratios; and, (3) a range of ratios using O_{90}, O_{50}, O_{15} in combination with d_{15}, d_{50} or d_{85}. The third range of ratios have become more popular in recent years due to advances in the determination of Opening Size Distribution (OSD) curves of synthetic materials. The second series of ratios are common in the US and France but have one main disadvantage and that is the determination of the O_{95}.

European standards recommend O_{90} as the characterising property for filtration by synthetic materials, because the O_{95} can vary considerably as it is generally located on the gentle sloping part of the semi-logarithmic OSD curve (implying that a small change vertically results in large changes in opening size). The type 1 ratios (O_{90}/d_{90}) seem to be the most practical criterion at present.

Hydraulic criteria. There are three criteria that are most frequently mentioned in the literature: (1) $K_e > 0.1 K_s$, (2) $K_e > K_s$, and (3) $K_e > 10 K_s$. Criterion 1 has little or no safety built in. The others have to different degrees. For agricultural criteria, it would seem that criterion 3, which is similar to expectations with granular envelopes, is best applied to most field conditions.

Anti-clogging. Clogging is a decrease of permeability in the long-term of synthetic fabrics by particles of the base soil. This is different from blocking of synthetic envelopes, which is the immediate and near total loss of permeability of the envelope by a layer of fine particles (commonly caused when smearing under wet construction conditions takes place). Dierickx (Table 8.6) proposed to use $O_{90} > 200$ μm as a guideline both from the hydraulic as well as the anti-clogging point of view. His criterion is to prevent clogging of the fabric itself. Christopher and Holtz remark that when $C_u > 3$ use $O_{95} > 3d_{15}$ and when $C_u < 3$ use the maximum opening size allowed from the retention criteria (for C_u see Eq. 10.5).

Mechanical strength properties are important for handling and installation. Drainpipes wrapped with geotextiles are sometimes exposed to natural weathering and chemical deterioration, which may affect their functioning. Therefore, information on the following properties of geotextiles, used as envelopes for drainpipes, are needed: thickness and mass per unit area; strength of the material (tensile-, grab-); strength of the joints (seam strength); static puncture resistance; compressibility; abrasion resistance; and, resistance to material deterioration. For sample specifications see Koerner (1994) and Vlotman et al. (2000).

From the foregoing, it is clear that there is a plethora of guidelines as well as criteria. For agricultural drainage conditions Vlotman et al. (2000) found the criteria in Table 8.6c the most practical.

Example of synthetic envelope design

From the base soil bandwidth for Pakistan and the UK (Figure 8.10) find that for Pakistan the d_{90} varies between 0.08 and 0.4 mm or 80–400 μm. The UK soil ranges between 93 and 290 μm. It is common to express characteristic opening sizes of geotextiles in μm, or 'microns'. Manufacturers publish specifications for geotextiles, e.g., the annual Specifier's Guide of the Journal Geotechnical Fabrics Report, Engineer's Guide to Geosynthetics publishes the Characteristic Opening Size (AOS) which is the O_{90}.

Suppose from the thickness determination (see section 8.4.2), it was found that the envelope should be $T_g > 4$ mm and that there is a ready supply available of nonwoven thick geotextiles. With compression taken into account, the material should have a thickness range of 4/0.58 to 4/0.7 or 6.9–5.7 mm. Manufacturers often specify thickness in mils (Box 8.3): the material should be between 225 and 270 mils.

Box 8.3   Common terms used with synthetic and geotextiles applications in civil engineering

Geosynthetic	materials manufactured from various types of polymers used to enhance, augment and make possible cost effective environmental, transportation and geotechnical engineering construction projects. They are used to provide one or more of the following functions; separation; reinforcement; filtration; drainage; liquid barrier.
Geotextiles	flexible, textile-like fabrics of controlled permeability used to provide all of the above functions, except liquid barrier, in soil, rock and waste materials. Natural fibre geotextiles (e.g., jute, etc.) are also considered to fall within the geotextile classification.
Geomembranes	essentially impermeable polymeric sheets used as barriers for liquid or solid waste containment.
mil	a measure to indicate the thickness of the fabric. 1 mil = 0.001 inch = 0.0254 mm.
dtex, tex	the weight of a 10 000 m length of fibre in grams. tex is weight per 1 000 m. The d is for deci.
Dernier	the weight in grams of 9 000 m of a single fibre. Mixtures of fibres of different dernier are used to create desired characteristic opening sizes (COS). It gives the fibre size of woven materials.
COS	Characteristic Opening Size. The COS is subject to many definitions: AOS, EOS, FOS, O_{90}, O_{95}, O_{98} see Vlotman et al. 2000.

Next, check the characteristic opening requirement, AOS or O_{90}. From Table 8.6c it follows that for $T_g = 4$ mm the O_{90}/d_{90} ratio should be less than 4.3 and greater than 1. For the Pakistan soil the O_{90} should range between 80–345 μm for the fine soil boundary and 400–1 720 μm for the coarse soil boundary. In addition to prevent clogging $O_{90} > 200$ μm is recommended. Hence, consider materials that have a characteristic opening size of 300–1 800 μm assuming that due to compression the actual opening size will be somewhat less (a reduction of 10% was found to be typical for non-woven needle punched materials when thickness was reduced under pressure by 50%). Similar consideration for the UK soil bandwidth would result in O_{90} between 300 and 1 300 μm.

The permeability of the material should be considered as well. Manufacturers sometimes report permittivity and transmissivity values. Permittivity is permeability normal to the plane and values of 0.02–2.2 s^{-1} without a load (pressure) are typical. Permeability or hydraulic conductivity is obtained by multiplying permittivity by the material thickness T_g resulting in a possible range of 0.0005–1.1 cm/s ($T_g = 0.25$ mm – 5 mm). For heat bonded woven and non-woven materials the reduction of permeability under load is slight, while for non-woven needle punched geotextiles it was slight to moderate; no actual data are reported (Koerner 1994). Transmissivity is the permeability in the plane of the material and is not relevant for agricultural drainage but is highly relevant for vertical drainage on large civil works. There are no strict guidelines (Table 8.6c) for synthetic material permeability and common sense dictates that as long as the permeability of the material is greater than the surrounding saturated hydraulic conductivity of the soil, one should be alright.

Finally, various mechanical strength requirements should be considered. A material that can be torn apart with your hands is likely to deform during the pressure and stretching that occur during installation. Pre-wrapped drainpipes are likely to be lifted in the field by grabbing the envelope material; hence it should be strong enough. Sharp stones, roots, sticks, etc. that may be found in the sub-soil may puncture the material and hence puncture strength should be adequate. Also, for the mechanical strength of the geotextiles there are no precise criteria, but some guidelines may be found in various government specifications that may or may not be founded on actual research and testing (see Koerner 1994 for geotextiles in general, Vlotman et al. 2000 for agricultural drainage application).

When there are uncertainties about the quality of the material, laboratory testing in permeameters is recommended and detailed instructions for this may be found in Vlotman et al. 2000.

8.6   Structures in pipe drain systems

Most structures are of a simple type and occur mainly in the larger composite systems. Typical examples are shown in Figure 8.11.

8.6.1   Surface water inlets

Surface inlets are not often used for agricultural drainage, except perhaps the buried or blind inlet (sometimes also referred to as a French drain). Inlets by means of a grating or via a riser pipe are more commonly used in municipal park settings where incidental removal of surface water from localised areas is necessary. Provision of proper silt traps is essential for open inlets; an inlet located to the side of the drain line, providing a safeguard against poor maintenance, is preferred in this respect.

8.6.2   Inspection, junctions and control

Inspection access and junctions are often combined in manholes. With the advent of controlled drainage various types of control mechanism are built in the manhole. Manholes are used in composite pipe drain systems to connect selected lateral pipes with the collector drainpipes. Most connections are made using blind junctions, which cannot be inspected after installation (Figure 8.4 and Figure 8.5). In critical locations a combined junction/silt trap should be provided, with access for inspection, regulation (controlled drainage), and for cleaning the drain from time to time. This type of inspection chamber is expensive and so should only be used in vital locations, e.g., where the slope of a collector drain flattens out (liability to silting) or on long lengths of drain (every 250 m on long runs). Experience has shown that above ground inspection chambers are liable to be vandalised or damaged by farm implements and so there is an increasing tendency to install buried chambers with some marker (e.g., metal covers used in combination with a metal detector) to enable their location to be pinpointed later.

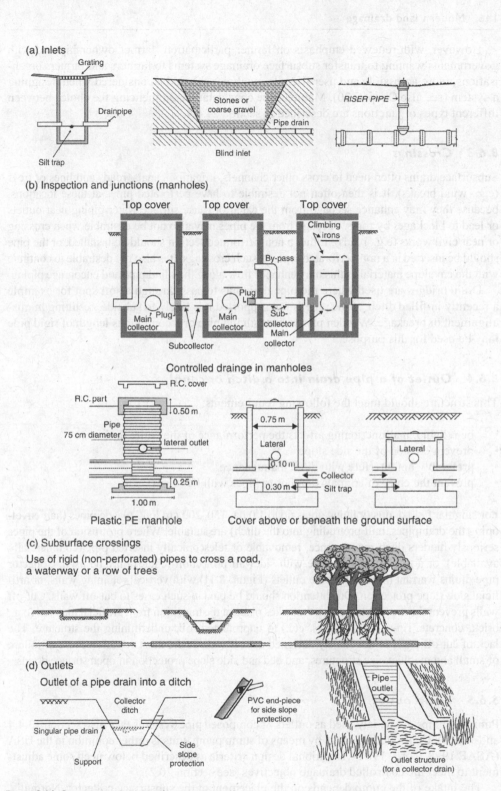

(a) Inlets

Grating

Drainpipe

Silt trap

Stones or coarse gravel

Pipe drain

Blind inlet

RISER PIPE →

(b) Inspection and junctions (manholes)

Top cover

Top cover

Top cover

Climbing irons

By-pass

Plug

Main collector

Plug

Subcollector

Main collector

Sub-collector

Main collector

Controlled drainge in manholes

R.C. cover

R.C. part

0.50 m

Pipe 75 cm diameter

lateral outlet

0.25 m

1.00 m

Plastic PE manhole

0.75 m

Lateral

0.10 m

Collector

0.30 m

Silt trap

Lateral

Cover above or beneath the ground surface

(c) Subsurface crossings

Use of rigid (non-perforated) pipes to cross a road, a waterway or a row of trees

(d) Outlets

Outlet of a pipe drain into a ditch

Collector ditch

Singular pipe drain

Support

Side slope protection

PVC end-piece for side slope protection

Pipe outlet

Outlet structure (for a collector drain)

Figure 8.11 **Structures in a pipe drain system**

However, with renewed emphasis on farmer participation, farmer ownership and with governments wanting to transfer subsurface drainage systems to farmers and farmers organisations, more accessible and user friendly manholes should be considered when designing a system (see also Chapter 20). Maintenance considerations influencing the choice between different types of junctions are described in section 24.4.4.

8.6.3 Crossings

Subsurface drains often need to cross other channels, irrigation canals, roads and lines of trees (e.g., wind breaks). It is then often not desirable to have perforated pipes at these locations, because they may enhance seepage from the canal or cause surface water piping near outlets or lead to blockages by roots. Seepage from the pipes may also not be desirable when crossing or near civil works (e.g., roads). Either a non-perforated section should be installed, or the pipe should be inserted in a non-perforated pipe. At such crossings, it is also not desirable to continue with the envelope material as this may enhance flow along the blind pipe and encourage piping.

Drain bridges are used where the pipe drain has to pass through a soft spot for example a recently in-filled ditch. If the pipe is not supported it is likely to subside, resulting in misalignment or breakage. Wooden planks beneath the pipe or a continuous length of rigid pipe may be used for this purpose.

8.6.4 Outlet of a pipe drain into a ditch or canal

This structure should meet the following requirements:

- be reliable: malfunctioning affects the performance of the entire drain
- prevent erosion of the side slope
- preferably not interfere with ditch maintenance
- prevent the entry of animals (applies to pipes with ID>5 cm).

For singular field drains draining areas up to 1.0 ha, 150–200 cm long plastic pipes (half enveloping the drainpipes, half protruding into the ditch) are suitable. Where protrusion of the pipes seriously hinders ditch maintenance, removable or telescopically inserted pipe (commercially available), or a non-protruding pipe with sidewall protection (chute) may be used. Collector pipe drains warrant more expensive outlets (Figure 8.11) with vertical retaining walls, or artificial side slope protection. Due attention should be paid in such cases to cut-off walls. Cut-off walls prevent the inevitable erosion that takes place at the transition from artificial material (e.g., brick, concrete, rip-rap, geotextiles, etc.) to unprotected soil, undermining the structure. The lack of cut-off walls or adequately designed cut-off walls is the most common cause of failure of small and medium size structures, and bed and side slope protection in open surface drains.

8.6.5 Sump outlet

Pumped sumps are widely used as outlets of composed pipe systems (see also section 13.4.4 and 16.3.3). Controlled drainage by means of sump pump units is rather common in the USA (ASAE 1998, Corey 1981). Traditional design criteria as described below need some adjustment to meet the controlled drainage objectives (see section 20.2).

The intake of the pump depends on the placement of the subsurface collector. Normally, the highest water level in the sump should not exceed the bottom of the collector (Figure 8.12). However, to store water in the soil and not mobilise solutes, water levels above

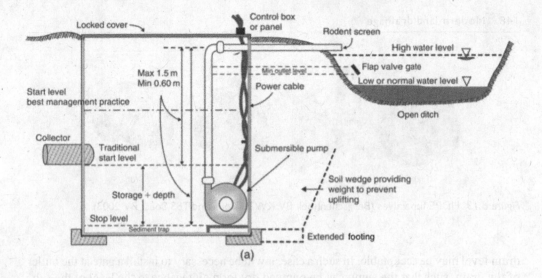

(a)

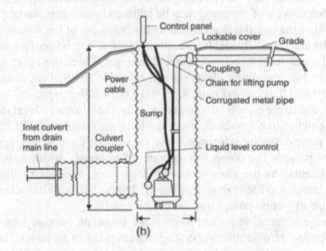

(b)

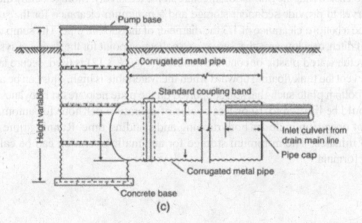

(c)

(a) Small agricultural sump unit, with fixed submersible electric pump: two start levels indicated
(b) Portable submersible pump in a corrugated steel manhole
(c) Details of a design of steel manhole and collector drain

Figure 8.12 Sump and submersible pump for small scale subsurface drainage (after ASAE 1998 and Corey 1981)

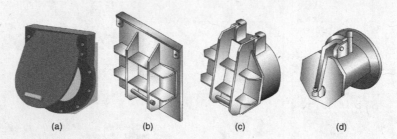

(a)   (b)   (c)   (d)

Figure 8.13 HDPE flap valves (Bergschenhoek BV, KWT Group and TBS Soest BV 2003)

drain level may be acceptable. In such a case, it will be necessary to install a gate at the outlet of the drain, such that the sump can be pumped dry for maintenance to the level of the submersible pump. Start levels of the pump may be adjusted based on seasonal drainage needs.

Discharge level of the pump is determined by the condition of the natural outlet. If water levels at the outlet are relatively stable, the discharge pipe can be set just above the maximum anticipated water level to provide a free discharge. Where the outlet water elevation fluctuates considerably, the outlet elevation can be made lower, thus reducing operating costs. To achieve this, a flap valve (Figure 8.13) must be installed to prevent backflow during periods when the discharge pipe is submerged. Finally, the discharge level may be selected to be deliberately higher to accommodate storage for reuse (see section 20.2.3).

The required pump capacity must match the capacity of the drainage system such that during peak flow periods the pump can operate continuously while satisfying drainage discharge requirements. As the growing season progresses, drainage requirements should decrease and the pump will then run intermittently. Temporary, portable submersible pumps or the transportable irrigation pumps may be used instead.

The sump storage volume, shape and position are important because together with pump capacity, they determine the intermittent operating characteristics of the pump. The sump must be large enough to ensure that the pump doesn't start/stop excessively. Storage below the minimum water level serves to provide sediment storage and a minimum clearance for the suction pipe. There should be a bottom clearance of 1/3 the diameter of the suction pipe. The sump can be a pit, tank, section of ditch, or a low area that serves as a collection point for the drainage system. When a tank (e.g., prefabricated plastic or concrete manhole (Figure 8.12) is used, secure anchoring is essential to prevent the tank floating upward when the watertable is high. This can be achieved by extending the bottom plate such that soil supported on the plate helps resist buoyancy.

Pumps should be limited to 10 or less cycles of operation per hour for automatic operation. A cycle of operation includes both running and standing time. Running time should not be less than 3 minutes. The minimum storage for automatic operation can be calculated by the following formula:

$$A = \frac{Q}{\Delta hn}$$

Eq. 8.7

where:
A = the minimum storage area (circular or square)
Δh = the height between start and stop level

n = the desired number of cycles per hour

Q = is the total pumping capacity (more than one pump may be installed, see also section 13.4.4.)

The storage area and depth (Δh) are then chosen so that their product is equal to or greater than Q. For an economical operation the sump should be large and shallow, not small and deep. Storage depths of 0.6–1 m are recommended for closed sumps and 0.3 m for open sumps.

8.7   Construction of pipe drain systems

The construction of a pipe drain line normally involves:

a)   setting out of alignments and levels
b)   installation of drainpipe in the soil.

The main technical aspects of these works are described in the following sections while relevant information and guidelines on their costing is presented in Chapter 2.

8.7.1   Setting out, depth and grade control

With advances in equipment for setting out, such as laser and Global Positioning Systems (GPS), setting out and depth control have evolved significantly, however, the principles remain the same (Figure 8.14a & 8.14b). The beginning and end of each drain line is marked by stakes (for long lines intermediate stakes are installed). The stake levels are adjusted to establish a line of sight parallel with the grade line (= trench bed slope) at a convenient height above the soil surface (sightline). Boning rods (also called travellers) are used as necessary to establish the correct level of the bed of the trench. Most times these days the staking is primarily for the alignment, while elevation control is achieved by laser level (Figure 8.14c). Installation depth and grade may be controlled manually by the machine operator maintaining a reference mark on the digging part of the machine (Figure 8.14b). With hand installation, boning rods are used for this purpose.

Modern methods employ laser equipment and GPS. A laser control unit consists of an emitter, located at the edge of the field, which establishes a sloping reference plane over the field by means of a rotating laser ray. A receiver on the machine visualizes this plane of reference and transmits the signal via the regulating system to the machine's control hydraulics (Figure 8.14c). The reference plane can be set to a grade with an accuracy of 0.5 cm per 100 m (0.005%). About five depth checks per second are normally made (rotation speed of the laser beam). The GPS units provides the necessary x-y coordinates. Elevation can also be obtained from a GPS when sufficient satellites are available at the same time. However, accuracy is nowhere near that of the laser unit and may not be sufficient. More elevation accuracy can be obtained with a differential GPS unit (DGPS), which uses a ground base station for elevation accuracy.

Standards of depth control vary somewhat between different countries though there is general agreement that the vertical position of the drainpipe should not be allowed to deviate by more than half the pipe diameter from its designed depth. Wavy vertical alignments may lead to the development of airlocks which reduce the flow in the pipe, and in extreme cases even

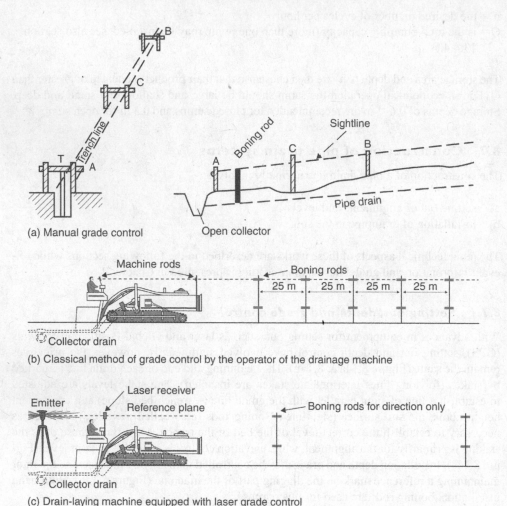

(a) Manual grade control

(b) Classical method of grade control by the operator of the drainage machine

(c) Drain-laying machine equipped with laser grade control

Figure 8.14 Alignment and elevation of a pipe drain line (after EPADP/RWS 2000)

block the pipe. For grades, the maximum permitted deviation is normally half the pipe diameter per 100 m. These standards can be achieved by skilled operators using manual control while machines fitted for laser control can easily lay claim to even more exacting standards. Inaccuracies in alignment and elevation are often caused by factors that have nothing to do with the method of setting out, but rather with the method of installation and the conditions under which the installation takes place. Stability of the trench, the bedding of the pipe and the position of the watertable at the time of installation are also major factors.

To maintain alignment in the trench, pipes are usually embedded in a V-notch or other form-fitting groove excavated in the bottom of the trench (most machines form such a groove). A gravel envelope provides very good bedding conditions for the pipe, although this is seldom the sole reason for it being used. In order to maintain the proper level, it is sometimes necessary to use a gravel envelope to weigh down the pipe and prevent uplifting when construction takes place below the water level. This may be an additional reason for using a gravel envelope when easily deflecting plastic corrugated pipe is used. The special problems encountered with pipe installation in soft, un-ripened soils, are considered in section 18.2. For further details on installation methods, reference is made to the ILRI Publication no 60 (Nijland et al. 2005) and to the FAO Irrigation and Drainage Paper no 60 (FAO 2005).

8.7.2 Installation methods and machinery

Pipes may be installed in a trench, excavated by hand or by machine, or they may be installed directly in the soil without first excavating a trench. The principal types of machine used for installation are illustrated in Figure 8.15. They are the backhoe, the trench- and trenchless machines.

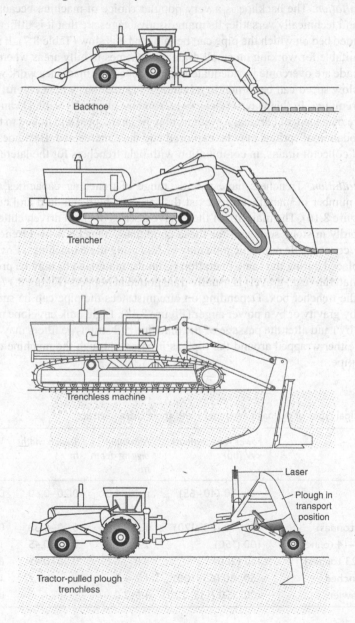

Figure 8.15 Principal types of drainage machines

Manual installation. Manual installation is justified when a very small area needs to be drained, when there is no machine locally available, or when external (socio-political) conditions favour labour intensive work. Considerable skill, however, is required to match the accuracy and overall performance of the available machines. In Pakistan it was found that hand installation (incl. excavating the trench by hand) was ten times more expensive than with trenching machines, even though cheap labour was available. This was caused primarily by the time involved due to the difficulty of installation with high watertables. Manual installation is often done in combination with a backhoe, in particularly for the larger pipe sizes of collector drains.

Backhoe installation. The backhoe is a very popular choice of machine because it is relatively cheap and technically versatile. Its main disadvantages are that it is difficult to form a uniformly graded bed on which the pipe can be laid, and it is slow (Table 8.7). It is, however, particularly suitable for working on small drainage schemes in hilly areas where the limitations of the grade are overcome by adequate natural slope. The machines work well in stony soils and its slow speed can be turned to advantage in situations where careful exploration of the soil is required. In this way services (water mains, electricity cables) can be exposed before they are damaged; old drainage systems may be intercepted and linked to the new system and the location of springs may be explored and thus intercepted. Backhoes are widely used to install collector drains, in combination with light trenchers for the laterals.

Trencher installation. Trenchers have a wide range of trenching capacities (Table 8.7). Worldwide a number of manufacturers exist that produce both standard and custom-made machines (Figure 8.16). The majority of these have a series of chain driven cutters but these wear very rapidly in stony soils and for these conditions, a bucket wheel trencher is more suitable. Clay/concrete tile or plastic pipe may be laid using these machines. Envelope material is either placed during the same operation (granular materials) or may be pre-wrapped.

Granular material (e.g., gravel) is usually transported in hoppers (Figure 8.17) to the field and fed into the trencher box. Depending on circumstances the pipe can be surrounded by gravel either by gravity or by a power auger (Figure 8.18). Light bulk envelope materials are often applied by hand after the passage of the machine and strip type filters may be fed from a roll and be either wrapped around the pipe as it passes through the machine or be placed on top of the pipe.

Table 8.7 Principal types of drainage machinery and their characteristics

Machine type	Power requirements kW (hp)	Maximum digging depth m	Trench widths m	Working speed km/hr
Backhoe	30–50 (40–65)	up to 4.0	0.20–0.60	0.2
Trenchers				
Light (8–10 tonnes)	60–90 (80–120)	1.5–2.0	0.20–0.30	0.5–1.0
Medium (13–14 tonnes)	100 (150)	2.5	0.20–0.45	1.0–2.0
Heavy (16–23 tonnes)	150 (200)	3.0–3.5	0.20–0.45	up to 3.0
Trenchless, winched	50–80 (65–100)	1.2–1.7		1.7–2.0
Trenchless, crawler	100–150 (135–200)	1.6–2.0		up to 5.0

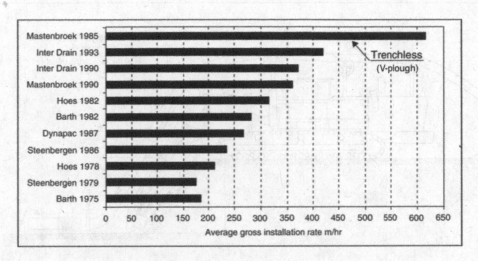

Figure 8.16 Gross production of trenchers and a V-plough in Egypt in irrigated lands (DRI staff 2001)

Figure 8.17 Trencher and gravel hopper (photographs W.F. Vlotman)

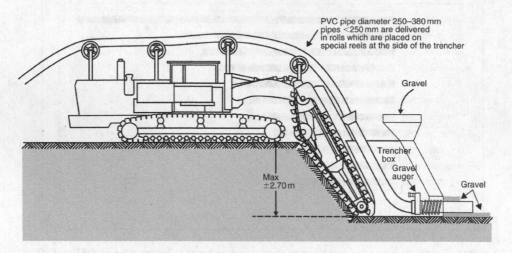

Figure. 8.18 Trencher with optional power feed of gravel

The speed of operation of drainage machines for lateral laying can be up to 1 000 m/hr although 600 m/hr is more typical (Figure 8.16), and over time this reduces considerably. The net output depends on installation depth, soil type, field dimensions, project size, weather conditions and work organisation. For medium textured soil, 1.0–1.2 m installation depth and 200 m long drain lines, a normal output would be 300 m per gross workable hour, reducing to 200 m/hr in heavy clay soils. For large machines, laying collectors, working at 2.5–3.0 m depths, the output is much less.

Trenchers normally require a crew of five men, comprising the operator, the assistant-operator (who also does the staking) and three labourers who supply the pipe, put in the envelope material, install the outlet pipe and backfill the trench (Figure 8.17). For normal output, this involves some 10–20 manhours/1000 m drain length (compared to 250–300 manhours/1000 m for manual installation). With pre-wrapped plastic pipe, the number of labourers can be reduced to two. With gravel envelopes additional equipment and labourers are required.

Trenchless installation. The trenchless method of installation was introduced around 1960 and has since developed into a viable alternative to installation in a trench, in particularly for difficult soil conditions (i.e., high watertables, unstable soil) and conditions when trenches are undesirable (pastures in the Netherlands). Trenchless machines have the potential to work 2–3 times faster than trenchers (Figure 8.16). They are best suited for the installation of pre-wrapped corrugated plastic pipe.

There are two types: a vertical plough and a v-shaped plough (Figure 8.19). To avoid compression of the soil around the pipe, leading to high entry resistances, it is essential that the opening in the soil where the drainpipe is to be placed be created mainly by lifting the overlying soil rather than by forcing the soil to flow around the implement. The V-plough is better suited for this than the vertical plough. With the vertical plough smearing of the soil immediately adjacent to the pipe was found to be a problem in the UK (Spoor 1995).

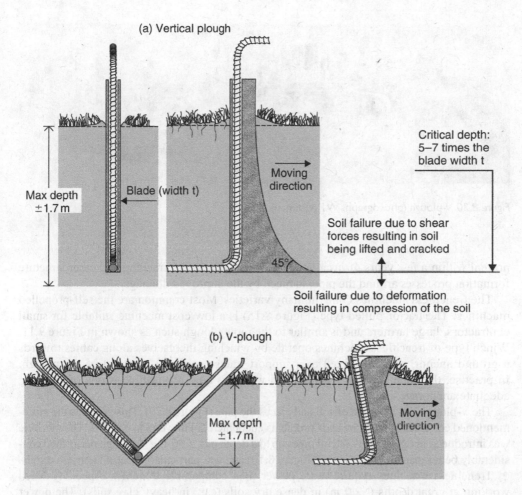

Figure 8.19 V-plough vs. vertical plough and the concept of critical depth

The principles of soil failure by the passage of a vertical blade as described in some detail in section 9.3 (mole drains) indicate that there is a critical depth for installation. The soil below the critical depth fails by plastic deformation, which compresses the soil and reduces its hydraulic conductivity. It therefore follows that the pipe should be installed at, or above this critical depth. The critical depth depends mainly upon the aspect ratio (working depth of implement/width of base of implement) as well as upon soil conditions, especially the soil's moisture content. In wet soils critical depth is always much shallower than in dry soils. The geometry of the foot of the blade used for opening up the soil (or the foot of the plough if the blade is of the plough type) also has an influence. The foot should be inclined and be rather wide, although the latter of course increases the draught requirement. High installation speeds are also not favourable. All in all, these considerations limit safe installation depth to 1.5–2.0 m in dry soils and much less in wet soils. It has, however, been observed that initially high entry resistances, due to installation below critical depth, often reduce to

Figure 8.20 V-plough (photographs W.F. Vlotman)

normal within a few years of installation, probably due to the cracking and other structure formation processes around the pipe, induced by the improved drainage.

The trenchless machines come in many varieties. Most common are the self-propelled machines. The tractor pulled type (Figure 8.15) is a low cost machine suitable for small contractors, large farmers and is similar to the mole plough such as shown in Figure 9.11. Winch type of trenchless machines operate by winching themselves along cables towards a ground anchor, normally fixed to a support tractor some 50 m ahead of the plough. In practise, the main problem encountered with this type of machine is one of ensuring adequate anchorage.

The V-plough lifts a wedge of soil and places the pipe (Figure 8.20). This prevents the afore-mentioned compaction and smearing problems encountered with vertical ploughs. The machine was introduced in Egypt to lay lateral pipes up to a depth of 1.5 m. The V-plough performed considerably better than trenchers in use (Figure 8.16) and was particularly useful in unstable soils.

Trenchless machines require more power than comparable trenchers especially when working at great depths (>2.0 m) in dense dry soils (e.g., in heavy clay soils). The power requirements (P) roughly increase with depth to the third power, $P = f(d^3)$, and a major problem experienced with the powerful crawler mounted machines is one of securing adequate traction under wet surface conditions. Special cleats may be mounted on the track, while triangular shaped blades of the tracks may also provide the necessary extra traction. For soils with low bearing capacity machines can be equipped with special wide tracks.

8.7.3   *Construction and quality control*

During construction the tender documents and specifications provide instruction on quality control during installation, as well as that of the materials used.

Quality control of materials may be achieved by using only certified suppliers or material that has been certified by material/branch specific organisations. Still material needs to be inspected when it arrives on site and when in doubt samples may be sent to laboratories for testing. Sampling procedures, as well as prescribed frequency may or should be given in the tender document/specification. There are usually also guidelines for storage of the material to protect it from weathering, or inappropriate stockpiling.

Pipe material needs to comply with appropriate ISO norms, and if not available with similar national norms. Perforations need to be selected based on the guidelines shown in Box 8.1.

Strength criteria should be determined based on expected soil pressure. During transport and handling in the field care should be taken that tie ropes do not damage the pipe. Visual inspection should be done before coils of material are inserted on the machine. Final inspection is possible just before the pipe enters the trencher box or the blade of the trenchless machine.

Granular envelope material requires testing during various stages from material acquisition to placement around the subsurface drain. First material needs to be selected, appropriately graded if necessary, and then after each storage and transport phase, it needs to be checked that no segregation of fine and coarse material has taken place.

With geotextiles the pore size distribution needs to be known, while strength properties of the material should be supplied by the manufacturer, with or without certification. Wrapping of the material needs be done in the factory (pre-wrapped), which potentially allows better quality control, or in the field. For large pipe diameters material may be sewn on site with special handheld sewing machines.

Installation below the watertable and back fill procedures that avoid formation of sinkholes during the first irrigation or rainfall, require special attention and details are described below. To check the integrity of the pipe after installation and back fill, an inflatable rubber ball or a metal wire cage may be pulled through the pipe after completion of a section of the drain. During construction a rope is inserted in the pipe, that will be attached to the wire cage or ball. A rope of similar length is on the surface and both are pulled simultaneously. If there is any blockage, the metal wire cage will get stuck and the rope on the surface immediately pinpoints the location.

Alternatively, pipes may be inspected with video inspection equipment, or by simple rodding equipment which is normally used for maintenance (see also section 24.4).

8.7.4   Timing of installation

Ideally, drain installation should be carried out during the dry season when the ground is able to support the heavy machines and the continual passage of heavy wheeled tractors and gravel trailers (NB special trailers on tracks are available, Figure 8.17). Drains should be installed even through a growing crop in preference to installation under wet surface conditions. Crop loss is at the most ~5%. The serious compaction of the soil by the passage of heavy machinery on wet land should not be underestimated. The greatest damage results from the passage of wheeled vehicles. The provision of crawlers leads to a lower and more even distribution of pressure (0.20–3 kg/cm^2) although the larger trenchless machines often have higher ground pressures which do not allow them to operate on soft soils unless fitted with special wide tracks.

8.7.5   Installation below the watertable

Modern pipe laying machines are capable of installing drainpipe beneath the watertable, but the risk is that the soil around the pipe will become puddled during excavation or backfilling, resulting in high entry resistance. Use of good, thick envelopes reduces the effect of low hydraulic conductivity adjacent to the pipe due to smearing and puddling, and for these conditions gravel is often most suitable since it also helps with preventing uplifting of the pipe before backfilling could have taken place.

Figure 8.21 Manual blinding, trench back fill with bulldozer to prevent pipe uplift, and method 3a of Figure 8.22 (photographs W.F. Vlotman)

In certain saturated soils, the soil may not remain in the claw of the digging chain and the excavator only puddles the soil in the trench, without excavating the trench. In such case a trenchless method is preferred.

To prevent uplifting of the perforated drainpipe, either a bulldozer immediately backfills the trench by pushing soil on the trencher box (Figure 8.21) or one makes sure that the pipe is filled with water the moment it is at drain depth when it leaves the trencher box. This situation occurs when construction of the pipe takes place below the watertable and the drainpipe is not yet filled with water. Filling the pipe with water before it leaves the trencher box will prevent uplifting (i.e., prevent mis-alignment) and also prevents initial high gradients at the soil-pipe, or soil-envelope interface, and thus reduces also the risk of excessive initial sediment entry in the pipe, resulting in sinkholes.

A special problem occurs with the installation of drainpipe under quick (sand) conditions (occurrence of unstable fine sandy/silty or sodic material at drain depth, below the watertable), a situation which is occasionally encountered in the deep drainage for salinity control of alluvial soils. The provision of suitable guard plates on the trencher will ensure that the trench will remain sufficiently open to allow machine installation to be used. The problems are most acute when installing collector pipes at depths > 2.5–3.0 m. Such installation by machine requires highly skilled operators. Extra gravel may help to weigh down the pipe and prevent flotation. Manual installation requires extensive dewatering to keep the trench dry and prevent it from collapsing. This may be done by conventional well-pointing along the trench[3] (on one or both sides, depending upon the severity of the caving in problem) although horizontal dewatering provides a suitable alternative system. This latter method involves pumping water from a pipe (usually an ordinary corrugated plastic drainpipe) installed lengthwise below the bottom of the trench. Special machines have been developed which are able to install these horizontal dewatering pipes at great depths (5 m and more; some types of conventional drainage machines can also be adapted for the installation of horizontal dewatering pipes up to a depth of 3 m).

8.7.6 Backfill

With open trenches the initial backfilling (20 cm on top of the pipe) should be done with topsoil (blinding) since this is generally more stable than the subsoil. This is to prevent macropores with a direct link to the pipe perforations, that may cause excessive pipe siltation. Blinding is only possible by hand. Back fill may be done by hand or mechanically

(Figure 8.21). In some areas it is common practice to run the tracks of the trencher back along the trench line to consolidate the apparent excess fill.

Also, in the case of trenchless construction, some form of after construction surface consolidation may be necessary to eradicate large surface cracks that may pose a problem for livestock. To prevent livestock being injured, a special roller can be attached to the plough (i.e., in the front of the plough) to also run back along the drain line and press excess soil back into position.

In irrigated areas serious problems can arise when water moves rapidly through the unconsolidated trench fill causing severe erosion (piping) and in extreme cases washing much of the trench backfill into the collector ditches. This problem can be overcome by prescribing the aforementioned blinding procedure, or, if not possible due to trench collapse, by prescribing a combination of trench backfill consolidation procedures, preventing irrigation water coming into direct contact with the trench, and by commissioning the work after the first few irrigation applications so that any voids appearing may immediately be backfilled with soil. Any remaining fill should finally be formed into a low bund to prevent water ponding above the recently filled trench. Piping normally ceases to be a problem after one season, and irrigation may then proceed without regard to the position of the drain.

To prevent delayed sinkhole formation, and possible envelope failure, backfill and consolidation procedures should be prescribed in the specifications, using one of the various methods available for trench backfill and consolidation as shown in Figure 8.22. Sinkhole development should be stopped immediately. Methods 3, 4, 5, and 6, or combinations of the methods were deemed to have most chance of success, while method 8 has been applied successfully at the Fourth Drainage Project (FDP) in Pakistan. Method 3b has been used successfully in the Mardan project in Pakistan. This was possible because trench collapse due to unstable soil conditions was generally not a problem. At FDP the method would most likely not have worked because the problematic macropores and unconsolidated backfill would still be below the reach of the roller with drains being laid at depths of 1.8–3.6 m depth. Controlled backfilling is not possible when trenches collapse.

When the outlet is into an open drain, the level of consolidation may be varied between the critical 10 m of trench adjacent to its outlet and the main length of the drain. This first 10 m of trench should contain a non-perforated drainpipe and the trench should be backfilled in 0.3 m deep layers. Each layer should be sprinkled with water to raise its moisture content to the optimum for maximum compaction by hand tamping. The final backfilling of the trench may be done to less exacting standards.

Similar problems may occur with trenchless installation. The (below surface) crack formation, which is considered a desirable feature in humid zones, can lead to piping under irrigated conditions. The passage of the plough will also cause the surface to heave and whilst this settles over a period, it presents problems for the mowers used on grassland/pastures.

Notes

1 Coir, or coconut fibre, is a natural fibre extracted from the husk of coconut and generally used for floor mats, doormats, brushes, mattresses and as wrapping material around drainpipes.
2 The true thickness of the synthetic envelope at drain depth is the actual thickness as specified by the manufacturer times the compressibility factor.
3 NB also essential for manual installation, and one of the reasons why manual installation was much more expensive than machine installation in Pakistan.

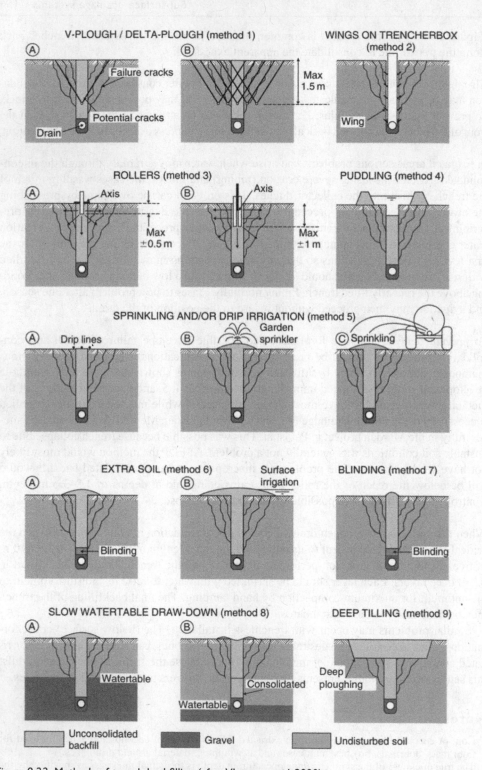

Figure 8.22 Methods of trench backfilling (after Vlotman et al. 2000)

Chapter 9

Surface/shallow drainage systems

These systems are conventionally referred to as surface drainage systems but may, on the basis of their functioning, actually be better referred to as shallow drainage systems. They apply to land in which subsurface drainage systems cannot be used because of the inadequate infiltration or percolation characteristics of the upper soil layers and/or inadequate hydraulic conductivity in the substrata. This situation is typical in so-called *heavy land (heavy soil)*. Representative examples are the basin clay soils of the alluvial deposits, the pseudo-gley soils of Mid- and Eastern Europe, the planosols of the semi-humid/semi-arid tropics, the vertisols of the semi-arid tropics (black cotton soils) and the glacial till soils, (boulder clays) covering much of northern America, United Kingdom and northern Europe. The prevailing profile characteristics of these soils are shown schematically in Figure 9.1.

Whereas in these soils the excess water is unable to drain downwards through the profile, it should be drained by lateral flow. Thus, water ponded on the surface will be drained by lateral overland flow whilst water impeded at some depth in the rootzone, will be drained by lateral interflow over the impeding subsoil layer. The depth of the drained soil will extend down to the impeding layer and because hardly any roots will penetrate into this layer, it will coincide with the main rootzone depth. As this type of subsoil normally has a low drainable pore space ($\mu < 2$–3%), there is little to be gained by deep drainage unless it is combined with subsoil improvement.

Surface ponding is also caused by flatness and unevenness of the land surface, for under these conditions the water will only run off slowly. The excess water will collect in depressions leading to an unequally distributed infiltration load. Moreover, conditions for infiltration in depressions are poor as the soil structure deteriorates under the frequent ponding and the pores become clogged by sediments carried into them by water. The ponding thus becomes self-promoting. Better grading and smoothing of the land may solve part of the drainage problem.

The rainfall intensity and magnitude also have an influence. Under conditions of high, intense, rainfall, subsurface drainage may well fail to provide adequate drainage in moderately permeable soils which could otherwise be drained under less severe climatic conditions (relevant to much of the humid tropics).

Finally, surface ponding and impeded percolation are also frequently caused by poor soil management. Modern mechanised arable cropping systems severely tax the soil structure, on the one hand by adding very little organic matter to the soil and, on the other hand by working the soil with heavy machines, frequently under adversely wet conditions. As described

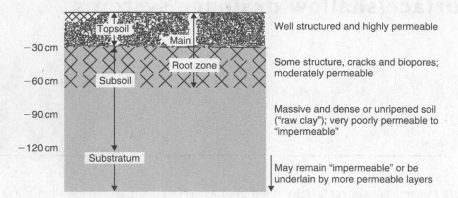

Figure 9.1 Heavy land soil profile

in section 3.6, soils with a low organic matter content are rather sensitive to the compacting, puddling and smearing effects of cultivation, resulting in poor soil structure and in impermeable layers (*pans*) at some depth below the seedbed. Of course, where pans occur in a soil having otherwise adequate hydraulic conductivity they should be broken up by sub-soiling, and their re-establishment prevented by improved soil management so that normal subsurface drainage may be used.

As a rough guide, shallow drainage systems are applicable when the impeding subsoil lies within some 50 cm below the soil surface, and the hydraulic conductivity of the subsoil is less than 0.1 m/d. When the impermeable subsoil occurs deeper, say at 80–100 cm depth, normal subsurface drainage may be used. The drains in this case should be laid just on top of the impermeable subsoil since all flow towards the drains will take place above this layer. These drains would have to be installed at rather narrow spacings in view of the limited flow zone and the small head (mid-spacing watertable height) that can be permitted. Nevertheless, reasonable rootzone drainage can be achieved.

Of course, it is no coincidence that much of the heavy land in the temperate zones is used as grassland and in the warmer zones as rice land. Rather, it is a good example of adapted farming in the sense described in section 1.9. The heavy lands as a class, however, are far from uniform and many can be drained sufficiently well using shallow drainage systems to allow a number of other field crops to be grown successfully.

9.1 Bedding systems

Bedding is the classical drainage method for flat, heavy land in humid climates. The excess surface water is drained by lateral overland flow towards the field drains. However, where permeable topsoil exists, much of this water will also drain by lateral flow through the perched groundwater zone forming itself in the topsoil on top of the impermeable subsoil (interflow see Figure 9.2). This *topsoil drainage* helps to remove ponded water and also

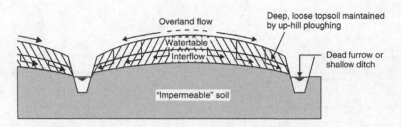

Figure 9.2 Drainage by overland flow and perched groundwater flow in a cambered bedding system

promotes early restoration of the aeration of the upper rootzone (usually the main rootzone in this type of soil). Topsoil drainage is enhanced by maintaining a deep plough layer on the beds (30–35 cm).

For the typical conditions under which bedding is used, flat beds should generally be no wider than 10 m to obtain good drainage. However, narrow beds are not conducive to mechanised farming. Crowning of the beds promotes overland flow and interflow, enabling wider beds to be used (up to 20–30 m), commonly being referred to as cambered beds. The crowning may be achieved in the course of time by deliberate uphill ploughing towards the centre of the beds or be constructed by grader and/or scraper.

Bedding is most appropriate for crops grown on the flat like grassland, other fodder crops and various grain crops. Ridge cropping, although in itself a desirable practice on heavy soils, has its drawbacks when used on beds. Ridges aligned along the length of the beds, parallel with the field drains, prevent lateral surface runoff and should only be used when there is a good deep topsoil capable of allowing considerable interflow to pass beneath the ridges. Small furrow drains across the ridges may be needed to permit ponded water, standing in the furrows between the ridges, to flow to the field drains (these small drains, usually made by hand, are termed *quarter drains* or *ridge cuts*).

The field drains separating the beds, may vary in form from simple dead furrows left by the uphill ploughing of the beds, to small ditches constructed by hand or with a ditcher. For unobstructed topsoil drainage, the field drains should be incised well into the subsoil so that water levels at normal flow remain below the topsoil (a total depth of some 40–50 cm will generally suffice). They should have a continuous gradient towards the outlet. Many field drains are constructed without staking and suffer from adverse gradients. This also applies to machine constructed field drains, as the ditchers used for this type of work are generally not provided with sighting equipment. Ideally the land should have some slope so that field drains can run down the slope, guaranteeing good discharge (this applies especially to the bedding system shown in Figure 9.3b).

Correct maintenance of the field drains is a time-consuming task due to their great number, the required frequency, and the fact that an established crop may only permit handwork. In practice, the poor construction and/or maintenance of the field drains is usually the main constraint to the proper functioning of the bedding drainage system.

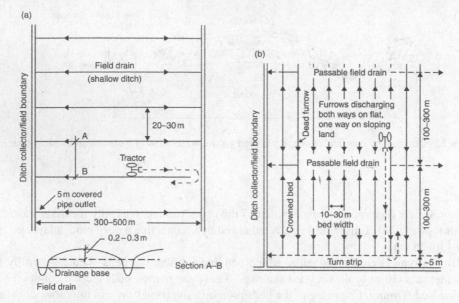

(a) Bedding system designed to achieve good topsoil drainage. The beds are formed by ploughing and grading, and are then maintained by uphill ploughing. The dead furrows between the beds are excavated well into the poorly permeable subsoil (depth about 50 cm; bottom width about 30 cm). The non-passable field drains discharge both ways and can only be crossed at the ends. The farm operations are carried out parallel to the field drains. The system is suitable for flat or gently sloping land; in the latter case field drains are best laid out parallel to the contours (in which case the field drain can discharge both ways).

(b) Bedding system with a shallower drainage base than the first system, achieving less topsoil drainage. The crowned beds are established and maintained by ploughing. Crown height is about similar to the first system although the dead furrows are shallower. The furrows discharge into passable field drains, the latter having side slopes of not more than 1:6.

Figure 9.3 Examples of modern bedding systems, suitable for mechanised farming

9.2   Shallow ditch systems

These systems comprise shallow field ditches laid out in certain patterns. The ditches are usually too shallow and the subsoil too impermeable to achieve much subsurface drainage. Their main function is to collect surface runoff and provide shallow profile drainage, mostly through interflow.

9.2.1   Types of shallow ditch systems

Based on the applied layout pattern and on the type and main function of the installed ditches, the following four types of shallow ditch systems may be distinguished.

The depression ditch system

This system (Figure 9.4) is applicable to fields in which a limited number of pronounced, elongated depressions exist, lending themselves to drainage by means of ditches. The depressions may be drained individually or be linked. The ditches discharge toward the field boundary into the main system. In the example of Figure 9.4, the soil and hydrological conditions are such as to require surface drainage only in the upper half of the field while both shallow and subsurface drainage are required in the lower part. For the latter area a parallel non-passable ditch system (described further on) has been planned. The ditches are made 0.75 m deep to make them reasonably effective as subsurface drains and their spacing is 30 m. Surface water will collect in the ditches partly by lateral overland flow and partly by lateral interflow drainage. Small cross drains (passable) may be installed in depressions (quarter drains/ridge cuts). The ditches discharge into (deep) collector drains which either follow pronounced depressions or field boundaries. In the upper part of the field surface water collects in depressions or border ditches. These ditches may be rather shallow and are made passable where desirable. The land is carefully graded towards the ditches to provide good conditions for overland flow. No watertable control is required in this part of the field (watertables kept adequately low naturally).

Only the main depressions should be provided with ditches as too many infield ditches hinder farm operations. Small and/or inconveniently located depressions should be filled in with, for example, the spoil from the excavated ditches. An infield ditch system connecting a number of isolated, widely distributed depressions is commonly referred to as a *random ditch system* which is comparable to the natural system for subsurface drainage depicted in Figure 8.2.

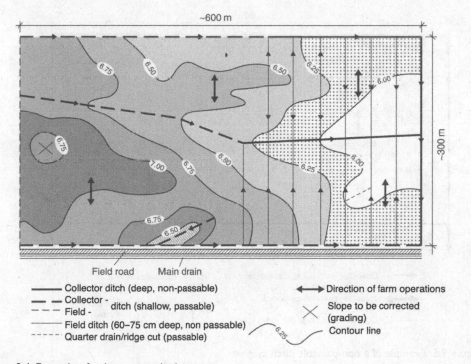

Figure 9.4 Example of a depression ditch system

Because of their low location, depression ditches are effective in drawing excess water from the surrounding, higher land. Where the overland flow rate towards these ditches is too slow, smoothing of the land, land forming, row drainage, etc., may be used to promote the flow (section 9.2.2).

The parallel non-passable ditch system

Two alternative drainage solutions have been presented for a field with a rather irregular topography i.e., the parallel non-passable ditch system (Figure 9.5) and the parallel passable ditch system (Figure 9.6). The parallel non-passable ditch system would be applicable when:

- the field requires both shallow and subsurface drainage
- very little earth movement in the field can be tolerated
- the field will be cropped in such a manner that the ditches will not interfere unacceptably with the required field work.

The planned system of Figure 9.5 requires only minimal earth movement. Only a few depressions have to be filled-in (using the spoil from the ditch excavations). To be effective subsurface drains, the ditches should be at least 75 cm deep. Culvert-outlets may be used to enable the ditches to be crossed by tractors. Portable bridges may also be used, or a short length of the ditch may be temporarily filled in during farm operations. Ponded water can enter the ditch by lateral overland flow and/or by lateral interflow. Under ridge cropping, a good number of ridge cuts should be installed to allow water collecting in the depressions within furrows to escape to the nearest ditch.

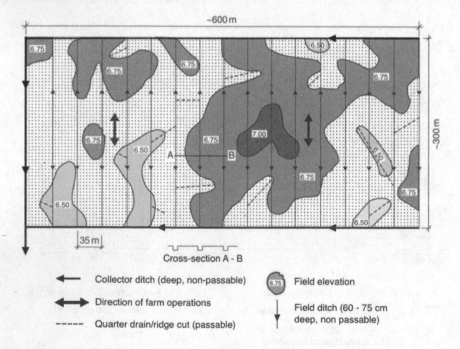

Figure 9.5 Example of a non-passable ditch system

Non-passable ditches are often a good solution for fairly regular, even-surfaced, flat land and also for land with a highly uneven (pocketed) surface. The internal field topography of this type of land does not indicate a preferential alignment of the ditches and so a parallel alignment which least hinders farm operations, is often the most appropriate. The ditch spacing may vary within a field although where conditions do not differ greatly, the spacing is best kept uniform.

The ditches should be well incised and definitely aim to achieve shallow profile drainage as well as to collect excess surface water. The system differs from the flat bedding system in that the drains are deeper and wider spaced and, to be successful, the soils should allow somewhat better and deeper profile drainage than is possible on the poorly drainable soils which need bedding. By promoting lateral interflow, e.g., through sub-soiling, and by promoting surface flow through smoothing and careful land preparation, bedding may often be replaced by a parallel ditch system.

As with bedding, ridge cropping (parallel with the ditches) is on the one hand desirable whilst on the other hand it prevents lateral overland flow and so should only be used where there is good profile drainage. Ridge cuts should be made to evacuate any ponded water from within the furrows.

The parallel passable ditch system

The parallel passable ditch system in Figure 9.6 presents an alternative to the system of Figure 9.5 which is more compatible with the field accessibility and manoeuvrability requirements of modern mechanised farming. The system, however, provides surface

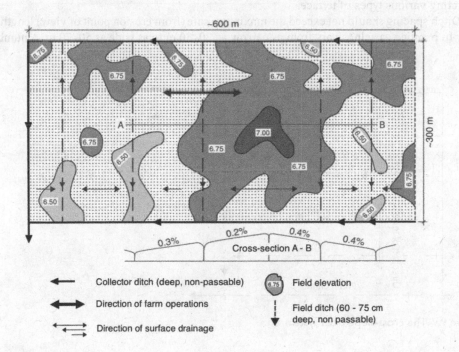

Figure 9.6 Example of a parallel passable ditch systems

drainage only; no water is drained from the soil profile and fields may be expected to dry up slowly whilst crops may suffer from poor aeration unless there is adequate natural drainage.

Ridge cropping is desirable in this situation as it provides the means for good row drainage while at least the soil in the ridge bodies is reasonably well drained. High quality grading of the land towards the ditches is required which may involve considerable earth movement and locally may leave very little (valuable) topsoil.

By making the ditches passable by machine, the earlier noted dilemma with respect to ridge cropping can be overcome. Ridges are laid out, and indeed all farm operations are done, across the ditches. The land between them is carefully graded towards one or both of the ditches. With ridge cropping, good row drainage can thus be achieved, enabling the intercepting ditches to be spaced rather widely. Hardly any profile drainage is, however, achieved (not even topsoil drainage).

The cross-slope ditch system

This system (Figure 9.7) is applicable on gently sloping land where on the one hand, sufficient runoff may occur to require control but on the other hand, the water may also remain ponded on the land. Under this system, cultivations are carried out with the slope (straight down or at an angle to the slope) with runoff being intercepted at regular, short intervals. Slopes should generally be not less than 0.5% and not more than 2% for row crops and 4% for close growing uniformly broadcast crops. On steeper slopes, more emphasis should be placed on detaining/retaining the excess water by practising contour farming and by constructing various types of terrace.

Ditch spacing should not exceed the maximum safe (from erosion point of view) length of run. In practice, spacings vary from as narrow as 20–30 m to as wide as 50–70 m. A number

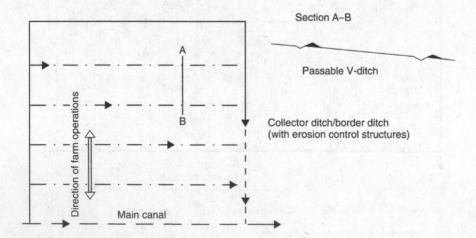

Figure 9.7 The cross-slope ditch system

of factors influence the spacing, the effect of the most important ones being indicated qualitatively below:

	Narrow spacing	Wide spacing
- Soil type: infiltrability	low	high
Erodibility	high	low
- slope of the land'	moderate	gentle
- intensity of the rainfall	high	low
- crops to be grown	row type	close growing

Required spacings are, however, difficult to predict because the influencing factors interact and therefore are best determined by trial and error.

9.2.2 Some technical aspects of shallow ditch systems

The field and collector ditches may be made passable or non-passable, examples of the first type being presented in Figure 9.8. The prevailing conditions will generally dictate which type is most suitable. Depression ditches which for example divide up the field rather badly, should preferably be made passable. This also applies to parallel ditches aligned across the direction of farm operations (as in Figure 9.6). However, where the ditches run parallel to the direction in which the farm operations are carried out non-passable ditches, occupying less land, may be used instead (as in Figure 9.5). The function of the ditch also plays a role: ditches which only need to collect surface water can be made passable rather more easily than the deeper ditches which are designed to function as (shallow) subsurface drains as well. Not only are the latter deeper and thus more expensive to make passable but they also dry up later.

Passable ditches are normally constructed using commonly available earth movement machinery such as: bulldozers, graders and scrapers. Self-propelling graders and scrapers may be used for large collector ditches whereas small ditches may be constructed using farm-type equipment (blades, drags, small scrapers, etc., drawn by farm tractors). Various types of ditching equipment are also available for constructing non-passable ditches such as the rotary ditcher shown in Figure 9.9.

Overland flow

Without good overland flow conditions, the excess surface water will not reach the ditches, rendering them ineffective. Good overland flow requires primarily that the land draining towards a ditch has a continuous grade towards that ditch. Where the topographical conditions are unfavourable, earth movement should be undertaken to establish such grades artificially. This may involve correcting minor irregularities in the slope (*smoothing*) or extensively changing the natural lie of the land (*land forming*). Under favourable conditions, field ditches may be widely spaced with travelling distances for the overland flow up to 100–200 m (Figure 9.4–Figure 9.6).

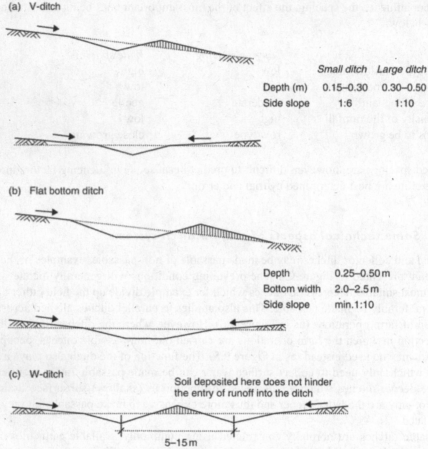

(a) V-ditch

	Small ditch	Large ditch
Depth (m)	0.15–0.30	0.30–0.50
Side slope	1:6	1:10

(b) Flat bottom ditch

Depth	0.25–0.50 m
Bottom width	2.0–2.5 m
Side slope	min.1:10

(c) W-ditch

Soil deposited here does not hinder
the entry of runoff into the ditch

5–15 m

These ditches are generally farmed through and the upper side slopes may well be planted. They should be cleaned before the drainage season, by hand or using a V-drag. A small furrow drain is often installed in the centre to ensure that the ditch is dry in sufficient time for tractors to pass through.

Figure 9.8 Types of passable field and collector ditches

Figure 9.9 Rotary ditchers (Courtesy Dondi, www.dondi.net.it)

Row drainage

This refers to a special form of overland flow, viz the run-off through the furrows formed by two crop ridges. Given a well-maintained furrow of a good grade, row drainage can be very effective; moreover, as mentioned earlier, ridge cropping is in itself a beneficial practice on poorly drained land.

Furrows should be aligned in relation to the slope of the land so as to achieve the most desirable furrow grade. For clean, well maintained furrows a minimum grade of 0.10–0.20% is required, but rough, trashy furrows require more slope to achieve a good discharge (minimum 0.20–0.25%).

Long furrows are, of course, desirable from the farming point of view but are more liable to become blocked. A length of 150–200 m is normally a good compromise. The erodibility of the soil should also be considered. Few soils tolerate furrow grades in excess of 2% while on erosive soils the grades should generally not exceed 0.5% at normal furrow lengths. On steeper slopes, row length should be adapted to the degree of slope and the erodibility of the soil for which some general guidelines have been given in Figure 9.10.

The water draining from rows is best intercepted by field drains running across the slope or through depressions as for example in Figure 9.4 and Figure 9.6. Where the ridges run parallel with the field drains as in Figure 9.5, the ridges are usually made more for agronomic reasons than for the row drainage. Drainage in this situation is assumed to be more lateral (topsoil and shallow profile drainage) than longitudinal (row drainage). To provide an outlet for any flow that may occur along the rows, cross-drains (ridge cuts) should be made. These may be pre-planned and installed at planting or they may be installed on an ad-hoc basis during the rainy season, when and wherever the need arises.

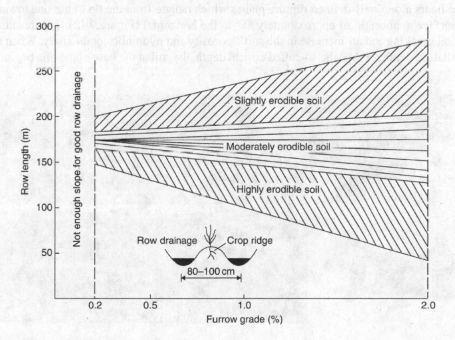

Figure 9.10 Furrow grade/length relationships for row drainage

Spoil

A common cause of malfunctioning of ditch systems is that the spoil from excavation or from maintenance is not disposed of adequately. Often the overland flow into the ditches is blocked by spoil banks.

The spoil from ditch construction should be used to fill in depressions on the slope or be feathered out by carefully grading it away from the ditch. Where satisfactory disposal of spoil from maintenance is difficult due for example to maintenance during the crop-season, the installation of a W-ditch, as depicted in Figure 9.8, offers a solution.

9.3 Mole drainage systems

Here the lateral shallow flow through the soil is promoted by mole channels. These are installed by a mole plough (Figure 9.11) whereby a bullet attached to a vertical tine/blade is drawn through the soil, leaving a tunnel (mole channel). The bullet is followed by an expander which expands and strengthens the channels somewhat. The mole channels thus formed normally have a diameter of some 5–10 cm and are spaced every 1.5–3.0 m. Installation depth is usually between 0.40–0.60 m below soil surface, well into the impermeable subsoil. Normally, heavy and powerful farm tractors can be used for moling. The installation capacity under normal conditions is around 5 km/hr (gross output) or some 5–10 ha per workable day.

Installation

During installation, the soil is displaced forwards, sideways and upwards and in the process fails/shears along well-defined rupture planes which radiate from the tip of the tine towards the surface at an angle of approximately 45° to the horizontal (Figure 9.12). This results in the soil cracking and an increase in the soil's porosity and hydraulic conductivity. When the installation depth reaches the so-called critical depth, the soil at the base of the tine begins to

Figure 9.11 Moling and sub-soiling equipment (Photos www.greatplainsint.com_)

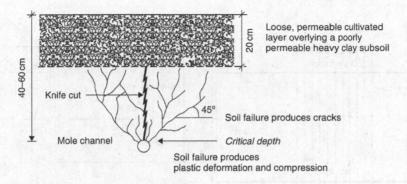

Figure 9.12 Water movement to mole drains

flow forwards and sideways, leading to compaction and a decrease in porosity and hydraulic conductivity (Spoor 1979, 1995).

The critical depth generally occurs at a depth corresponding to an aspect ratio (tine depth/ tine thickness) of the order of 5–7. It becomes shallower as the soils become more plastic (i.e., wetter) or the upper layers become harder and drier. Ideally, moling should form a channel just below the critical depth, linked to a zone of cracked soil which extends up into the topsoil. These cracks provide direct access for impeded water to the mole channels. The channels also establish a groundwater sink but when the surrounding soil is nearly impermeable, hardly any water will be drawn into the mole drains by normal subsurface drainage flow.

Moling is best suited to clay soils with a minimum clay content of ~ 30%. When the moisture content at mole depth is near the lower plastic limit, not too much smearing (hindering water entry) will occur during installation. Channels installed under much drier conditions than the lower plastic limit are liable to deteriorate rapidly. Best installation conditions will often occur shortly after harvest, when the upper part of the soil profile has dried out.

Length, slope and outlet

The effective lifetime of mole channels depends on the inherent stability of the soil to repeated wetting, on the soil uniformity and on the conditions at the time of installation. Liability to blocking increases with increasing channel length. Safe lengths vary with soil type between 20 m and 80 m and should be established on the basis of experience. Channels have been known to function for well over 20 years in stable calcareous clays, although it is sound practice to re-mole every three to seven years.

It is also important to prevent water stagnating in the mole channels since this will weaken the walls and lead to premature collapse. Some slope and a good outlet are therefore important. Too much slope, however, may cause erosion of the channel, or lead to blow-outs when combined with long lengths. Safe gradients are in the range of 0.2–3.0%.

Mole drains are usually aligned with the slope of the land where the latter is within the acceptable range. Few mole ploughs allow a very accurate grade to be maintained and when the land surface is rather uneven, some irregularities in the grades of the channels are to be expected. Mole channels may discharge directly into a ditch, either with an unprotected or with a protected outlet (the latter e.g., in the form of a 1-2 m long pipe inserted into the channel).

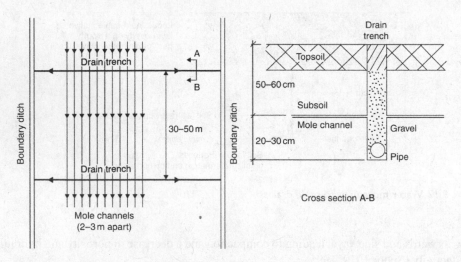

Figure 9.13 Mole drains discharging into drain trenches

A better system is to intercept the flow in the mole channels at suitable intervals by means of drain trenches installed across the mole channels (Figure 9.13). This provides a dependable outlet and allows the use of short length mole channels in large fields. The drain trenches are installed first and then the mole channels are drawn across and through the trenches. The trenches may be parallel and regularly spaced or be laid out in close accordance with the topography, with the trenches mostly following the depressions in the field. The trenches are provided with a pipe drain and are backfilled to well above mole depth with gravel to allow the discharge from the mole channels to flow readily down to the pipe.

Mole drains generally respond rapidly to rainfall and their discharge rates can be substantial. Therefore, it is essential that the hydraulic conductivity of the backfilled gravel should be high in order to minimise head losses at the discrete points where water cascades from the mole channel into the gravel-filled trench. The gravel should be clean and have a minimum size of between 3 and 5 mm. The pipes should also be designed to cope with this rapid response to rainfall.

Soil improvement

Mole drainage seeks to establish a medium deep drainage base in the soil and although not all of the soil above this depth is well drained, opportunities for water movement and root development are enhanced by the cracking of the soil. Further improvements may occur in time as the improved drainage induces more cracking and deeper, more thorough exploration of the soil by roots. This process of rootzone deepening may be assisted by the measures described in section 9.5. By gradually increasing the installation depth up to a maximum of 50–60 cm over a 10–20 years period, a parallel deepening of the rootzone may also be achieved. Mole drainage should not be considered as a cheap version of pipe drainage: rather it is a soil and drainage improvement measure, whereby the latter supports the former and vice versa, based on cracking and subsequent improvement of the subsoil and on the provision of lateral outflow facilities for impeded water through the mole channel.

9.4   Pipe drainage systems

Pipe drainage systems are seldom feasible in heavy land as these derive their value (and justify their high costs) from their subsurface drainage function. Owing to the immobility of the
groundwater in the subsoil and the inhospitable nature of these strata for root development,
subsurface drainage is in-effective in this case. Pipe drains thus serve mainly as (expensive)
underground ditches (Figure 9.14), achieving essentially only the same kind of shallow drainage as the bedding and the shallow ditch systems described in sections 9.1 and 9.2, and this
only for so long as the backfill material in the trench retains adequate hydraulic conductivity.

Pipe drainage may however be used effectively in heavy land in the situation where
the impermeable subsoil is underlain within normal pipe drainage depth (1.0–1.5 m) by
a reasonably permeable substratum, as sometimes occurs with clay soils in alluvial basins
(Figure 9.15). Pipe drain trenches installed in this substratum perform the shallow drainage

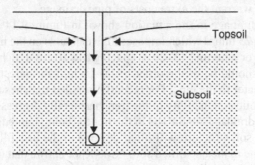

Figure 9.14 The functioning of a pipe drain in heavy land

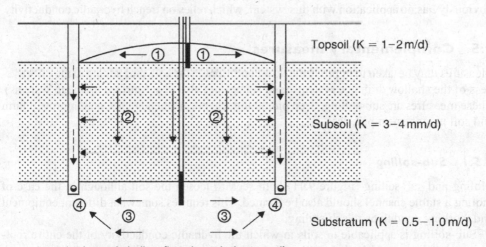

1   Lateral shallow flow through the topsoil
2   Vertical flow through the subsoil
3   Horizontal flow through the substratum
4   Radial flow through the substratum

Figure 9.15 Functioning of pipe drainage in heavy land with a well permeable substratum (after van
Hoorn 1974)

function described above and, in addition, establish a low hydraulic potential in the permeable substratum which induces downward flow in the overlying profile. This flow reaches its maximum value when the watertable is in the topsoil and the soil water potential in the substratum at drain depth is about zero. This condition is met when the drains are rather narrowly spaced and little head is lost in horizontal and radial flow in the permeable substratum. Under these conditions virtually all head is available for vertical flow through the impeding layer. This vertical flow will occur under a hydraulic gradient approaching its maximum attainable value of one, with the flow rate equal to the hydraulic conductivity of the subsoil (e.g., in Figure 9.15 about 3–4 mm/d). This rate, although not very high, is similar in magnitude to the normal rates of evaporative drying of the soil and thus contributes as much to the early workability of the soil after rain as the latter. The flow also continues for a longer period than the shallow drainage and in time achieves a more thorough, deeper drainage of the soil than shallow drainage alone. Plant growth and workability conditions generally improve sufficiently to warrant the extra costs of pipe drainage.

The lateral topsoil drainage in the situation shown in Figure 9.14 and Figure 9.15 may reach peak flows of 20–25 mm/d which imposes high demands on the hydraulic conductivity of the trench backfill. For a normal 15–20 m spacing and 20 cm trench width, the minimum K-value in the trench should be of the order of 1.0–1.5 m/d for unrestricted conveyance of the shallow drainage water to the pipe. For recently installed trenches this is generally not a problem, but the hydraulic conductivity of the trench generally decreases as the backfill consolidates. However, as drainage conditions in the trench (being immediately above the pipe) are favourable, a good soil structure will often develop in the backfill, which may even be promoted by adding lime to the backfilled soil. Experience shows that trenches are more permeable than the undisturbed soil. At wide spacings, however, the trench conductivity may well act as a bottleneck to the drainage of the topsoil. Trenchless installation of pipe drains obviously has no application with this system, which relies on trench hydraulic conductivity.

9.5   Complementary measures

Measures may be taken to improve the flow conditions in the soil, thus adding to the effectiveness of the shallow drainage systems (and sometimes also of subsurface drainage systems). These measures are sub-soiling, deep ploughing, chemical amendments, and improved farm and soil management practices.

9.5.1   Sub-soiling

Moling and sub-soiling (Figure 9.11) both seek to loosen the soil although in the case of moling a stable channel should also be formed. This requires somewhat different equipment and working conditions to sub-soiling.

Sub-soiling is applicable to soils in which the hydraulic conductivity of the entire rootzone, or of layers within the profile, is very low. It involves pulling tined shanks through the soil, lifting, shattering and loosening the treated soil layers in the process (Figure 9.16). These implements can be mounted on a frame or on a toolbar. For shallow sub-soiling, for example, to break up a plough pan, 4–5 tines spaced 50 cm apart may be used; for a working depth of 60–80 cm (~max. depth) 2–3 tines only, about 75 cm apart is normal. As shown in Figure 9.16, the base (taken at soil surface) of the loosened triangular area has about the same width as the working depth. The tine spacing is usually about equal to the working depth which provides a considerable overlap of soil disturbance in the upper soil layer and adequate disturbance in the lower layer.

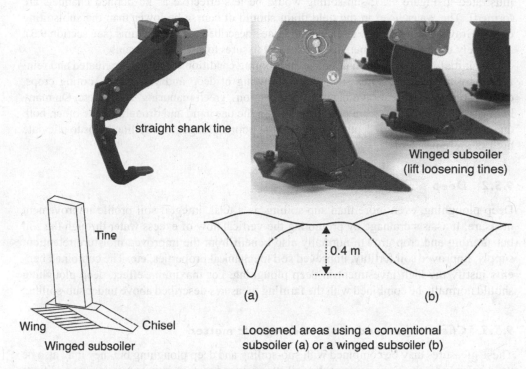

straight shank tine

Winged subsoiler
(lift loosening tines)

Tine

0.4 m

(a)   (b)

Wing   Chisel   Loosened areas using a conventional
Winged subsoiler   subsoiler (a) or a winged subsoiler (b)

Figure 9.16 Patterns of soil failure created by conventional and winged sub-soiler (Spoor 1979) and Great Plains International (www.greatplainsint.com/en-gb/products/709/sub-soiler)

Sub-soiling may be done instead of moling in heavy textured soils which have such a variable nature (e.g., containing pockets of sandy soil) that the integrity of the mole channels cannot be assured.

For sub-soiling to be effective, the working depth should always be less than the critical depth (section 9.3) which depends upon the width of the tine and the installation conditions. The tine should be rather wide and have an inclination of not less than 25–30° to produce a good lifting and loosening of the soil. Thin, shallowly inclined tines are less effective although they may be used for light work. As the tines become larger and more steeply inclined, power requirements increase. Power requirements also increase with the addition of side wings to the conventional chisel shaped foot of a subsoil tine (Figure 9.16) although this is generally more than compensated by the increased loosening which occurs.

Sub-soiling without adequate drainage may create a 'bathtub situation': the excess water readily sinks into the loosened soil but is unable to drain away. Both subsurface drainage and shallow drainage systems may be used in combination with sub-soiling. The first system would for example be applicable where the soil below drain depth was reasonably permeable and the sub-soiling would improve the vertical drainage in the soil all the way down to drain depth. Shallow drainage would be applicable where sub-soiling has loosened the upper part of the soil while below the sub-soiling depth the soil remains virtually impermeable.

Sub-soiling should, as a rule, be aligned across the field drains since, normally, flow conditions for water in the sub-soiling direction may be expected to be improved by the treatment. Sub-soiling for example may be done across the pipe drains such as in Figure 9.14 which will improve the lateral shallow flow (compare to moling across the pipe trenches illustrated in Figure 9.13; sub-soiling would be less effective as no defined channels are formed). The water level in the field drain should of course be lower than the sub-soiling depth. Gravel backfills quite similar to the ones described for mole drains (see section 9.3), are widely used by UK farmers to link subsoil fissures to subsurface drains.

The initial improvement in drainage and rooting conditions may be perpetuated and reinforced by farming measures such as the growing of deep and vigorously rooting crops, careful soil management (avoidance of compaction), green manuring, liming, etc. On many heavy soils, crops suffer from poor drainage on the one hand and drought on the other, both due to the limited rooting depth imposed by the poor drainage. Sub-soiling helps to alleviate both of these problems.

9.5.2 Deep ploughing

Deep ploughing, even more than sub-soiling, is a total, integral soil profile improvement measure. It assists drainage by promoting the vertical flow of excess water through the soil but farming and crop growth normally also benefit from the improved moisture retention/ supply, improved soil fertility, improved soil mechanical properties, etc. The combined benefits justify the high investment in deep ploughing. For maximum effect, deep ploughing should normally be combined with the farming measures described above under sub-soiling.

9.5.3 Chemical amendments and organic matter

These measures may be combined with sub-soiling and deep ploughing but they may also be used independently. They improve the soil structure and as such improve drainage conditions although their effect is most pronounced in the topsoil. The beneficial effects of liming and of organic manuring on the structure of clay soils are widely recognised. Easily dispersed soils become less prone to slaking and surface sealing. The structure of sodic soils may be improved by gypsum application. Lime and organic matter also have a favourable influence on the FC/PL relationship and as such on the soil workability (section 6.5). It generally follows that well-limed soils with a good percentage of organic matter are less liable to compaction. In each of these cases infiltration and percolation of excess water into and through the soil improves. Drainage is assisted, and the harmful effects of imperfect drainage are partly alleviated by farming geared towards the maintenance of adequate organic matter in the soil. Soil management can enhance or cause drainage to deteriorate (Smedema 1993).

9.5.4 Land levelling

Appropriate land levelling will assist the uniform distribution of irrigation water and therefore can reduce (deep) percolation which traditionally occurs at the upstream parts of furrows and basin/border irrigation. It will also reduce local runoff and enhance uniform infiltration of irrigation water. Surge flow irrigation systems are known to enhance the uniformity of irrigation water application and reduce deep percolation of irrigation water.

Since the 1970s laser levelling has been developed and a tower- or tripod-mounted laser level is used in combination with a sensor on a wheel tractor-scraper to level the agricultural land to near-flatness or to a slight grade for irrigation water distribution and drainage. The wheel tractor-scraper can be equipped with Digital GPS, which will enhance the control of elevations at any point in the field.

Chapter 10

Main drainage systems

A main drainage system receives water from the field drainage systems. In a small system, the field systems may be of a uniform type. Large systems receive water from a variety of field systems and receive surface runoff, interflow as well as subsurface water flow. The system is also likely to receive natural drainage flow and it may have to convey the discharge from urban drainage systems.

The principal function of the main drains is to convey all this water to the outlet point. Main drains may, however, also collect a fair amount of excess water directly from the fields themselves (act as direct drainage sinks), especially as these drains tend as a rule to follow the depressions or run along the lower sides of the fields.

10.1   Main features

The situation shown in Figure 10.1 is fairly typical of the layout and functioning of a main drainage system. The area encompassed (*drainage basin*) is about 8 000 ha of farmland which includes two small urban areas. The hierarchical composition of the canal network with small lower order canals combining into larger higher order canals in the process of centralisation of the flow and convergence towards the outlet, corresponds to the arrangement of most natural drainage systems. Such a composition of the drainage network emerges naturally in a design based upon topography and is compatible with the aim of arriving at a well ordered, lucid system which can readily be managed. No hard and fast rules apply, however, and where significant savings may be made, by for example allowing tertiary drains to discharge directly into a primary canal rather than via a secondary canal, this should be done.

Designation of the different canal reaches and auxiliary structures may follow the format used in Figure 13.11. A systematic designation greatly facilitates the design of the canal system as well as the subsequent operation and maintenance of the system.

Agricultural fields (plots) may vary in size from < 1 ha up to some 25–30 ha. These are drained internally by the field drainage systems described in Chapters 8 and 9. They will normally be bounded on one or more sides by a lower order main drainage canal (collector ditch or tertiary canal) into which the field system discharges. In Figure 10.1 only the secondary drains are shown which typically serve areas from a few hundred up to several thousand hectares.

In planning the layout of the drainage system, the basin should be perceived at different levels of generalisation. Initially, perception at a high level of generalisation is required to determine the division of the basin into sub-basins, the alignment of the primary drains and the outlet location. Next, at a lower level of generalisation, the secondary networks

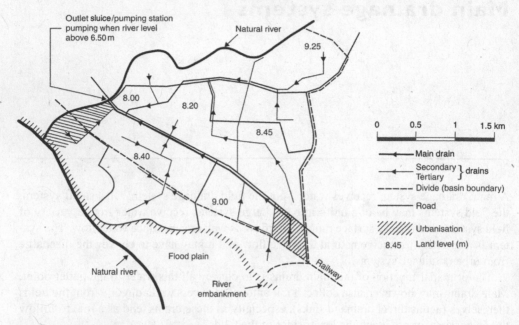

Figure 10.1 Example of a main drainage system

and finally the tertiary networks should be planned. In this process various factors must be weighed and integrated: the topography of the area, the present and planned infrastructure, property and administrative boundaries, soil conditions, costs, operation of the system, etc. The layout in Figure 10.1 is based mostly on the geography and topography of the area, although historical and infrastructural developments have obviously also had a considerable influence. The latter is for example apparent in the interrelationships between road and canal systems and in the division between the two sub-basins formed by the main road crossing the area.

10.1.1 Drainage basin (watershed, catchment)

The area served by a main drainage system should preferably constitute an independent hydrological unit. If the system only serves a part of such a unit, remedial or compensatory arrangements will normally be required to safeguard the drainage of the part outside. Furthermore, the drainage of the area within may not be quite so straightforward as if the entire area had been included. For example, a system that is only concerned with the upper part of a natural basin will usually have to pass through the lower part to the outlet and so special provisions will have to be taken in order to ensure that the drainage of the lower area is not impeded. Main drainage systems, however, seldom cover an entire large river basin; rather it is more likely that the river will serve as the receiving stream for a number of drainage systems organised on a sub-unit basis. In river and coastal plains and in delta areas, hydrological divides are often not very pronounced and hydrological sub-divisions can be changed or established relatively easily as desired. These divisions will often result in the formation of polders (further described in section 10.2).

Drainage basins may vary in extent from less than 100 ha (covering one large or a number of small farms) to more than 100 000 ha. They are generally largest in lowland areas, but it also depends on how much of the natural drainage system has been incorporated. Technically there are no strict limits to the basin size provided the area constitutes a suitable hydrological unit. Most large sized main drainage systems, however, are communal or public undertakings and limits arise from the consideration that the management of too small an area is inefficient while of too large an area it becomes unwieldy. Optimum sizes therefore range from a minimum 5 000/10 000 ha to a maximum 20 000/50 000 ha. Very large basins may be sub-divided into smaller management units under the overall direction of an authority/board (for details see Chapter 22).

10.1.2 Types and alignment of drainage canals

Main drains will usually be (open) canals, as these normally constitute the most economical way to convey the relatively large quantities of water involved in drainage discharge. The storage potential of canals is also valued under certain conditions. Closed conduits (e.g., pipelines) are only used in exceptional circumstances, and then for lower order drains only. This may, for example, apply to a highly mechanised large-scale farming area where collector ditches may be replaced by collector pipes to form larger fields and to make the field more accessible from the adjoining roads. Buried main drains may also be feasible in an irrigated area to avoid the complicated crossing of two open canal systems (irrigation and drainage systems). Such drains will usually only serve the subsurface drainage system and so a relatively small diameter pipe may be used since discharges will be rather low. A separate system for the discharge of excess surface water will then be required. Unfavourable soil conditions causing by for example sloughing of side slopes may also be a factor in favour of using pipelines. For various reasons (aesthetic, costs, right of way, safety), closed conduits may also be appropriate for drains passing through urbanised areas.

Flow velocities in drains can generally be kept sufficiently low to allow the use of earthen canals, where necessary with bank protection. Smooth rectangular bends may well be used in the smaller tertiary canals (capacity up to $1-2$ m³/s). For larger canals, curves should be used. Recommended radii of curvature generally vary from 5 m for canals with a capacity less than 5 m³/s to about 5 to 10 times the bed width for large main canals with a capacity of over 10 m³/s. Transitional sections as commonly used in road curves, are generally not needed for drainage canals.

Two main considerations apply with respect to the alignment and layout of drainage canals:

a) main drains should, wherever possible, follow the depressions and downslope sides of the fields. Higher order canals should preferably follow the valley lines or otherwise follow the main depressions in the area, the next lower order of canal following the next lower sites and so on. This alignment establishes the best gradients for the discharge of field and collector drains into the tertiary canals, tertiary into secondary, then into primary and onwards to the outlet. In addition, this approach will make the main drains themselves effective sinks and collectors of excess water. So, on the basis of this consideration, the general layout of the main system, as well as the alignment of the individual drains, will be largely dictated by the macro/meso relief of the land

b) field and farm boundaries will be largely formed by the lower order drainage canals towards which the field drainage systems discharge, and therefore these should preferably be laid out in a manner compatible with efficient farming. This calls for some regular field pattern, (rectangular or trapezoidal) of fair dimension (Figure 10.2).

These two considerations are obviously not always in harmony. However, an acceptable compromise is often possible where the relief of the land is regular and even. A small number of depressions in a field can always be filled-in, using the spoil from the newly excavated ditches and canals.

However, where the macro/meso topography of the land is irregular the choice must be made between either forming regular shaped, good sized fields (but, containing a number of in-field depressions) or adapting the layout to conform to the relief of the land, resulting in an irregular layout, comprising a large number of odd shaped, small to medium sized fields (Figure 10.2). These considerations apply mainly to the lower order canals; higher order canals should generally be aligned solely on the basis of topography.

Main drain alignments may also be dictated by the existing infrastructure (roads, railways, existing bridges and culverts, etc.) and/or by administrative divisions (property boundaries, district boundaries, etc.). There may be a preference to plan the new drains along existing roads as this facilitates inspection and maintenance and also avoids unnecessary further division of the land. In complete contrast, they may be sited away from the roads for reasons of safety, to avoid abuse, etc. The latter considerations also apply with respect to the alignment

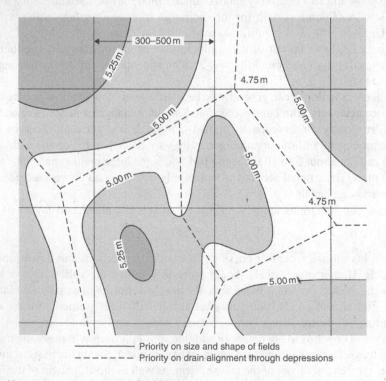

Figure 10.2 Alternative main drain and field layouts

of main drains relative to settlements. Alignment of main drains along property and administrative boundaries divides the land loss evenly between the two parties and emphasises the communal character of these drains.

Box 10.1   Drain density

For drainage systems with an irregular layout, the drain density is a useful system characteristic. It gives the drain length per unit area (usually expressed in m/ha); for example, when a basin of 5 000 ha has a total drain length of 125 km, the drain density is 25 m/ha. Drain density may interpreted as the spacing L of an equivalent parallel system, whereby L = 10 000/drain density (so for the 25 m/ha drain density, the equivalent L = 400 m). Drain densities are mostly used for main systems (although they could equally well be applied to a random field system as shown in Figure 8.2). Main system densities range from 40–60 m/ha in polder areas in the Netherlands to 2–5 m/ha for irrigation commands in India and Pakistan.

Wherever feasible, use should be made of the existing drains (excavated or natural). Natural drains as a rule follow the predominant gradient of the land, often through the valley bottom and depressions, and as such can normally be integrated rather well into the main drainage plan. Some minor canalisation and/or re-alignment can always be done to improve the fit of these natural drains into the desired layout. In general, the scope for incorporating the natural drainage system into the new plan will increase towards the outlet.

The location of the outlet is of course also a determining factor in the planning of the main drainage system. It determines the overall orientation of the system, in particular of the primary canals. The shortest route to the outlet point should be sought.

10.1.3   Outlet and water levels

The outlet point of a drainage system will normally be a conveniently low point on a river, a lake, the sea or any other component of the hydrological system which may be suitable to act as a recipient of drainage water, e.g., a nearby (marshy) depression.

Site selection

The outlet into a river will often be located downstream to create an acceptable gradient in the main drainage system and to allow gravity discharge (section 10.2.2). Prevailing wind directions, foundation conditions, liability to scouring/siltation, accessibility, proximity to power lines, etc., should also be considered. In humid climates, the natural drainage system is usually well developed and so finding an outlet does not generally pose a problem. In arid/semi-arid areas where the runoff is much less and the natural drains are fewer and less well developed, outfall drains for irrigation projects may have to be extended far outside the area to end up at a suitable outlet point. In some cases, the discharge may be led into an evaporation-pond or even a vertical outlet may have to be used i.e., discharge into the underlying substrata. The latter solution is, for example, suitable where an isolated small, enclosed, area is underlain by a well-drained aquifer. Disposal problems and solutions of irrigation projects have been described in more detail in section 16.5.

The functioning and maintenance costs of outlets can be seriously affected by siltation and this should be duly considered in the site selection. Outlets should wherever possible, be located at sites with a good frontal sediment flushing capacity. Siltation problems can to some extent be reduced by the concentration of the outlet flows in order to create a more powerful flushing flow. Serious siltation problems in front of outlets are almost always to be expected when the outlets are not discharging during significant parts of the year and when the closed outlet structures do not allow for natural scouring. The siltation in front of outlet structures installed in tidal creeks with small dry season flows is notorious. Dredging and river training are often too costly for land drainage while results are often quite unpredictable.

Environmental considerations should also be considered in the outlet site selection. Low quality drainage water should only be disposed in water bodies which have sufficient assimilative capacity or can otherwise accept such water. These and other water quality problems are described in more detail in sections 1.8 and 16.5.

Water levels

The level difference between the regional drainage base (= controlling outer water level at the outlet point) and the field drainage base (= water level in the field drains) constitutes the total available head for drainage flow out of the area (P_1–P_3 in Figure 10.3).

The field drainage base should be low enough to provide an efficient sink for the collection of the excess water in the field. Depending on the field drainage system used, these base levels vary from 30–40 cm (shallow drainage) to 120–200 cm (subsurface drainage) below the soil surface. Starting from this level (P_1 in Figure 10.3), the inner water level at the outlet (P_2 in Figure 10.3) is determined by the hydraulic grade line adopted in the main canal system (see also Figure 13.4):

$$P_2 = P_1 - H$$

where:
$H = h_1 + h_2 + \ldots = \Sigma$ headloss in the primary, secondary, etc., canals (including the structures).

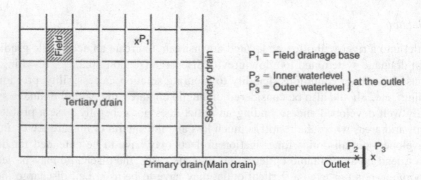

Figure 10.3 Important water levels in a drainage system

When $P_2>P_3$, a gravity outlet may be used, but pumping is required in situations where $P_2<P_3$. Often the outer water level will vary in time. This is for example the case with outlets into the sea or into the lower reaches of rivers subjected to tidal influences. River levels also depend on the discharge which will often vary seasonally. In all these cases the frequency and duration of high outer levels should be investigated, specifically for those periods when the drainage discharge from the area occurs. These periods may or may not coincide with high outer water levels. On the basis of these investigations it may then be decided that gravity or a pump outlet is most appropriate. Various combinations should also be investigated, as well as the scope for storing water in the area during short periods when gravity discharge is not possible (section 12.7).

10.1.4   Outlet structures

Various outlet structures may be used. Where the outer level is always well below the inner level, an open free fall structure is suitable. Various types of gated structures are used to stop the intrusion of water due to high outer levels. Automatic gates are suitable for small tidal outlets. Where the outlet is on a river with a mainly seasonally fluctuating water level, a manually operated gate (Figure 10.4) would be preferable since the operation of the gate is not too much of a burden and the infrequent operation of the gate makes automatic operation risky. Automatic gates may still be the best solution, however, for remote, inaccessible outlet points.

Large outlets may comprise single or multiple barrel, reinforced concrete box culverts or take the form of a sluice. In the case of tidal outlets, structures may function automatically by providing the culverts with top-hung flap gates or the sluice with side-hung self-moving doors (Figure 10.5 and Figure 10.6). Very large outlets may have specific operating rules or otherwise, warrant non-automatic operation.

The top-hung flap gate is popular as this allows water to flow out of the area when the outer water level is below the inner level but closes automatically when the outer level rises above the inner level. Flap gates may be placed in a small outlet sluice or be incorporated into a culvert passing through a dike (e.g., could be fitted in the culvert shown

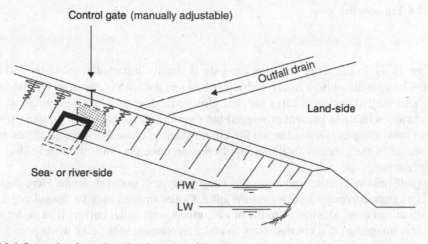

Figure 10.4 Principle of gated outlet through a dike

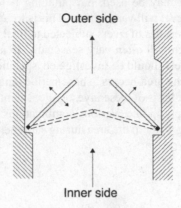

Figure 10.5 Small outlet sluice with side-hung vertical doors

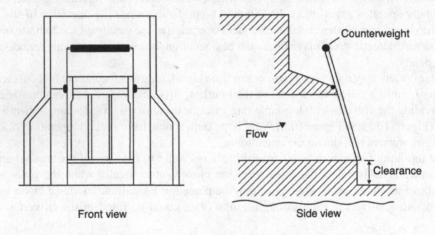

Figure 10.6 Top-hung flap gate

in Figure 10.4). In the latter situation the gate is usually installed at the outer end of the culvert. The gate may hinge freely with a vertical rest position or rest slightly slanted outwards. The slanted design ensures the best closure (and as such is often preferred where salt intrusion is liable to present problems) but requires, of course, more pressure from the inside to open the gate. Depending on the circumstances, counterbalanced weights may be incorporated in the gates to facilitate rapid opening once the outer water level has fallen below the inner level.

The (self-priming) outlet siphon is a less common type of outlet structure. Here the excess water is siphoned through a pipe over the dike. Outlet siphons may be considered in tidal outlet situations as an alternative to the more conventional outlet culvert. The outlet siphon has the advantage that it is not necessary to make an opening in the dike while poor foundation conditions for the culvert may also make the siphon an attractive solution.

Pump outlets may vary from small sump-installations with a capacity of 20–50 l/s to large stations occupied by several pump units, each with a capacity of some 10–20 m³/s. Often pump outlet and gravity outlet are combined in one outlet structure, with the pumps mainly coming into operation during prolonged periods of high outer water levels. The main characteristics of drainage pumps are described in section 13.4.

The structural design of the outlet structures, especially of the gates, largely determines their proper functioning. For tidal outlets, the experience in Malaysia are of special relevance (DID Manual 1977). For other specifications see the pertinent NRCS (formerly SCS) manuals.

10.2 Lowland and upland drainage

Different drainage problems are encountered in the different reaches of a drainage basin. In the upper to middle reaches, most of the land has a good elevation above the local drainage base and gradients are rather steep. In the uplands in particular, emphasis has often to be placed upon retarding rather than promoting the discharge. In the lowlands, the land is usually rather flat, and the natural drainage base is at a shallow depth or even above the land level (e.g., during high river discharges or high tide when the land is flooded). Here the classic polder situation exists.

10.2.1 Lowland polder

Broadly defined, a polder is an area in which the open water level can be artificially controlled at a level which deviates from the prevailing regional open water level. In the classical conception of a polder as developed in the Netherlands, polders maintain a lower water level in the area. This is still the main aim in most cases where polders are established (although during summer or dry periods, the inner level may well be maintained above the outer level).

Polders are used in areas where the regional drainage base is too high for the intended land use. This situation may exist permanently (swamps, sea and lake bottoms) or only temporarily (tidal lowlands, seasonally flooded river plains). Polder formation involves separating the area to be developed hydrologically from the surrounding area by constructing dams or weirs in the connecting canals or by 'endiking' the area to protect it from flooding. When the outer water level is permanently above the desired inner level, the latter can only be maintained by pumping the excess water out of the polder. Where the regional drainage base is controlled by the sea or river, excess water may be discharged from the polder by gravity during periods of low tide and during low river flow periods when the outer level falls below the inner level (Figure 10.7).

Figure 10.8 presents an example of the layout of a modern polder. The outlet is conventionally located in the topographically lowest corner (SE corner in this case). The land has a slight slope (0.01–0.02%) towards the outlet. The target open water level to be maintained in the polder is termed the *polderlevel*. It serves as a reference for the operation of the outlet. The polderlevel is established based on agricultural considerations (it should be deep enough to allow unobstructed discharge of field drains) although other considerations sometimes also play a role (e.g., navigation requirements). During periods of discharge, actual canal levels in the upper part of the polder will generally somewhat exceed the polderlevel (+10 to +20 cm) while near the outlet the water level may be lowered well below the polderlevel

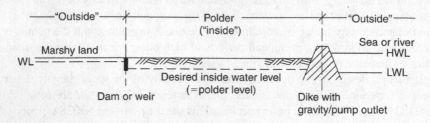

Figure 10.7 Main characteristics of the classical polder

(−30 to −40 cm), thus creating a gradient for flow towards the outlet. During dry periods, continued evaporation will cause the water level to drop below the polderlevel unless water is brought in from outside the polder. A high polderlevel is sometimes maintained during summer/dry seasons to encourage the lateral flow of water from the canals into the adjoining land, for example via the drainpipes (sub-irrigation).

Polders may be divided into sections having different polder water levels. This applies for example to a polder with an upper and a lower part which would probably be better served by maintaining a higher polder water level in the upper part and a lower water level in the lower part, rather than one water level for the entire polder. Different polder water levels are also advisable where soil conditions within differ greatly or where different parts have different functions (e.g., urban versus agricultural use).

During prolonged dry periods all water tends to collect in the lowest part of the polder whilst the canals in the upper part become (nearly) dry. Where this is undesirable (e.g., because the soil may over drain or ditches need to contain water for livestock water supply), water in the upper part may be conserved by installing weirs in the canals. These will also retard the discharge from the upper part somewhat during flood periods. The weir in Figure 10.8 conveniently divides the polder into an upper and a lower section. The width and crest level of the weir should be such that when the design discharge passes over the weir, the upstream water level is not too high (must still allow free discharge from the field drains) while during dry periods, when the canal levels in the upper section become horizontal at the crest level, a substantial water depth is maintained in all canals (for more information see section 13.2.2).

10.2.2   River polders

Drainage discharge problems in the middle to lower reaches of a river are illustrated in Figure 10.9. Gradients along the river are usually of the order of 20–30 cm/km. Natural raised levees are formed adjacent to the river and low-lying basins occur further away. The basins are often semi-enclosed by the levees, the terrace land bordering the valley and higher ground on the valley floor (often natural levees related to a former river course), leaving only a narrow outlet at the downstream end.

Tributaries entering the valley from the upland areas typically change direction in the basin area. With a slight gradient (less than the river gradient), they follow a course parallel to the river, centrally through the basin area, until sufficient head has been generated to break through the levee into the river. For tributary I this occurs 50 km downstream of its point of entry into the valley.

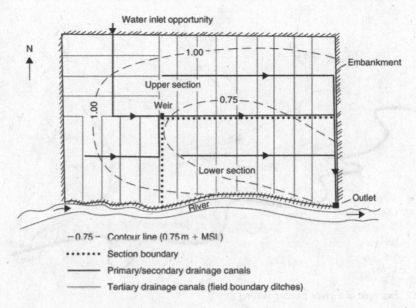

- 0.75 - Contour line (0.75 m + MSL)
••••••• Section boundary
━━━━━ Primary/secondary drainage canals
─────── Tertiary drainage canals (field boundary ditches)

Figure 10.8 Typical layout of a modern polder

Box 10.2   Field dimension of polders in the Netherlands for drainage

In a rather flat area, the hydraulic requirements of the canal layout can usually be accommodated to harmonise rather well with a modern field and farm layout. Canals and roads may be arranged to form fields of 400–500 m width and 800–1 200 m length, giving field sizes of 32–60 ha. If the field drainage is based on singular pipe drainage, as depicted below, field widths should generally not exceed 500–600 m because:

- commercial drainpipe flushing equipment generally only reaches effectively up to 250–300 m
- with long drain lines and for normal gradients, drain depth at the beginning and end of the line will differ too much
- with long lines, deep ditch levels have to be maintained.

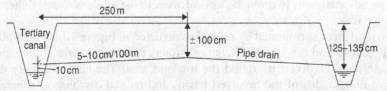

Long fields save on the required number of secondary canals although this advantage is generally outweighed by various farm management disadvantages when the length exceeds 1 000–1 500 m.

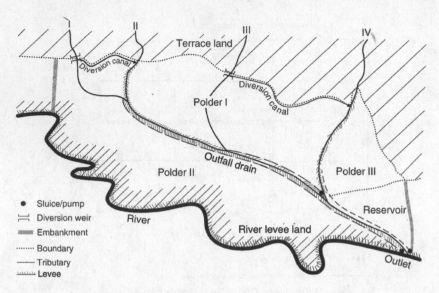

Figure 10.9 Example of a river polder setting

This principle is used to provide gravity discharge for the polders shown in Figure 10.9. The tributaries entering the valley from the side are diverted to join a neighbouring tributary. Within the alluvial plain they are normally 'endiked' so as to convey high discharges safely (without flooding) to the main river. The polders discharge by gravity into the tributary when levels permit (flap gate), otherwise pumping is used. When floods on the tributary coincide with floods on the main river, the outfall drain is allowed to spill over near its outlet into a low-lying reservoir area. Where a suitable reservoir area exists (low lying, low value land), this may be preferable to pumping (see section 13.4).

10.2.3   Upland discharges

Where upland and lowland occur in the same drainage basin and share the same outlet, it is often advantageous to keep the drainage flow of these two areas separate. Typical is e.g., the situation in some parts of the Po Valley (Italy) with two different main drainage systems: a high-level system for the upland areas, discharging by gravity and a low-level system for the lowland areas, discharging by pumping. For small upland areas, it might however be advantageous to drain the upland areas to the lowland areas, rather than keeping them separate.

For lowland areas surrounded by upland as depicted in Figure 10.10 an important decision is whether to lead the upland discharge through the lowland area (following the natural drainage path) or divert it around the lowland area. The choice usually depends on the relative magnitudes of the involved basins and related discharges. Where the existing river serves a large catchment and follows a distinct, well incised bed, the first solution may be the preferred one, in spite of the need for establishing two separate drainage infrastructures.

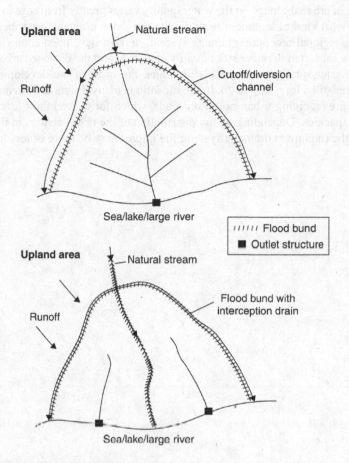

Figure 10.10 Alternatives for dealing with upland flow

10.2.4   Drainage of urban areas

Although in most catchments, compared to the rural area, the urban area is quite small, urban drainage often has a disproportional impact on its hydrological regime. The impact of urban drainage is generally most pronounced on the high stream flows and on the water quality. Low stream flows and groundwater regimes are generally far less affected.

Due to the high percentage of built-up, paved and other land with low infiltration and retention properties and its generally lack of significant storage facilities, streamflow hydrographs for urban areas are often much more peaked than those in rural areas. In small highly urbanized catchments, the peak flows at the outlet may be easily two to three times higher than for comparable rural catchments. Due to normal hydrograph attenuation and falling urbanization percentages, the impact of urban drainage on design discharges usually becomes smaller as the catchment size increases and often becomes quite insignificant when catchments reach sizes of < 10 000–25 000 ha.

The impact of urban drainage on the water quality varies greatly from case to case, mainly depending on what kind of treatment is applied to the urban drainage water before it is disposed onto the regional receiving drainage system. In most developed countries, the urban disposal is now subjected to rather strict quality standards and the disposed urban water may well be of the same quality as the receiving water. For most of the developing countries, there is however still a long way to go. In the meantime, many drains and streams in the arid climatic zone are receiving urban pollution loads, which far exceed their low flow season assimilation capacities. Depending on the distribution of the urban centres in the catchment and layout of the catchment drainage system, the impact may be more or less localised.

Part IV

Design

Chapter 11

Design of pipe drainage systems

Subsurface drainage by means of buried pipe systems is probably the most comprehensively studied subject in land drainage. As a result, the relationships between the variables and parameters are well established. The remaining design problems are mostly due to the great variability in the geometry and in the hydraulic properties of the soil, and due to the many other interactions between the soil and the system (affecting the choice of materials, methods of installation, etc.).

The descriptions in this Chapter are almost exclusively focussed on pipe systems with a parallel layout, which is by far the most common way in which pipe drains are laid out, both in the temperate and in the arid zone. Typical parallel layouts have been indicated in section 8.1. Layouts of pipe drainage in the arid zone (mostly for salinity control of irrigated land) often have some particular features which warrant special consideration. These are described in Chapter 16.

11.1 Flow patterns

The streamlines for the drainage flow towards two parallel pipe drains typically show a pattern as in Figure 11.1. The recharge percolates vertically downward through the unsaturated zone towards the watertable. In the saturated zone below the watertable the flow continues more or less in a *vertical*, downward direction, but soon turns into a lateral, *horizontal* flow towards the drains. Towards the end of its path the flow converges *radially* onto the drain. The extent of the three flow zones differs from case to case, depending especially upon the relative magnitude of h, L and D. When L is large in comparison to both h and D, the flow is predominantly horizontal[1]. An extensive radial flow sector is to be expected when L and D are of the same order of magnitude. A distinct vertical flow zone only occurs when h is relatively large.

The horizontal flow may extend to depths down to ¼L below the drainage base while the radial flow zone is roughly confined to a circle with radius ½ D√2 ≈ 0.7D around the drain.

The total headloss (h) may be visualised as being made up of the headloss due to vertical flow (h_v), horizontal flow (h_h), radial flow (h_r) and entry flow (h_e). This break down of the headloss components constitutes the basic concept underlying the Ernst drain spacing equations; for details see Ernst 1962 and ILRI 1994:

$$h = h_v + h_h + h_r + h_e$$

The different component head losses are indicated in Figure 11.2, although in this case h_e is assumed zero (ideal drain case).

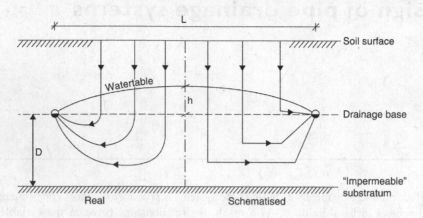

Figure 11.1 Typical pipe drainage flow pattern[1]

Vertical flow

The headloss due to a vertical flow of q (m/d) through a soil layer of thickness D_v (m) and a hydraulic conductivity of K (m/d) is according to Darcy's Law:

$$h_v = \frac{qD_v}{K}$$
Eq. 11.1

In situations such as in Figure 11.2, the vertical flow zone is usually assumed to extend from the watertable to the drainage base depth (although in reality it will often go deeper). The headloss h_v may then be determined as the difference between the readings of the piezometers (1) and (2). Its value is usually rather small. For example for q = 10 mm/d, K = 0.5 m/d and D_v = 1.0 m, h_v is only 2 cm. The headloss due to vertical flow becomes significant only when it passes through very poorly permeable soil (heavy clay, impeding layer, etc.).

Horizontal flow

The headloss over the horizontal flow zone equals the difference between the readings of the piezometers (2) and (3) in Figure 11.2. An analytical expression for this headloss may be derived by considering the (horizontal) flow Q_x through a vertical plane of unit width, which may be described by the following two expressions:

(a) $Q_x = q(L_h/2 - x)$ and (b) $Q_x^2 = KD_h \frac{dh_x}{dx}$

where D_h is the conceptual average thickness of the horizontal flow zone ($D_h \cong D + \frac{1}{2}h$). Equating these two expressions and solving the resulting differential equation for $\times = 0$, $h_x = 0$ and $\times = L_h/2$, $h_x = h_h$:

$$q \int_0^{L_h/2} (L_h/2 - x)\, dx = KD_h \int_0^{h_h} dh_x \rightarrow h_h = \frac{qL_h^2}{8KD_h}$$
Eq. 11.2

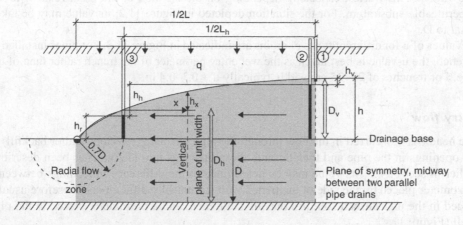

Figure 11.2 Head losses in pipe drainage flow

This implies a parabolic curvature[3] of the watertable over the distance L_h. The gradient dh/dx becomes increasingly steeper in the direction of the flow as is to be expected in view of the fact that an increasing amount of water passes through a decreasing cross-sectional area (Q_x increases and h_x decreases in the flow direction).

Box 11.1 Alternative derivation of Eq. 11.2

Eq. 12.2 may also be derived by noting that Q_x increases over the distance $L_h/2$ from $Q_x = 0$ for $x = L_h/2$ to $Q_x = \frac{1}{2}qL_h$ for $x = 0$, giving an average flow of $\frac{1}{4}qL_h$ through an average flow cross-section D_h. Applying Darcy's Law $Q = KiA$ gives:

$$\frac{1}{4}q_h = K\frac{h_h}{\frac{1}{2}L_h}D_h \text{ or } h_h = \frac{qL_h^2}{8KD_h}$$

This formula which assumes the flow to the drain to be fully horizontal is known as the Donnan formula

---

Radial flow

The headloss h_r may be measured as the difference between the reading in piezometer (3) and the water level in the drain (Figure 11.2). It may be expressed as (Ernst 1962, see also ILRI 1994):

$$h_r = q\frac{L}{\pi K}\ln\frac{aD_r}{u}$$
Eq. 11.3

where:
aD_r = an indicative geometric parameter for the radial flow zone
u = the wet entry perimeter of the drain

The value of aD_r varies, especially depending on the location of the drain relative to the impermeable substratum. For the situation depicted in Figure 11.2, its value may be taken equal to D.

Values of u for different types of drains are indicated in Figure 11.3. For pipes installed in a trench, the u-value is best taken as the wet entry perimeter of the trench rather than of the pipe. For trenches of 20–25 cm width, typically u = 0.3–0.4 m.

Entry flow

The headloss h_e incurred in the flow through the pipe surround (envelope and/or backfill) to the openings in the pipe and then through these openings into the pipe has been described earlier in section 8.3. Its value may be determined as the difference in readings between a piezometer placed by the side of the trench wall (technically difficult and therefore usually placed in the undisturbed soil, just outside the trench) and a piezometer placed in the pipe itself (Figure 11.4).

In general, the aim is to minimise the value of h_e (the *ideal drain* case with $h_e \sim 0$) so that the total head is available for vertical, horizontal and radial flow through the soil. The open entry areas of drainpipes in common use are sufficiently large to allow water to enter with negligible headloss during normal design discharges. The flow through the surround to the entry points may constitute a bottleneck. To be an ideal drain, the hydraulic conductivity of the surround should be at least 10 times higher than that of the undisturbed soil outside the trench. In practice this requirement is not often met due to inadequate or clogged envelope. Some entry headloss usually occurs. For details on appropriate envelope design see section 8.4.

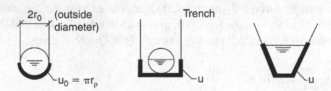

Figure 11.3 The wet entry perimeter for different types of drain[4]

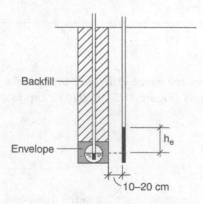

Figure 11.4 Measurement of the headloss at entry

11.2   Drain spacing formulae

The basic design criterion for a pipe system for watertable control specifies the recharge (q) that the system should be able to cope with while maintaining a desired watertable depth (H). Suitable values for W (= field drainage base depth) may be selected based on considerations such as those described in section 11.6. This determines the watertable head h = W – H (Figure 11.5). Required drain spacing (L) may then be calculated using a drain spacing formula.

Drain spacing formulae may be categorised as either *steady state formulae* or *non- steady state formulae*. Steady state formulae are based upon the assumption that a steady constant flow occurs through the soil to the drains. Discharge equals recharge and the watertable head (h) is constant. In the *non-steady state formulae* all these parameters vary in time and the watertable fluctuates during the drainage process.

Soundly based and tested procedures and criteria generally only exist for steady state design (partly due to the fact that these procedures were developed earlier in time and the underlying models and tests are so much simpler than for the non-steady case). Consequently, although it is recognised that the drainage process is in fact non-steady, design is normally based on steady state formulae and criteria. The latter criteria reflect a fictitious average recharge (q) and average watertable depth (H) during the critical period(s). Systems designed according to these criteria control the high watertable occurrences to within acceptable limits as regards frequency and magnitude, indirectly taking into account the non-steady nature of the drainage process.

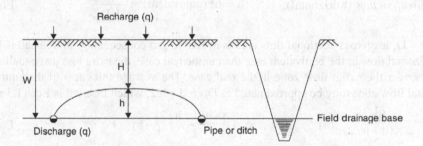

Figure 11.5 Procedure for the determination of the drain spacing for parallel pipe drainage systems (for steady state conditions)

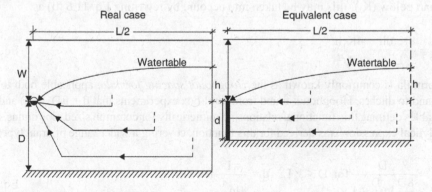

Figure 11.6 Transformation underlying the Hooghoudt formula

A number of steady state formulae are available although the principles will only be described in relation to the *Hooghoudt formula*. This formula has a wide applicability and a relatively simple structure. There are other formulae which may be superior in some cases but far worse in others, while some are quite unwieldy in use. In any event the accuracy with which drain spacing can be determined is limited by the accuracy of the soil parameters (due especially to the high variability of the hydraulic conductivity of the soil) rather than by the formula adopted.

11.3 Hooghoudt formula

A steady state drain spacing formula for pipe drainage was developed in 1940 by the Netherlands drainage researcher *Hooghoudt*. In this formula, only the head losses due to horizontal and radial flow to the pipe are considered (losses due to vertical flow usually being insignificant). As shown in Figure 11.6 Hooghoudt conceived that the horizontal/radial flow to pipe drains (described by Eq. 11.2 and 11.3) could be treated as a flow to trenches with the impermeable base at a reduced depth (d). This equivalent flow is essentially horizontal and may be described by the simpler Eq. 11.5.

$$\text{\textit{real flow} (horizontal + radial):} \quad h = h_h + h_r = \frac{qL_h^2}{8KD_h} + q\frac{L}{\pi K}\ln\frac{aD_r}{u} \qquad \text{Eq. 11.4}$$

$$\text{\textit{equivalent flow} (horizontal):} \quad h = h_h^* \text{(equivalent)} = \frac{qL^2}{8KD_{h*}} \qquad \text{Eq. 11.5}$$

Since d < D, less cross-sectional flow area is available and consequently more head is lost in the horizontal flow in the equivalent case than in the real case, the extra loss just equalling the headloss over the radial flow zone in the real case. The average thickness of the equivalent horizontal flow zone may be approximated as $D_{h*} = d + h/2$, which inserted in Eq. 11.5 gives:

$$h = \frac{qL^2}{8K(d+h/2)} \text{ (a)} \qquad \text{or} \qquad q = \frac{8Kh(d+h/2)}{L^2} \text{ (b)} \qquad \text{Eq. 11.6}$$

The equivalent horizontal flow takes place partly below the drainage base (the thickness of this flow zone being d) and partly above the drainage base (the average thickness of this flow zone being h/2). When the soil above drainage base has a different hydraulic conductivity (K_1) than below (K_2), this may be taken into account by rewriting Eq. 11.6 (b) as:

$$q = \frac{8K_2dh}{L^2} + \frac{4K_1h^2}{L^2} \qquad \text{Eq. 11.7}$$

This formula is commonly known as the *Hooghoudt spacing formula*, applicable both to pipe drains and to ditches. Hooghoudt found from sand-box experiments that d = f(D, L, u) and prepared tables defining this functional relationship numerically for common sized pipe drains. Later an analytical expression was derived for this function, covering a wider range of drain types:

$$d = \frac{D}{\dfrac{8D}{\pi L}\ln\dfrac{D}{u} + 1} \text{ for } D < \tfrac{1}{4}L; \quad d = \frac{\pi L}{8\ln\dfrac{L}{u}} \text{ for } D > \tfrac{1}{4}L \qquad \text{Eq. 11.8}$$

In situations where the soil becomes less permeable with depth but there is no distinct impermeable substratum, the depth D may be taken equal to the depth at which the K-value has decreased to 1/10 of the (average) K-value of the layer(s) above. As shown in Eq. 11.8, the equivalent depth d becomes independent of D when $D > \frac{1}{4}L$.

11.3.1 Use of the Hooghoudt formula

The normal procedure for the determination of the drain spacing with the Hooghoudt formula involves the following steps:

1) formulation of the basic design criteria (q and H)
2) establishment of the field drainage base W and the available head $h = W - H$
3) establishment of the soil parameters K (or K_1 and K_2) and D
4) selection of drain type (pipe or ditch) and determination of u
5) determination of the drain spacing (L) by solving the Hooghoudt formula.

The final step, solution of the Hooghoudt formula, may be done by trial and error, by graphical means or by computer.

Trial and error solution: since L depends on d and d depends on L, the Hooghoudt formula is not explicit in L, which can only be found by trial and error (Box 11.2):

- assume a value for L and determine d from Eq. 11.8
- solve the Hooghoudt formula for L and compare this value with the assumed value
- modify the value of L and repeat until the calculated and assumed values are equal.

Graphical solution: a number of nomo-graphical solutions of the Hooghoudt formula have been developed in the fifties/sixties (pre-computer age). The nomograph in Figure 11.7 covers most conditions normally encountered in pipe drainage design (h = 0.1–1.2 m, q = 1–10 mm/d). For pipe drains with $r_0 = 0.04 \sim 0.10$ m, u = 0.30 m.

For all practical purposes, the graph is valid for all pipes installed in 20–25 cm wide trenches (u ≈ 0.30 m).

Computer solution: computer programs are available not only for the solution of Eq. 11.8 but also for the integrated solution of the full Hooghoudt formula. These programs have been described in FAO Irrigation and Drainage Paper No 62 (FAO 2007).

11.3.2 Notes on the Hooghoudt formula

The Hooghoudt formula shows that (all other variables remaining constant), the spacing L increases when:

- K increases (especially when K_2 increases; the value of K_1 has much less influence)
- q decreases ($L \approx q^{-\frac{1}{2}}$)
- D increases (has less influence when L is small than when L is large)
- h increases (implies increase of W or decrease of H).

Box 11.2 Example of drain spacing determination

Determine the required drain spacing for the basic design criteria q = 7 mm/d, H = 0.6 m.

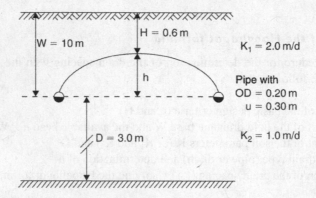

Trial and error solution:

For first trial assume L = 40 m. With D = 3.0, u = 0.3 m and L = 40 m, calculate with Eq. 11.8 that d = 2.15 m.

$$L^2 = \frac{4K_1h^2}{q} + \frac{8K_2dh}{q} = \frac{4 \times 2 \times 0.4^2}{0.007} + \frac{8 \times 1 \times 2.15 \times 0.4}{0.007}; L^2 = 1165; L = 34 \text{ m}$$

For second trial assume L = 32 m; from Eq. 11.8 → d = 2.0 m.

$$L^2 = \frac{4 \times 2 \times 0.4^2}{0.007} + \frac{8 \times 1 \times 2.0 \times 0.4}{0.007} + 1097 = 33 \text{ m}$$

Final solution L = 33 m.

Graphical solution:

Calculation:

$$\frac{4h^2}{q} = 91.5; K_1 = 2.0; \frac{4K_1h^2}{q} = 183$$

$$\frac{8h}{q} = 457; K_2 = 1.0; \frac{8K_2h}{q} = 457$$

In Figure 11.7 (graph B) connect 183 (on right hand scale) with 457 (on left hand scale); where the connecting line intersects with D = 3.0, read L = 33 m.

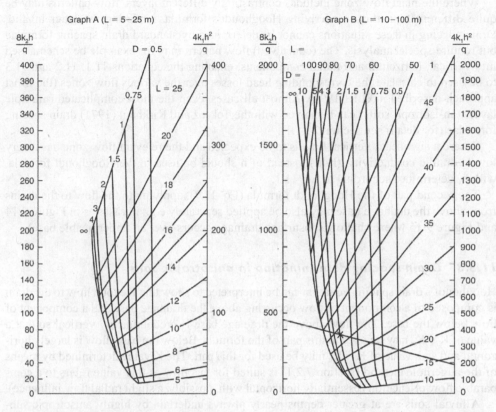

Figure 11.7 Nomograph for the solution of the Hooghoudt drain spacing formula (van Beers 1965)

When the drainage flow above the drainage base can be neglected, the Hooghoudt formula reduces to (simple Hooghoudt formula):

$$L^2 = \frac{8Kdh}{q}$$

Eq. 11.9

This formula may, for example, be used in situations where the watertable head (h) is small or where flow conditions below the drainage base are much more favourable than above it.

A change in hydraulic conductivity at about drainage base depth is quite common in non-stratified soils in the temperate zone. Here drains are commonly installed at 1.0–1.2 m, which corresponds to the depth of the most active part of the soil profile in terms of physical and chemical soil formation, soil biological life, soil moisture regime, root development, etc., all of which affect the pore geometry and distribution and thus the hydraulic conductivity of the soil. In stratified soils, however, the drainage base depth will rarely coincide exactly with a boundary between two soil layers. Slight deviations, especially when the hydraulic conductivities of the involved layers are similar, are of little concern. Otherwise, composed K values based on horizontal flow should be used (see section 7.4 for details). This is only acceptable when the K-values of the layers involved are similar in magnitude (differences of not more than 100%).

Where the main flow zone includes contrastingly different layers, flow patterns may be quite different from those underlying Hooghoudt's formula, rendering the latter invalid. Drain spacing in these situations cannot be determined by standard drain spacing formulae but require special analysis. The (conceived) flow pattern may for example be schematised into vertical, horizontal and radial components, enabling the equations 11.1, 11.2 and 11.3 to be used to calculate the corresponding head losses over the various flow zones (the Ernst approach described in ILRI 1994). Almost all cases, even the most complicated (multiple layers, an-isotropic soils) can be solved with the Toksöz and Kirkham (1971) drain spacing formulae (for details see section 21.1.1).

Where a significant vertical flow is to be expected and the relevant flow zone has a very low hydraulic conductivity, $(h-h_v)$ instead of h should be used in the Hooghoudt formula, with h_v determined according to Eq. 11.1.

The second part of the Hooghoudt formula (Eq. 11.7), applying to the flow to the drains from above the drainage base may also be applied separately e.g., in cases as in Figure 9.14 and Figure 9.15 where shallow subsurface drainage occurs over an impermeable base).

11.3.3   Drain spacing determination in anisotropic soils

Hooghoudt's drain spacing equation can be interpreted to show that the total flow to the drain is composed of a component of flow occurring above the drainage base and a component of flow below the base. The flow above the drainage base is predominantly vertical so that a value of $K = K_v$ may be used in this part of the formula. Below the base, flow is largely horizontal and so a value of $K = K_h$ may be used for this part. The K value determined by means of the augerhole method (section 7.2.1) is suited for this flow as this value refers to a comparable flow orientation (essentially horizontal with, possibly, a slight radial flow influence).

Alluvial soils are at greater depths nearly always underlain by highly anisotropic substrata with K_h exceeding K_v by a factor of 20 to 30 (Boumans 1979). Anisotropy has a considerable influence on drain spacing. It may be dealt with by transforming scales in such a way that the soil may be considered to be isotropic. This transformation process can be done either by multiplying the vertical dimensions by a factor $R^{1/2}$, where $R = K_h/K_v$, or by dividing the horizontal dimensions by $R^{1/2}$. The equivalent isotropic hydraulic conductivity value is calculated as $K'=(K_hK_v)^{1/2}$. Transformation of the vertical scale is generally the most straightforward and will be illustrated for the case of Figure 11.8.

Example: Basic design criteria q = 2.5 mm/d and h = 1.0 m. The drain spacing is calculated using the simple Hooghoudt's equation (Eq. 11.9) applying the following steps:

1)   Calculate the ratio $R = \dfrac{K_h}{K_v} = 25$, and $R^{1/2} = 5$.

2)   Adjust the vertical scales by multiplying by $R^{1/2}$. Note that this transformation does not apply to the head h.

3)   Calculate the isotropic K-value and apply Hooghoudt's formula:

$$K' = (K_hK_v)^{1/2} = (0.75\times0.03)^{1/2} = 0.15\,\text{m/d}; L^2 = \frac{8K'd'h}{q} \text{ where } d' = \frac{d}{\dfrac{8D'}{\pi L}\ln\dfrac{D'}{u'}+1}$$

$$L^2 = \frac{8(0.15)\,d'(1)}{0.0025} = 480\,d'; \text{ by trial and error } L = 50$$

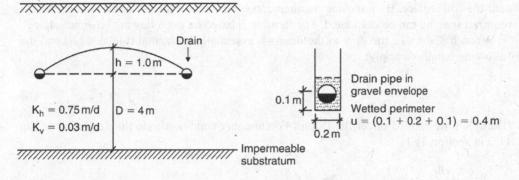

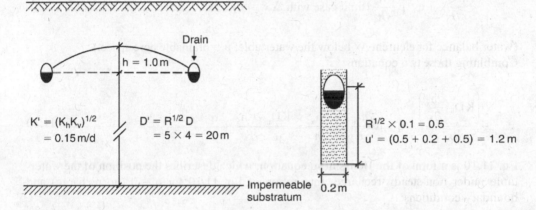

Figure 11.8 Principal dimensions before and after transformation

Complicated cases involving a number of layers with different anisotropic permeability can be solved with the aforementioned Toksöz and Kirkham equation.

11.4   Non-steady state drainage formulae

Steady state drainage criteria are not available or are not satisfactory in all cases. The first limitation applies to the many countries in which subsurface drainage has only recently been introduced or where there is no history of drainage research. The second applies to climates in which the rainfall is generally distinctly non-steady in nature (coming in discrete storms rather than as prolonged periods with rather uniform, medium intensity rainfall). The same applies to subsurface drainage of irrigated land with the draining events generated by distinct periodic irrigation applications. Steady state criteria are also unsatisfactory in situations where the soil hydrological conditions are such that watertables and drain-outflow respond directly and quickly to (rainfall) recharge, rather than the response being moderated. In all these cases, the non-steady drainage formulae as described below, are of value.

11.4.1   Falling watertable (Glover-Dumm formula)

This formula describes the fall of the watertable after it has risen almost instantaneously to near the soil surface. By imposing requirements on the rate of the fall of the watertable, the required spacing can be calculated. The formula is based on the following schematisation.

When $h < d \ll L$, the flow to the drains is essentially horizontal (Figure 11.9) and the following equations apply:

a) $Q_x = KD_h \dfrac{\partial h_x}{\partial x}$

(Darcy's Law applied under the Dupuit-Forchheimer conditions; see the derivation of Eq. 11.2 in section 11.1).

b) $\mu \dfrac{\partial h_x}{\partial t} \Delta x = Q_{x+\Delta x} + q\Delta x - Q_x$

$\dfrac{\partial Q_x}{\partial x} + q = \mu \dfrac{\partial h_x}{\partial t}$ (limit case with $\Delta x \to 0$)

(water balance for element Δx below the watertable; μ = drainable pore space.)
Combining these two equations:

$$\frac{\partial \left(KD_h \dfrac{\partial h_x}{\partial x} \right)}{\partial x} + q = \mu \frac{\partial h_x}{\partial t} \to \frac{\partial h_x}{\partial t} = \frac{KD_h}{\mu} \frac{\partial^2 h_x}{\partial x^2} + \frac{q}{\mu} \qquad \text{Eq. 11.10}$$

Eq. 11.10 is a form of the Boussinesq equation, which describes the position of the watertable under non-steady recharge. Integration of Eq. 11.10 for $q = 0$ (no recharge) and boundary conditions:

- $t = 0$; $h_x = h_o$ for all x values (an initially horizontal watertable at a height h_0 above the drainage base)
- $t > 0$; $h = 0$ at $x = 0$ and $x = L$ (an instantaneous lowering of the potential to zero in the drains)

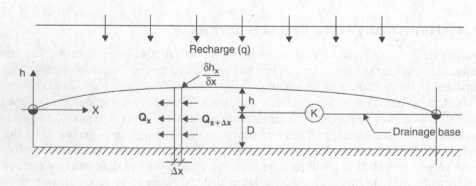

Figure 11.9 Horizontal drainage flow to parallel drains

yields (after some elaboration) the *Glover-Dumm* drainage formula which describes the rate of fall of the watertable after the latter has risen instantaneously (at t = 0) to a height h_0 above the drainage base (typical of the behaviour of shallow watertables in irrigated land which often rise sharply during a water application and then recede more slowly).

According to Glover-Dumm the mid-spacing watertable head h_t at t = t relates to the head h_0 (at t = 0, Figure 11.10) as:

$$\frac{h_t}{h_o} = 1.16e^{-\alpha t} \qquad\qquad \text{Eq. 11.11}$$

and

$$\alpha = \frac{\pi^2 Kd}{\mu L^2} \qquad\qquad \text{Eq. 11.12}$$

where:
α = reaction factor, d^{-1}
t = time (days)
h_0 = initial watertable head, at $t = t_0$ (m)
h_t = watertable head, at t = t (m)
μ = drainable pore space (m³/m³)
L = drain spacing (m)
d = equivalent depth to the impermeable substratum (m)
K = hydraulic conductivity (m/d)

Combining Eqs. 11.11 and 11.12 gives (rounding $\pi^2 = 10$):

$$L^2 = \frac{10Kdt}{\mu}\left[\ln 1.16\frac{h_0}{h_t}\right]^{-1} \qquad\qquad \text{Eq. 11.13}$$

In these formulae it is assumed that the shape of the falling watertable conforms to a fourth-degree parabola (explains the factor 1.16 in Eq. 11.11). Radial resistances are taken into account by replacing the depth D to the impermeable substratum by Hooghoudt's equivalent depth d.

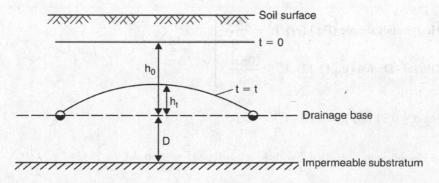

Figure 11.10 Notations and boundary conditions used with Glover-Dumm non-steady state formula

Example of solution of the Glover-Dumm equation

Given:

$W = 1.2$ m, $D = 4.0$ m, $K = 2.0$ m/d, $\mu = 5\%$ (= 0.05 m³/m³), pipe drain with $u = 0.2$ m

$t = 0$; $H_0 = 0$ m (watertable at soil surface)

$t = 4$; $H_4 = 0.8$ m

The basic design criteria thus state that the system should be able to lower the watertable from the soil surface to a depth of 0.8 m below the soil surface within a period of 4 days.

a)   *Direct solution*

$$\left.\begin{array}{l} h_0 = W - H_0 = 1.2m \\ h_4 = W - H_4 = 0.4m \end{array}\right] \frac{h_4}{h_0} = \frac{0.4}{1.2} = 0.33$$

From Eq. 11.11 $\alpha t = -\ln\dfrac{0.33}{1.16} = 1.24$ and with $t = 4 \rightarrow \alpha = \dfrac{1.24}{4} = 0.31$

Eq. 11.12: $\alpha = \dfrac{10Kd}{\mu L^2} = 0.31$ or $L^2 = \dfrac{10Kd}{0.31\mu}$; for

$\mu = 0.05$ and $K = 2.0 \rightarrow L^2 = 1290 \times d$

Since $L = f(d)$ and $d = f(L)$, a trial and error procedure is followed to solve this equation for L:

First trial: $L = 30$ m. . . . $d = 2.20$ m (from Eq. 11.8)
$L = \sqrt{1290 \times 2.20} = 53.3$ m

Second trial: $L = 60$ m $d = 2.84$ (from Eq. 11.8)
$L = \sqrt{1290 \times 2.84} = 60.5$ m

Solution is: $L = 60$ m

b)   *Indirect solution*

The Glover-Dumm formula and the Hooghoudt formula can be interrelated, enabling the non-steady basic design criteria to be translated into steady-state criteria:

$$\left.\begin{array}{l} \text{Hooghoudt simple (Eq.11.9)}\ \ L^2 = \dfrac{8Kdh}{q} \\[3mm] \text{Glover - Dumm (Eq.11.12)}\ \ L^2 = \dfrac{10Kd}{\alpha\mu} \end{array}\right] \frac{h}{q} = \frac{10}{8\alpha\mu}$$

For $\alpha = 0.31$ and $\mu = 0.05 \rightarrow \dfrac{h}{q} = 80.65$

In Hooghoudt: $L^2 = 8Kd\dfrac{h}{q}$ with $K = 2$, gives $L^2 = 1290 \times d$

By trial and error, as above, it is found that $L = 60$ m.

11.4.2   Fluctuating watertable (de Zeeuw and Hellinga formula)

This is not a design formula but a formula which can be used to simulate the watertable and drain discharge fluctuations under a non-steady recharge regime. The formula is based on the simple Hooghoudt's formula, in which the drain discharge (q) is linearly related to the mid-spacing watertable head (h):

$$\text{(i.e., } q = \frac{8Kd}{L^2}h \qquad\qquad \text{Eq. 11.9)}$$

The variation of the drain discharge with time is thus also linearly related to the variation in time of the watertable head:

$$\frac{dq}{dt} = \frac{8Kd}{L^2}\frac{dh}{dt} \qquad\qquad \text{Eq. 11.14}$$

When the groundwater body is recharged by rainfall or by another source (R) and is depleted by drain discharge (q), it follows that the watertable will rise when $(R - q) > 0$ and fall when $(R - q) < 0$. The watertable fluctuations may be described by:

$$\frac{dh}{dt} = \frac{(R-q)}{C\mu} \qquad\qquad \text{Eq. 11.15}$$

where C is a correction factor accounting for the fact that the watertable during its rise/fall will not be horizontal.

Combining Eq. 11.14 and Eq. 11.15 and taking C = 0.8, gives:

$$\frac{dq}{dt} = \frac{10Kd}{\mu L^2}(R-q) = \alpha(R-q) \qquad\qquad \text{Eq. 11.16}$$

So the change in drain discharge $\left(\dfrac{dq}{dt}\right)$ is proportional to the excess recharge (R – q), the constant of proportionality being a (reaction factor, see Eq. 11.12). Integration of Eq. 11.16 between the limits $t = t$, $q = q_t$ and $t = t - 1$, $q = q_{t-1}$ gives:

$$\int_{q_{t-1}}^{q}\frac{dq}{(R-q)} = \int_{t-1}^{t}\alpha dt \rightarrow \frac{(R-q)_t}{(R-q)_{t-1}} = e^{-\alpha\Delta t}$$

which may be elaborated into:

$$q_t = q_{t-1}e^{-\alpha\Delta t} + R_{\Delta t}(1-e^{-\alpha\Delta t}) \qquad\qquad \text{Eq. 11.17}$$

where $R_{\Delta t}$ is the mean value of R during the time interval Δt between $t = t - 1$ and $t = t$.

Using the linear relationship between q and h of Eq. 11.9 ($q = \dfrac{8Kd}{L^2}$ h = 0.8 αμh) it also follows that:

$$h_t = h_{t-1}e^{-\alpha\Delta t} + \frac{R_{\Delta t}}{0.8\mu\alpha}(1-e^{-\alpha\Delta t}) \qquad\qquad \text{Eq. 11.18}$$

Equations 11.17 and 11.18, first derived by de Zeeuw and Hellinga (1960), may be used to simulate drain discharge and watertable depth fluctuations on the basis of (historical) weather records, given the value of the reaction factor α. The non-steady processes are approximated by a series of intervals over which the recharge $R_{\Delta t}$ is assumed to be uniform. This requirement can generally be met by adopting intervals of length $\Delta t = 1$ day (expressing t in days makes the coefficient $\alpha\Delta t$ dimensionless since α has the dimension d^{-1}).

The first rains after a dry period are normally mostly retained within the soil. In general, the watertable will not rise above the drainage base until the overlying soil is at or above field capacity. Soil moisture balance calculations as described in section 5.1.1 can help to establish when field capacity is reached and the watertable response to rainfall commences.

Reaction factor (α)

The reaction factor $\alpha = 10Kd/mL^2$ is a direct indicator of the intensity with which the drain system responds to changes in the recharge. Values generally vary from $\alpha = 0.1-0.3$ for land with a slow response (low KD value, wide drain spacing, high drainable pore space) to $\alpha = 2.0-5.0$ for rapidly responsive land (high KD value, narrow drain spacing, low drainable pore space). Its value may be calculated according to Eq. 11.12 (from known KD, L and μ values) but since the μ-value is especially difficult to determine, best estimates of α are obtained by observing the actual response of the system in the field. From equations 11.17 and 11.18, it follows that in periods during which there is no recharge (R = 0):

$$\alpha = \frac{\ln q_{t-1} - \ln q_t}{\Delta t} = 2.30 \frac{\log q_{t-1} - \log q_t}{\Delta t} \qquad \text{Eq. 11.19a}$$

$$\alpha = \frac{\ln h_{t-1} - \ln h_t}{\Delta t} = 2.30 \frac{\log h_{t-1} - \log h_t}{\Delta t} \qquad \text{Eq. 11.19b}$$

Observed q_t or h_t values may be plotted on log normal paper as in Figure 11.11. If the system obeys the assumptions underlying the basic formulae (equations 11.9, 11.14 and 11.15), the observed values more or less fit to a straight line with a slope equal to α. Observations are best made during periods of low evaporation shortly after the end of a few good rainy days when the recharge to the groundwater has ceased and the watertable starts receding. The recession sections of the watertable/drain flow vs. time graphs, see Figure 12.6; the observations from day 6 to day 8 in Table 11.1 may for example be used to estimate α.

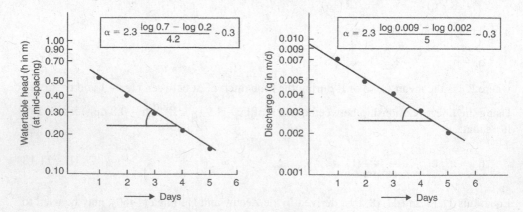

Figure 11.11 Determination of the reaction factor (α) from observed watertable heads and drain discharges during recession

Table 11.1 Watertable head and drain outflow calculations on the basis of the de Zeeuw-Hellinga
formula

Day	Rainfall (P in m)	Evapotranspiration (ET in m)	Recharge (R in m)	Watertable head[1] (h in m)	Drain outflow[2] (q in m/d)
0				0.10	0.001
1	0.020	0.001	0.004	0.16	0.002
2	0.020	0.001	0.019	0.54	0.007
3	0.010	0.002	0.008	0.57	0.007
4	0.025	0.001	0.024	0.96	0.012
5	0.005	0.002	0.003	0.77	0.010
6		0.002		0.56	0.007
7		0.002		0.41	0.005
8		0.002		0.30	0.004
9		0.002		0.22	0.003
10		0.002		0.16	0.002[3]

1) $h_t = h_{t-1} e^{-\alpha\Delta t} + \dfrac{R_{\Delta t}}{0.8\mu\alpha}(1 - e^{-\alpha\Delta t})$ (Eq. 11.18)

for $t = 1$: $h_1 = (0.10 \times 0.73) + \left(\dfrac{0.004}{0.012} \times 0.27\right) = 0.073 + 0.090 = 0.16$

for $t = 2$: $h_2 = (0.16 \times 0.73) + \left(\dfrac{0.019}{0.012} \times 0.27\right) = 0.117 + 0.427 = 0.54$

for $t = 6$: $h_6 = 0.77 \times 0.73 = 0.56$; for $t = 7$: $h_7 = 0.56 \times 0.73 = 0.41$ etc.

2) $q_t = q_{t-1} e^{-\alpha\Delta t} + R_{\Delta t}(1 - e^{-\alpha\Delta t})$ (Eq. 11.17)

for $t = 1$: $q_1 = (0.001 \times 0.73) + (0.004 \times 0.27) = 0.002$
for $t = 2$: $q_2 = (0.002 \times 0.73) + (0.019 \times 0.27) = 0.007$
for $t = 6$: $q_6 = 0.010 \times 0.73 = 0.007$; for $t = 7$: $q_7 = 0.007 \times 0.73 = 0.005$. etc.

3) evapotranspiration may be expected to start interacting.

Application of de Zeeuw-Hellinga formula

Parallel pipe drainage system with W = 1.20 m, L = 40 m, KD = 2.5 m²/d and μ = 0.05 m³/m³

$$\alpha = \frac{10Kd}{\mu L^2} = \frac{10 \times 2.5}{0.05 \times 1600} = 0.31; \ \Delta t = 1.0 \text{ day}, \rightarrow e^{-\alpha\Delta t} = e^{-0.31} = 0.73; \ (1 - e^{-\alpha\Delta t}) = 0.27;$$

$0.8\mu\alpha = 0.012$

At the start of the rain (day 0), the watertable head h = 0.10 m, so the mid-spacing
watertable depth H = W – h = 1.20 – 0.10 = 1.10 m below the soil surface. The corresponding drain outflow q = 8Kdh/L² = (8 × 2.5 × 0.10)/1 600 = 0.001 m/d. The overlying soil profile can store 15 mm before reaching field capacity. The evapotranspiration
is low (1–2 mm/d). All rainfall infiltrates and excess water in the profile readily percolates through to the ground watertable.

11.5   Basic design criteria

The criteria to be used depend upon the objective(s) to be served by the drainage system. Broadly, the objective of subsurface drainage systems is to control the watertable in the soil in order to create favourable soil water conditions for crop growth and farm operations (see Chapter 1). Criteria for subsurface drainage for soil salinity control are treated separately in section 16.3.2.

Most of the criteria described are formulated in steady state terms although it will be shown that they can equally well be expressed in the non-steady state form. A number of specific objectives may be distinguished (for details see section 1.4):

- improvement of the rootzone aeration
- early soil workability after rain
- early warming up of the soil in spring
- prevention of soil structural deterioration
- promotion of useful biological, microbiological and biochemical processes (especially related to nitrogen availability to plants).

Criteria generally express the integrated requirements based on one or more of these objectives although the weight given to individual objectives varies from case to case. This is why different criteria are used for off season and crop-season drainage, reflecting the different objectives of the drainage in these two seasons. The most critical season determines the criteria to be used for the design.

11.5.1   Criteria for off-season drainage

Figure 11.12a illustrates the subsurface drainage needs in the moderate, humid climates of NW Europe. The rainfall in this area is mostly well distributed throughout the year, totalling some 700–900 mm per annum. During the winter, P (precipitation) >> ET (evapotranspiration) and watertables tend to be high. However, this is of little concern since there are very few farm operations which have to be carried out on the land during this period, and crop growth is limited anyway by the prevailing low temperatures. Some control, however, is desirable to prevent continuous, severe waterlogging of the main rootzone as this has an adverse effect on the structure of the soil and on the nutrient availability in the following spring and summer, the main farming season. Excess water in the soil during the late winter/early spring also delays planting, retards the warming up of the soil and creates poor soil-water-air conditions during the critical germination and seedling growth periods.

During the summer ET > P and watertables tend to be low. They may rise occasionally well into the rootzone after heavy rain, but even low intensity drainage assisted by the considerable bio-drainage provided by evapotranspiration, will generally suffice to lower the watertables in sufficient time to prevent crop damage. In the autumn as P once again overtakes ET and watertables start rising, harvesting and other farm operations may be hindered by excess water conditions although these are generally somewhat less critical than in the spring.

To ensure the desired excess water control during the critical winter and early spring period, subsurface drainage systems in NW Europe are commonly designed on the basis of criteria such as detailed in Table 11.2. Under the prevailing climatic conditions in NW Europe, systems with a drainage intensity (see Box 11.3) h/q = 55 to 100 generally suffice. Such systems are able to control the watertable during the winter/off-season mostly below the topsoil and ensure early return of the upper rootzone to FC (Field Capacity) in the spring, an essential condition for effective evaporative drying of the land. In this way, drainage contributes to good seedbed preparation, timely planting, good germination and seedling growth.

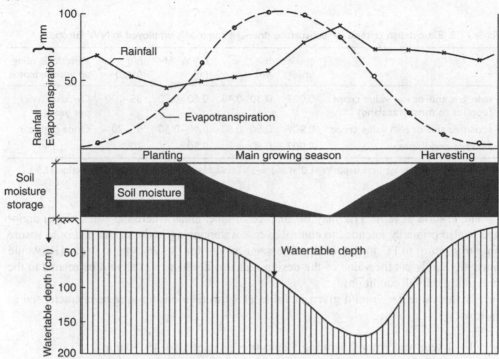

(a) Netherlands

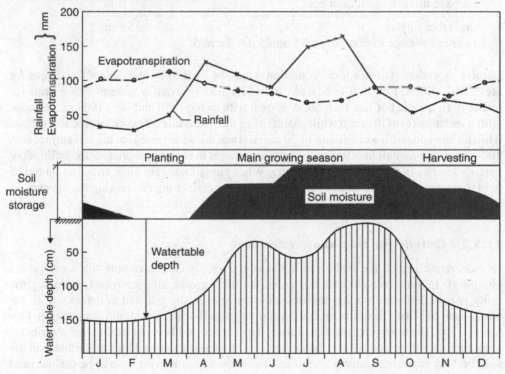

(b) Nyandarua District (Kenya)

Figure 11.12 Illustration of prevailing agro-hydrological conditions during the year and in the various seasons

Table 11.2 Basic design criteria for subsurface drainage commonly employed in NW Europe

	q (m/d)	h (m)	h = W–H (m)	h/q (days)	*Watertable rising up into the topsoil*
• tolerant and/or low value crops (applies to most grassland)	0.007	0.30–0.40	0.60–0.70	85–100	Once or twice per year
• sensitive and/or high value crops	0.007	0.50–0.60	0.40–0.50	55–70	Once per year
• average conditions	0.007	0.50	0.50	70	

NB: valid for field drainage base depth W=1.0 m and μ~5% (drainable pore space), see further section 11.5.3.

The criteria in Table 11.2 may be adapted to other areas where the (subsurface) drainage is also primarily intended to control excess waterlogging in the winter and/or to ensure an early return to FC in the spring. The same values for the watertable depth (H) would normally apply but the value of the design recharge/discharge (q) should be related to the prevailing rainfall conditions.

For the value q = 7mm/d given in Table 11.2 this relationship may be characterised as follows:

• 1 × 1 year winter rainfall during 5 days (design rainfall)	45–50 mm
• storage (under unfavourable conditions)	10–15 mm
• evapotranspiration	– mm
• seepage minus natural drainage.	– mm
• drainable surplus	35 mm
• required average discharge q = 35 mm/5 d = 7 mm/d	

Similar procedures (to suit local conditions) may be used to derive design discharges for areas with different rainfall conditions. Few situations warrant a system with a drainage intensity (h/q-value) of less than 50. A system with an h/q = 50 and W = 100 cm, can cope with a recharge (q) of 10 mm/d while maintaining the watertable 50 cm below the soil surface whilst it can cope with a maximum of 20 mm/d when the watertable is at the soil surface (h = 100 cm). Daily rainfall in NW Europe seldom exceeds these rates, moreover the infiltration/percolation rate of the soil is often limiting. When rainfall rates are high, attention should be paid to improving surface/shallow drainage conditions rather than increasing the intensity of the subsurface drainage system.

11.5.2   *Criteria for crop-season drainage*

In moderate climates, the main farming season normally coincides with the warm period (summer). In (sub) tropical climates crops are often grown all year round, including the rainy season. Indeed, cropping seasons are often specifically planned to coincide with the wet season to benefit from the rain (applies to most semi-arid/semi-humid climates, see Figure 11.12b). The evapotranspiration during the warm period is considerable and contributes very significantly to controlling the occurrence of excess water and high watertables in the soil, both by removing water from the soil (which would otherwise have to be drained) and by depleting soil moisture during dry periods, thus creating storage for subsequent rainfall. Drainage requirements during warm growing seasons, therefore, are rather small, except for occasional heavy (summer) rains lasting several days which can create periods during which excess water occurs for sufficient length of time to cause harm.

Crop-season drainage aims primarily at a rapid restoration of the rootzone aeration, following such a heavy rain. Other considerations for controlling excess soil water (early workability, warming-up, nitrification) are generally less critical.

However, there are few well tested criteria for excess water/watertable control during this season. Values for H are often taken somewhat higher than for the off-season and could for example be taken equal to the depth of the main rootzone. Of course, steady state criteria depict a rather fictitious watertable situation, which is not even an average situation, and so no direct, straightforward relationship need exist between H and rootzone depth. The following depths serve as rough guidelines in selecting suitable values for H:

- shallow rooted and/or tolerant and/or average value crops: H = 0.50 m
- deep rooted and/or sensitive and/or high value crops: H = 0.75–1.00 m

---

Box 11.3   Drainage intensity

The simple Hooghoudt formula Eq. 11.9 $\left(L^2 = \dfrac{8Kdh}{q} \right)$ may be rewritten as $\dfrac{h}{q} = \dfrac{L^2}{8Kd}$ to show that for a given system h/q is a constant determined by the drain spacing L of the system and the KD values of the soil. For identical field drainage base depths, the parameter h/q is indicative of the drainage intensity of the system. A system with a low h/q-value is able to cope with higher recharges and/or maintain lower watertables than a system with a high h/q-value.

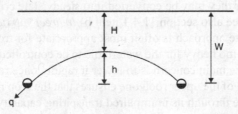

	Aeration drainage	Salinity drainage
q	5–10 mm/d	1–3 mm/d
H	0.5 m	1.0 m
W	1.0 m	2.0 m
h = W – H	0.5 m	1.0 m
h/q	50–100 days	~500 days

Aeration drainage requires a responsive system which can restore desired air-water conditions with the tolerance limits, while for salinity control it generally suffices to have a small seasonal downward flow in the rootzone (for more description see section 16.3.2).

The corresponding value for q may be derived from the 1–2 year return period, 5 days rainfall. This rainfall should of course relate to the critical period (growing season, or even a specific period within this season). Fairly high reductions for storage and evapotranspiration may be allowed for during the main growing season.

Example

Criteria for crop-season drainage of vegetables:

- H = 0.50 m (fairly sensitive and high value crops but mostly with a rather shallow root system)
- 1×1 year rainfall for 5 days   120 mm
- storage (mostly in the soil).   - 30 mm
- surface runoff (for the intense storms to be expected assuming that a
  properly functioning surface drainage system has been installed)   - 20 mm
- evapotranspiration (5 mm/d on average)   - 25 mm

- drainable surplus   45 mm
- q = 45 mm/5d = 9 mm/d

An alternative (and often better) approach towards establishing crop-season drainage criteria is to formulate them in non-steady terms i.e., a certain rate of fall of the watertable (with corresponding rate of restoration of the rootzone aeration) is prescribed, following a heavy rainfall which is assumed to have caused the watertable to rise to the soil surface. It may, for instance, be prescribed that within one to two days after the end of the rain the watertable should have fallen sufficiently so as to have restored the aeration of the upper rootzone, while within two to three days the entire main rootzone should be drained to field capacity. Relevant criteria are formulated in Table 11.3.

These criteria may be used directly in the non-steady state Glover-Dumm formula to calculate drain spacing or they may be converted into steady state criteria and then used in the Hooghoudt formula (see also section 11.4.1 item b) *Indirect solution*).

The non-steady state approach is often most appropriate for tropical/sub-tropical conditions where the rain is too heavy for the watertable to be controlled below the topsoil during the actual storm and the main concern is to lower it rapidly once rain has ceased. Rapid restoration of the aeration of the upper rootzone ensures that the crop remains vigorous, capable of further self-drainage through its unimpaired transpiring capacity.

Table 11.3 Non-steady state criteria for crop-season drainage

	Tolerant and/or average value crops	Sensitive and/or high value crops
Shallow Rootzone	t = 0 days H = 0.00 m t = 2 days H = 0.20 m t = 4 days H = 0.35 m	t = 0 days H = 0.00 m t = 1 day H = 0.20 m t = 2 days H = 0.35 m
Normal to deep Rootzone	t = 0 days H = 0.00 m t = 2 days H = 0.30 m t = 4 days H = 0.50 m	t = 0 days H = 0.00 m t = 1 day H = 0.30 m t = 2 days H = 0.50 m

Tropical storms can be intense and prolonged and much of the rain, which is unable to infiltrate into the soil, needs to be discharged over the surface, requiring the provision of surface drainage in addition to the installation of a subsurface drainage system. The ditch systems (especially the system described in section 9.2.1 sub 2) combine these two functions and as such are well suited to tropical/subtropical conditions.

11.5.3 The impact of drain depth and drainable pore space

The design criteria also depend to some extent upon on W (the selected drainage base depth) and on μ (drainable pore space of the soil). This is due to the influence of these two parameters on the watertable regime. During rain-free periods watertables will fall, eventually to drainage base depth or even below. When W is large, watertables will on average have fallen deeper by the time the next rain arrives than when W is small. As a result, more rain can be stored in the soil without watertables rising to harmful levels thus reducing the quantity to be discharged. Consequently, lower q-values should suffice when W is large.

Similar reasoning may be used to demonstrate that a lower q-value may be adopted when μ is large (>15%) since the rise of the watertable due to rainfall decreases as μ increases. The converse does not apply because, although watertables rise rapidly when μ is low (<5%), they also fall quickly and so there is generally no need to increase q when μ becomes very small.

Some rough guidelines for taking into account the effects of μ and W on q are suggested in Table 11.4. This table shows the variation in q relative to the standard case of W = 1.0 m and μ = 5%. It shows that for a non-standard case such as W = 1.50 m and μ = 10%, the design discharge should be taken as $0.65 \times q = 0.65 \times 7 = 4.6$ mm/d. Since according to the Hooghoudt formula $q \propto 1/L^2$, the effects of these adjustments of q on L remain rather small (a 2 × smaller value of q only results in a 1.4 × wider drain spacing).

11.5.4 Drainage criteria determined by simulation

Non-steady state drainage formulae such as Eq. 11.18 enable the watertable behaviour over a certain period to be simulated on the basis of the (infiltrated) rainfall and (actual) evapotranspiration data for that period (Figure 11.13). Watertable hydrographs may thus be developed for, say, a 20 year period, using the historical daily weather data for a range of basic design criteria (q, H packages). From the results frequency tables, as in Table 11.5, can be compiled on an annual or seasonal basis. For example, separate tables may be prepared for

Table 11.4 Guidelines for the adjustment of the design discharge (q) in ground-water drainage design for different values of the drainable pore space (μ) and drainage base depth (W)

	Drainage base depth W			
	0.70 m	1.00 m	1.50 m	2.00 m
μ = 5%	1.5 × q	q*	0.8 × q	0.65 × q
μ = 10%	1.2 × q	0.8 × q	0.65 × q	0.5 × q
μ = 15%	1.0 × q	0.65 × q	0.5 × q	0.5 × q

* Standard case with μ = 5% and W = 1.00 m

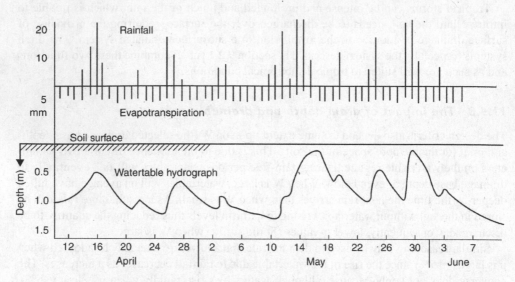

Figure 11.13 Watertable hydrograph determined by simulation

Table 11.5 Average number of days per annum with watertable at or within depth indicated from the soil surface during the main growth period (field drainage base depth W = 1.0 m)

Watertable depth below soil surface	q = 7mm/d H = 0.30m	q = 7 mm/d H = 0.50m	q = 10 mm/d H = 0.50 m
0.00 m	2 days	1 day	0 days
< 0.20 m	4 days	2 days	1 day
< 0.40 m	7 days	5 days	3 days
< 0.60 m	12 days	7 days	5 days

the off-season, for the crop-season, or for a particular period posing its own particular drainage requirements (planting-early growth period, main growth period, harvest period, etc.). The DRAINMOD and SWAP computer models (Chapter 21) are especially suitable for such watertable simulations.

For the case of Table 11.5 with q = 7 mm/d and H = 0.30 m, the watertable rises during the main growth period on average for two days per year to the surface, for four days per year to within 0.20 m from the surface, for seven days to within 0.40 m and for 12 days to within 0.60 m. If a crop is to be grown which can only tolerate the watertable in the upper rootzone for one to two days per year the more stringent criteria of q = 7 mm/d and H = 0.50 m should be adopted.

11.6　Drain depth

For a given value of watertable depth H the value of watertable head h increases with the value of drain depth W (Figure 11.5) and wider drain spacing may be used. This favourable effect of an increase of drain depth is reinforced by the fact that more excess water can on average be accommodated in the soil profile when W is large, allowing a lower value for q

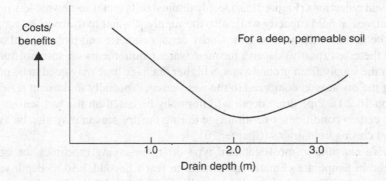

Figure 11.14 Normal relationship between costs/benefits and the field drainage base depth of a sub-surface drainage system

to be used which leads to a still larger drain spacing. Lowering the drainage base, however, raises the drainage costs since this requires deeper installation of pipe drains or deeper excavation of ditches and in addition, the main system will generally need to be deepened. Up to a depth of 2.0 m the savings resulting from the wider spacing normally more than compensates for these extra costs. Further lowering of the field drainage base is seldom profitable. Few machines are able to install drainpipe deeper than 2.5 m while deep ditches pose other problems (land loss, caving in, maintenance, etc.), resulting in pronounced increases in costs. Incremental benefits also tend to decrease at such low field drainage base depths, for a variety of reasons (points below). Figure 11.14 illustrates these trends.

The most important local conditions influencing the drainage depth selection are:

a) *The local/regional drainage base level(s)*: under gravity discharge, the field drainage base cannot be lower than the local drainage base (= water level maintained in the collector and lower order main system) while the latter cannot be lower than the regional drainage base (level in main outlet system). These local/regional levels are in their turn generally subjected to various other constraints (established rights, the use of the drainage canals for navigation, etc.). Where the natural drainage base is a constraint, pumped discharge may be considered although this of course involves extra costs

b) *Seepage*: a low local drainage base may constitute a sink, drawing excessive amounts of seepage water from the surrounding areas

c) *Soil conditions*: the stratification of the soil may favour or disfavour a certain depth. For example, pipe drains should not in general be installed in quicksand layers, in unconsolidated layers, in layers with a low hydraulic conductivity, etc. In situations where a rather poorly permeable upper soil profile is underlain (within the accepted maximum depth) by a more permeable substratum, advantage should be taken of the fact that drains are always much more effective when they are installed well into the more permeable layer. On the other hand, when a reasonably permeable upper profile is underlain by a very poorly permeable layer, little is to be gained by installing the drains into this latter layer. Establishment of a deep drainage base in unconsolidated or unripened soils may be undesirable as it enhances subsidence and/or the formation of acid sulphate soils (Chapter 16)

d) *Drought risk*: a deep drainage base may over-drain the soil causing potentially considerable yield reductions (Figure 1.5). Deeply drained soils retain somewhat less moisture in the rootzone at field capacity while also the supply of water to the roots by capillary rise from the groundwater will be less. Under certain climatic and hydrological conditions where the crops (partly) depend for their water requirements on soil moisture storage and on the supply from groundwater, a higher drainage base may need to be maintained during the dry season compared to the wet season, especially in drought sensitive soils (section 10.2.1). Pipe drain depth will normally be based on the wet season situation. Under certain conditions, over-drainage during the dry season may also be avoided by (partly) closing field drains (Chapter 20)

e) *Available machinery*: most standard type drainpipe-laying machines for agricultural drainage in temperate climates are unable to reach beyond 1.50 m depth while their most economical working depth is usually around 1.0–1.2 m. Heavier machinery is available for the installation of deeper drains for salinity control of irrigated land in the arid zone.

In NW Europe pipe drains are generally installed at 0.9-l.2m depth, deeper installation generally not being feasible due to considerations under (a) to (e) above. This is also a common depth in other areas where the drainage is installed to control excess water during the wet season. In the case of pipe systems installed for salinity control, much deeper drainage bases are normally maintained (Table 11.6).

Drainpipes installed at shallow depths risk blockage by root penetration and a minimum depth of some 0.75 m is generally advisable. Using the hydroluis drain (Bahceci et al. 2017) may avoid root penetration while a new system called capiphon drainage (IDW13 2017)

Table 11.6 Selection of applied drain depths (see also Smedema 2007 and Table 16.2)

Temperate (humid) zones mostly for aeration control and mechanisation	
Holland, Northern Europe	1.20 m
Canada (Ontario, Quebec)	1.00–1.40 m
France (Atlantic and Northern Zone)	1.20–1.50 m
Turkey (coastal and inland plain)	1.50–1.70 m
Spain (alluvial river plains and deltas)	1.00–2.00 m
China (South, humid tropics)	1.00–1.50 m

Arid/Semi-arid zone: mostly for salinity control but also some aeration control during wet season	
Egypt (Nile Delta)	1.50 m
Mexico (national standard)	1.50–1.60 m
India (Haryana, Punjab, Rajasthan)	1.20–2.00 m
China (Xinjiang and other arid zone projects)	1.50–2.20 m
Pakistan (NWFP, Punjab, Sindh)	2.00–2.20 m
Central Asia (Aral Sea Basin)	2.50–3.00 m
USA (ASAE-PE 463.1 guidelines)	Minimal 2.00 m

may also overcome root penetration. In cold climates drainpipes should be installed beyond the depth of frost penetration (see also section 24.4).

A review of drain depth by Smedema (2007) found that there is no convincing justification for >2.00 m depths. Normal drain depths range from 1.2 to 2.0 m. A deeper drain depth will result in wider spacings for the same drainage coefficient. It was found that the golden standard is a depth of 1.5 m with corresponding spacing; resulting in possibly the lowest cost. This assumes that deeper drain installation is more expensive and is not offset by lower costs of a wider spacing.

11.7 Pipe diameter

The hydraulic design of pipe drains is based on the standard pipe flow formulae, which relate the discharge (Q) to the hydraulic gradient (i), the pipe diameter (d) and wall roughness. These formulae differ for smooth and corrugated pipes, and between pipes that simply transport a fixed discharge along their length (*uniform flow*) and those transporting a discharge, which increases along the length of the pipe (*varied* or *non-uniform flow*). The latter condition applies to most field drains. The provided formulae are generally quite adequate for the design of most pipe drainage systems except for the cases where large diameter pipes (diameter larger than 200 mm) with large corrugations are used (recommended formulae for the latter cases are can be found in FAO 2005). Conventional electronic calculator and nomograph methods but also computer programs are now available to perform the necessary calculations (see Chapter 21).

Pipe drains are conventionally laid under a slope of about 10 cm/100m. Hydraulic gradients may also develop by back up of flow and the drains may in fact be laid horizontally. The concept that subsurface drains should have a slope to allow sediment to pass through is a misconception. Self-cleaning subsurface drains are not possible: water flow velocity will drop below sediment transport flow velocity when water backs up in the pipe system. This will happen regardless of the slope under which the drainpipe has been constructed. Providing slope is, however, still a prudent engineering practice.

Uniform flow

The formulae for this type of flow are derived in the case of smooth pipes from the Darcy-Weisbach equation and for corrugated pipes from the Manning equation:

smooth pipes:	$Q = 50 \ d^{2.71} \ i^{0.57}$	Eq. 11.20
corrugated pipes:	$Q = 22 \ d^{2.67} \ i^{0.50}$	Eq. 11.21

where:
Q = the discharge along the pipe (m³/s)
d = the pipes internal diameter (m)
i = the hydraulic gradient (m/m)

It is prudent to over-design the pipe to allow for partial siltation and for misalignment during construction. This may be conveniently done by assuming a fictitious discharge $Q_f = \beta Q$ where $\beta = 1.33$ for continuous plastic pipe installed in stable soil and $\beta = 2.0$ for tile drains laid in siltation prone situations.

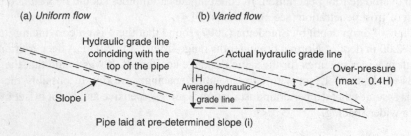

(a) *Uniform flow*

(b) *Varied flow*

Hydraulic grade line coinciding with the top of the pipe

Slope i

Actual hydraulic grade line

H

Average hydraulic grade line

Over-pressure (max ~ 0.4 H)

Pipe laid at pre-determined slope (i)

Figure 11.15 Hydraulic grade lines under uniform and non-uniform pipe flow

The hydraulic gradient is constant and it is assumed that for a pipe transporting its design discharge (Q) the hydraulic grade line will coincide with the top of the pipe (in other words, it is assumed that the pipe runs full of water at all points without over-pressure, see Figure 11.15a).

Example 1

- smooth concrete collector pipe;
- slope of the pipeline 0.1%;
- receiving water from 20 ha of land drained by lateral drains designed for q = 5 mm/d
- to be designed for a 25% reduction in transport capacity due to siltation ($\beta = 1/0.75 = 1.33$).

The discharge at the end of the line is:

$$Q = \frac{q \times A}{1000 \times 3600 \times 24} (m^3 / s)$$

where:
A = drained area in m²

In this case A = 200 000 m² and q = 5 mm/d, so Q = 0.0116 m³/s. Allowing for siltation Q_f = 1.33 (0.0116) = 0.0155 m³/s. Applying Eq. 11.20:

$$Q = 50d^{2.71}i^{0.57} \rightarrow d = \left[\frac{0.0155}{50(0.001)^{0.57}} \right]^{\frac{1}{2.71}} = 0.217\,m$$

Select a pipe with ID (inside diameter) of 22 cm or the nearest larger size available. Application of the nomograph of Figure 11.16 (smooth pipe, uniform flow) would have given the same result.

Varied flow

The formulae for varied, or non-uniform, flow derived from the same standard equations, are:

smooth pipes:	$Q = 89\ d^{2.71}\ \bar{i}^{\,0.57}$	Eq. 11.22
corrugated pipes:	$Q = 38\ d^{2.67}\ \bar{i}^{\,0.50}$	Eq. 11.23

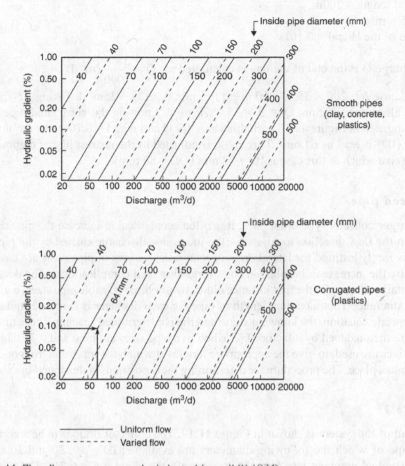

Figure 11.16 Pipe diameter nomographs (adapted from ILRI 1974)

in which Q and d are as previously described and $\bar{i}$ is the average slope of the pipe. The discharge in a lateral drain increases linearly from zero at the upstream end to the maximum value at the outlet. The slope of the hydraulic grade line also increases from zero at the upstream point to a maximum value at the outlet. For design it is assumed that the pipe runs full at all points and that the average hydraulic grade line coincides with the slope and position of the top of the pipe. As Figure 11.15b shows, this means that the water in the pipe is under a slight over-pressure for the whole of its length with the exception of the two ends.

Comparison of the equations for varied and uniform flow indicates that a varied flow field drain of a given size and slope can transport a discharge that is approximately 75% larger than a similar pipe, which conveys a fixed discharge along the length of the pipe.

Example 2

- parallel system of lateral pipe drains at 35 m spacing
- corrugated plastic drainpipe

- lateral length = 200m
- q = 5 mm/d
- slope of the lateral = 0.10%

The discharge Q at the end of the line is calculated as: $Q = \dfrac{q \times A}{1000} (m^3/d)$

In this case A = 200 × 35 = 7000 m²; q = 5 mm/d, so Q = 35m³/d. Taking a 50% safety factor to allow for siltation: $Q_f = 2 \times 35 = 70$ m³/d. Entering the corrugated pipe, varied flow nomograph of Figure 11.16 with this value for Q and with i = 0.10%, the required pipe diameter (ID) is read as 64 mm. This value is adjusted to the nearest higher commercially available size which in this case is ID = 72 mm (OD = 80 mm).

Composed pipe

For the larger collector type drain lines, it is often economical to increase the pipe diameter in steps in the flow direction to adjust for the increasing discharge carried by the pipe. This practice is rarely justified for lateral drains because the savings in pipe costs are easily outweighed by the increase in installation costs (organisational complications). The design procedure entails selecting in the first instance either two-or three possible pipe sizes, e.g., 10, 20 or 30 cm diameter. The maximum length of pipe for each diameter is then determined using the appropriate equation, the known q value and the predetermined spacing. These maximum lengths are then reduced by a factor[5] 0.85 when two pipe sizes are used and 0.75 when three or more sizes are used, to give the appropriate lengths at which the transition from one size to the next takes place. The procedure for determining these positions is illustrated in example 3.

Example 3

The layout of the system is shown in Figure 11.17. The collector drain is to be a corrugated plastic pipe of which the following diameters are available: ID = 10, 20 and 30 cm; q = 2 mm/d (typical drainage rate for salinity control); the slope is i = 0.2%. The maximum

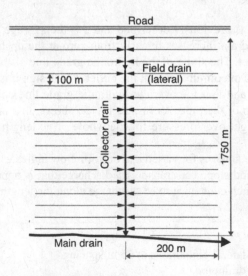

Figure 11.17 Layout of drainage system in example 3

Table 11.7 Determination of maximum drain length

	Pipe diameter (ID in cm)		
	10	20	30*
Maximum discharge capacity of corrugated pipe: $Q_{design} = 38d^{2.67} i^{0.50}$ in m³/d	313	1 997	4 189
Allowing for 25% reduction of capacity due to siltation Q = 0.75 x Q in m³/d	235	1 498	3 142
Drained area A = (1000 Q)/q in m²	117 500	749 000	1 571 000
Maximum drain length = A/400 in m	294	1 873	3 928

* for large diameter corrugated pipe (> 200 mm), it is generally recommended to use a somewhat smaller coefficient in the flow formulae (in this case the coefficient has been reduced from 38 to 27, see FAO 2005).

possible length of drainpipe is determined for each of the available diameters, using the equation for non-uniform flow (the discharge increases approximately linearly along the collector's length, Table 11.7).

It is checked whether a two-size collector composed of 10 cm and 20 cm diameter pipe would be acceptable. Applying the 0.85 reduction factor, the collector would be composed of:

• pipe with ID = 10 cm for section 0 – 250 m (85% of 294 m)
• pipe with ID = 20 cm for section 250 – 1 590 m (85% of 1 873 m)

The calculated max length of 1590 m is short of the required length (1 750 – 100 = 1 650 m, see Figure 11.17). The alternatives (a three-size collector with the last 60 m of 30 cm pipe, or a two-size collector with pipe sizes 10 cm and 30 cm) would fully meet the requirements but the 10/20 cm design would normally be close enough to be acceptable.

Notes

1 Flow patterns are inevitably presented in a distorted form in diagrams like Figure 11.1. In reality L is often 10 to 20 times larger than D and even 50–100 times larger than h. Due to the large difference in horizontal and vertical scale, the horizontal flow sector always becomes grossly under-represented.
2 Expression (b) is an application of Darcy's Law using the so-called Dupuit-Forchheimer assumptions i.e., it is assumed that the streamlines in the vertical plane considered, are horizontal and that the flow velocity in the plane at all depths is proportional to the slope of the watertable.
3 Applies only for D_h = constant as is assumed in the integration underlying Eq. 11.2. Integrating for $D_h = D + h_x$ yields the more precise equation of an ellipse.
4 For drain spacing calculations, it is generally assumed that the pipes run half full although the actual water depths in the pipe may vary from full to nearly empty.
5 These factors take into account the extra headloss incurred when part of the pipeline has a smaller diameter.

Chapter 12

Design discharges

Design discharges (Q_{design}) are generally based on rainfall since this is almost always the most critical source of excess water with which a drainage system has to contend. The designer should, however, be alert to exceptions (e.g., heavy seepage inflow into a polder or snowmelt).

Discharge generated by the intense storms normally taken for design, typically generates a wave in which the discharge in the drainage canals rises and then recedes. Hydrographs showing the rate of discharge in the basin as a function of time due to such storms have a peaked shape.

12.1 Discharge transformation

Drainage discharges are generally formed by a series process whereby the rainfall subsequently passes through the field drainage system and the main canals of various order, to the outlet (see also Box 12.1). In this process, the hydrograph is steadily transformed as schematically shown in Figure 12.1. From rainfall to outlet discharge, the peak rate of discharge expressed in mm/d or in l/s/ha is reduced (although the actual discharges in m³/s are of course likely to increase in the direction of the outlet in line with the increase in basin area).

Storage and flow resistance

Storage and flow resistance are the main factors causing transformation of the hydrograph. The head needed for the water to flow through the system is partly created by storing part of the inflow such that the upstream levels rise above those downstream (detention storage).

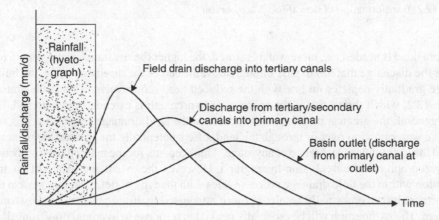

Figure 12.1 Transformation of the hydrograph along the drainage path

Box 12.1   Flood routing

Flood routing is a technique for analysing the movement of a flood/discharge wave through a channel system in order to predict the rise and fall in water levels at various downstream locations. The simplest flood routing technique only takes into account the increase/decrease in storage as depicted in Figure 12.2. Its application to the routing of a flood wave through an on-line reservoir is described in section 12.7.1 and its application to routing through a canal section is known as the Muskingum method. In the latter method, the storage in the section under consideration is assumed to include a combination of prism and wedge-shaped volumes whilst the transformation process is described by a set of mathematical equations which can readily be solved by hand or by computer calculation. Flood routing through entire systems should not only consider the impact of storage but also other wave transformation factors and processes such as longitudinal variations in stream bed characteristics and in flow resistance, junctions and local inflows, differences in time of travel and flow through structures. This type of routing (broadly referred to as distributed, kinematic routing) is generally based on solution of the Saint-Venant equations, the basic equations for non-uniform, non-steady canal flow. Examples include the flood routing routines of the HEC-RAS and HEC-HMS models described in Chapter 21.

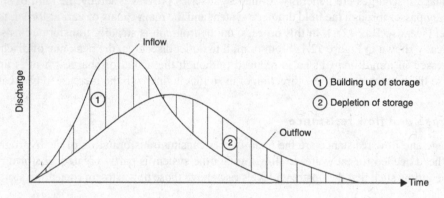

Figure 12.2 Transformation of flow through a reservoir

More head is needed i.e., more water is stored, the higher the resistance in the system or the higher the discharge that has to pass through it. At the end of the discharge wave, the built-up storage gradually depletes, in line with the reduced head requirement. This is illustrated by Figure 12.2, which shows that in the process the hydrograph is extended and flattened.

In general, the greatest transformation occurs at the field drainage stage since this stage, with the water moving over or through the land, offers potentially the greatest storage capacity and the highest flow resistance of any stage. This accounts for the marked contrast between the hyetograph and the field drain hydrograph. However, the extent and nature of the transformation within the field drainage process varies with the type of field drainage system used. Subsurface drainage normally results in more attenuated hydrographs than shallow/surface drainage. The hydrograph will be especially peaked in response to overland flow from sloping land, whereas on flat land it will be less peaked because more rain will be detained in depressions while the drainage, being mostly topsoil drainage/interflow will also be slower.

Time of travel

The field drains discharge into the collectors/tertiary canals at different points along the system whilst the collectors discharge at different points into the secondary system and so on into the primary canal. This spatial differentiation in inflow of lower order canals into higher order canals means that the flow in the latter canals is composed of flows having different times of travel. The effect this can have is illustrated in Figure 12.3 for the case of two identical tertiary canals, each draining a similar area of land, which discharge into a secondary canal at points distance L apart.

The hydrographs in the two tertiary canals will be more or less identical and synchronised. The discharge from tertiary 1 will pass though the junction point B with tertiary 2 a time Δt later than the discharge from tertiary 2. The travel time Δt between points A and B in the secondary canal equals L/v where v = the velocity of flow in this canal section. The combined hydrograph at point B may be developed as shown on Figure 12.3 by superposition

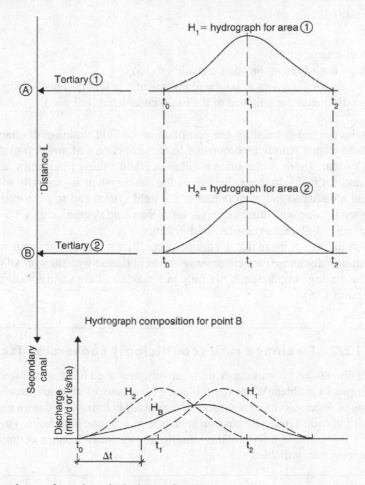

Figure 12.3 Effect of time of travel on the hydrograph

of the hydrographs from the tertiary canals 1 and 2 taking due account of the time of travel. This simple procedure ignores any transformation occurring in the discharge from tertiary 1 as it moves between points A and B.

The composed hydrograph at point B will be different from the component hydrographs. In the case described, the peak rate of discharge (mm/d) is less than the peak rates of the component hydrographs, although the actual discharge (m³/s) will be higher due to the increase in the drained area. The peak rate would be equal when the peak discharge from tertiary canal 2 occurs a time Δt later than the peak in tertiary canal 1, the two peaks arriving at point B simultaneously. However, it is more likely that the peaks will arrive at B at different times resulting in a reduced rate of discharge of the composed hydrograph, the more so for elongated basins as compared to fan shaped basins.

Basin vs field discharge; drainage coefficient

Design discharges at different points along the drainage canal system may in principle be calculated as:

$$Q = \frac{qA}{1000}$$
Eq. 12.1

where:
Q = basin discharge (m³/s)
A = drainage basin area at the point considered (ha)
q = the *drainage coefficient,* also termed the specific or unit design discharge i.e., the design discharge per unit area of the basin considered (in l/s/ha).

Basin discharges are formed by the combination of field drainage discharge although in general these cannot simply be summated because of the transformation of the discharge in the canal system. There is also often a difference in the design frequency adopted for field drainage and for main drainage. Failure of the main system is generally of greater consequence and more damaging than failure of the field system and so main systems are generally designed to cope with more extreme events than field systems, e.g., 1 × 5–10 year events compared to 1 × 1–2 year events for field systems.

All these influences mean that the drainage coefficient for main drainage normally differs from the design discharge for field drainage (also indicated with the symbol q in Chapter 11 and that the drainage coefficient is not fully independent of the basin area A (see also discussion in section 12.6).

Box 12.2   Drainage rate (coefficient) conversion factor

Conversion factor: various dimensions are routinely used in the quantification of agro-hydrological variables. While rainfall and evapo(transpi)ration are usually expressed in terms of water depths, discharges are expressed in terms of volumes per time interval. Both variables may also have an area as well as a time dimension. The conversion factor 1.0 l/sec/ha = 8.64 mm/day readily allows water balance components to be related to system variables.

12.2 Design considerations

Rainfall on the basin and the resulting drainage discharge are of course related. This relationship is, however, influenced and interacted upon by many factors limiting the applicability and reliability of the various established rainfall-discharge relationships. The reliability generally improves when they are calibrated for the type of rainfall and for the basin to which they are applied. Application of these methods may however require the prior recording of the passage of a discharge wave from the basin, generated under conditions comparable to the accepted design conditions. This of course limits the applicability of these methods. Other methods do not require such site discharge recordings but rely on more generally available basin characteristics, making these methods more universally applicable. The widely used conventional methods are presented in this chapter while for the available computer models, reference is made to Chapter 21.

Flat vs sloping land

As the slope of the land becomes steeper, an increasing proportion of the excess rain will discharge rapidly as overland flow or as another rapid type of shallow flow. As a result, hydrographs generally become more peaked with increasing land slope. This trend is often accentuated in the main system due to the reduction in travel time and the diminished canal storage of sloping basins. These features generally become apparent when the land slope is > 0.5% *(sloping land)* while they are normally insignificant when slopes are < 0.2% *(flat land)*. Other factors besides the degree of slope also influence the rapidity of the drainage discharge from the land (rainfall intensity, soil type, vegetative cover, cultivation method, etc.). These factors largely determine whether the land with slopes in-between 0.2–0.5% should be classified as sloping or as flat.

Another difference between flat and sloping basins is that over-topping of the canals, leading to flooding of the adjacent land, is generally less harmful on flat than on sloping land. On flat land, the floodwater will spread and retreat more calmly than on sloping land. On sloping land, floodwater will travel further and faster and be generally more devastating due to erosion and siltation.

These differences underlie the different approaches taken with respect to design discharges for flat and sloping drainage basins. Design discharges for sloping basins are generally taken equal to *the peak discharge* generated by fairly extreme, short duration storms to ensure that flooding is a rare event (e.g., 1×10–25 year peak discharges lasting only a few hours may be used in flood sensitive areas). Since hydrographs for these basins are in any event peaked, by implication the Q_{design} will also be quite high.

For flat basins design discharges are much lower because they can be based on more frequent events (recurrence interval < 5 years) and because they can be based on an *average discharge* during the period of high discharge considered. Figure 12.4 illustrates this last point. Periods of high discharge are likely to last from about half a day for basins with mainly shallow drainage to a period of two to three days for basins with mainly subsurface discharge. Taking Q_{design} equal to the average discharge implies that actual discharges will be higher than the Q_{design} during part of the discharge period. In view of the flat shape of the hydrograph, these discharges will not exceed the Q_{design} by much, so that in general they can be safely accommodated by the freeboard (as described in section 13.1.6). During this period the drainage flow will back up into

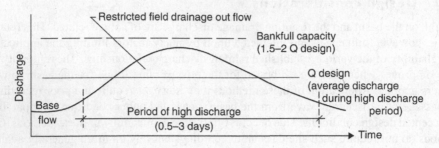

Figure 12.4 Design discharge approach for flat basins

the field drains, temporarily restricting the inflow into the main system, but for short periods this can generally be allowed.

Composed discharges

Most main systems will receive subsurface drainage as well as shallow drainage flow, but their contributions are likely to be quite different in magnitude and occur at different times. The design discharge should be based on the highest discharge. In sloping basins, the discharge due to shallow drainage (especially the overland flow) is almost always several times higher than the concurrently occurring subsurface drainage discharge (the peak of which comes later) so that the contribution of the latter may well be neglected in establishing design discharges. For flat basins in humid climates this may be different. In these areas the concurrently occurring discharges due to shallow drainage (mostly interflow and subsurface drainage) differ less and an approach taking their joint contribution into account may have to be adopted.

The problem is further complicated by differences in discharge between the different parts of the basin for the same type of drainage flow. In the Netherlands, for example, basins are often divided into two parts with respectively rapidly and slowly responding subsurface drainage. The rapidly responding part typically comprises the lower areas of the basin where the groundwater levels are always at shallow depth (within 1–2 m from the surface). The slowly responding part comprises the naturally better drained areas of land at a higher elevation in which groundwater levels fall to considerable depths during dry periods. Much of the rainfall on these latter areas will be detained/retained in the soil above the watertable and the groundwater recharge will be delayed. Moreover, as the subsurface drains are likely to be widely spaced (little need to improve the natural drainage), the outflow will be even more gradual and delayed, similar in character to base flow. In contrast, in the lower part of the drainage basin the drain outflow will increase soon after the start of the rain, reaching its peak within one to two days, and then recede rather rapidly. Design discharges for basins should obviously be based mostly on the discharge from the rapidly responding part of the basin with the contribution of the slowly responding part taken into account as a base flow only or even neglected.

There is very little authoritative guidance as to how composed design discharges should be determined. Analysis of land conditions noting such characteristics as the slope of the land, infiltrability, possibilities for retention/storage of water on the surface and in the soil, groundwater depths, KD-values and other characteristics of the substrata, etc. may provide some insight into the relative contributions of the different types of drainage discharge. For example, in the case described for the Netherlands, it was found that the two parts could be delineated on the basis of the (available) information on the annual groundwater regime.

Special consideration should also be given to the discharge from impervious surfaces within the basin (built-up areas, glasshouse complexes, etc.). The runoff from such surfaces is virtually 100% of rainfall and occurs rapidly, increasing the peak discharge of the relevant main drains very considerably (see also section 12.5.1).

12.3 Statistical analysis of observed discharges

Statistical analysis of historically observed discharges is the most straightforward and most reliable method for determining design discharges. The required series of discharges is, however, seldom available for agricultural basins.

The discharges may be measured directly but are more usually derived indirectly from so called *rating curves* which show the relationship between the discharge (Q) of a stream/canal and the stage level (h)[1]. Rating curves may be established by plotting measured Q-values, representing the entire range of expected discharges, against corresponding stage levels. They may subsequently be used to determine the discharge passing through the stream/canal by simply reading the stage level (example in Figure 12.5).

Note that antecedent moisture content in the side slopes and bed of the drain for which a rating curve is to be established will have a major effect on the correlation between head and discharge; this is because resistance to flow in the dry drain is higher than under wet conditions and loss of water through infiltration though the side slopes and bed is higher under dry conditions. Separate correlations for dry and wet conditions are recommended.

In most countries, hydrological services maintain a number of official measuring sites *(gauging stations)* on the main streams/canals. Discharge data on the drainage canals within an agricultural basin are, however, seldom available and will need to be specifically collected for the project. The observation series will then as a rule be short (1–2 years only). Various methods exist to extend the series (section 5.6.2), though generally at the expense of reliability.

Statistical analysis is applicable when a rather long series of discharge data is available (> 15–20 years). The observed annual maximum discharges are ranked in decreasing order

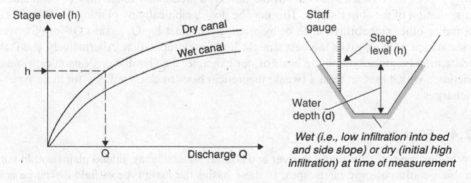

Figure 12.5 Typical rating curve for a small stream/drainage canal

Table 12.1 Frequency analysis of drainage discharges

Year	Maximum discharge (m^3/s)	Rank number (m)	Rank number (m)	Discharge (m^3/s)	Probability (P)	Recurrence interval (T years)
1958	85.1	4	1	98.3	0.05	20
1959	50.1	17	2	90.2	0.10	10
1960	48.2	18	3	85.3	0.15	
1961	68.3	10	4	85.1	0.20	5
1962	60.4	13	5	80.7	0.25	4
1963	55.2	14	6	80.6	0.30	
1964	80.7	5	7	78.4	0.35	
1965	90.2	2	8	78.3	0.40	
1966	85.3	3	9	76.7	0.45	
1967	61.3	12	10	68.3	0.50	2
1968	98.3	1	11	61.5	0.55	
1969	78.4	7	12	61.3	0.60	
1970	80.6	6	13	60.4	0.65	
1971	36.7	19	14	55.2	0.70	
1972	50.2	15	15	50.2	0.75	
1973	61.5	11	16	50.2	0.80	
1974	50.2	16	17	50.1	0.85	
1975	78.3	8	18	48.2	0.90	
1976	76.7	9	19(=N)	36.7	0.95	1

Note: $P = m/(N+1)$ is the annual probability of occurrence or of exceedance of the discharge indicated; $T = (N+1)/m$ is the recurrence interval (also called *return period*).

as in Table 12.1 (instead of annual maximums, the use of seasonal maximums per annum would be relevant if these are more critical).

For example, basing canal design on the 1×5 year event, it follows from this table, that $Q_{design} = 85.1$ m³/s. Discharges outside the observed probability range may be estimated by extrapolation of the above series. This may be done graphically by plotting the Q-values on normal or other probability paper, or by plotting P against log Q, against $Q^{1/2}$ or otherwise, to see which plotting gives the best straight line for extrapolation. Alternatively, the P and Q data may be processed on the basis of, for example, the Gumble or Pearson distribution functions, which have at least a (weak) theoretical basis to claim validity for these types of discharges.

12.4 Flat basins

Flat basins are typically found in river and coastal plains, deltas, inland plains and in comparable geomorphologic landscapes. In these basins the main type of field discharge is by subsurface flow and interflow and not by overland flow.

12.4.1 Subsurface drainage

For subsurface drainage, most of transformation of the discharge occurs in the field drainage phase, the transformation in the main system being negligible in comparison. Field discharge may in this case be directly equated to basin discharge. As the time of travel from field drain to basin outlet for normal basins (say basins up to 25 000 ha) is seldom more than half to one day and the peak rates of field drainage discharge are generally sustained throughout this period, the average basin discharge will be very similar to the average field drainage discharge.

As noted in Table 11.2, subsurface drainage systems in the temperate humid climates of NW Europe are typically designed on the basis of the steady state criteria. q = 7mm/d and H = 50 cm. The maximum possible discharge capacity of such a system (watertable at soil surface) is 2 to 2.5 × q = 14–18 mm/d (frequency of occurrence about 1 × 1 to 2 years). The corresponding average basin discharge, allowing for the effects of storage and time of travel, would normally be of the order of 1.25 to 1.50 × q = 9–11 mm/d.

Rule of thumb

The non-steady state formula, Eq. 11.17, enables the field drain discharge to be calculated for any rainfall event. As an example, drain discharges have been calculated in Table 12.2 for conditions normally encountered in NW Europe.

The calculations are done for daily periods, which period is of the same order as the travelling time in most medium sized basins. The reaction factors α = 0.22 and α = 0.69 represent respectively a slowly and a moderately rapidly responding field system. With respect to antecedent rainfall it is assumed that previous rainfall has wetted the soil profile to field capacity and that the field drains are discharging at a rate of 2 mm/d (tail end of the recession curve). The results in Table 12.2 are compared with those of an empirical rule according to which the daily rainfall discharges uniformly from the basin over a 3 day period starting with the onset of rain (1/3-1/3-1/3 rule, see Table 12.2). In the past this rule was widely applied for the determination of outlet capacities for polders in the Netherlands (now mostly replaced by the linear reservoir model).

Table 12.2 Subsurface drainage discharge calculations for the case of a 3 days, 1 × 1 year design rainfall

Design rainfall		Discharge according to Eq.11.17		Discharge according to the rule of thumb			
Days	(mm)	α = 0.22 (slow response) (mm)	α = 0.69 (rapid response) (mm)	day 1 rain (mm)	day 2 rain (mm)	day 3 rain (mm)	Total (mm)
0		2	2				
1	20	5.6	11.0	7			7
2	10	6.5	10.5	7	4		11
3	8	6.8	9.3	6	3	3	12
4		5.4	4.7		3	3	6
5		4.3	2.3			2	2
6		3.4	1.2				
7		2.7	0.6				

The rule of thumb applies to basin discharges and although it has been argued that sub-surface drainage discharges and basin discharges are more or less identical, some transformation in the main system does in fact occur. Results of Eq. 11.17 and the 1/3-1/3-1/3 rule therefore do not need to match.

Linear reservoir model

Experiences in the Netherlands indicate that the non-steady state formulae for subsurface field drainage can also be used to determine discharges at the basin level when the transformation of discharge in the main system is small compared to that in the field system (Kraijenhoff van der Leur, 1973 and de Zeeuw, 1973). By analogy with Eq. 11.17, basin discharges may be described as:

$$Q_t = Q_{t-1}e^{-\alpha\Delta t} + R_{\Delta t}(1 - e^{-\alpha\Delta t})$$ Eq. 12.2

This formula assumes the existence of a linear relationship between the discharge Q and the dynamic storage S (not yet discharged rainfall) in the basin i.e., Q = αS. Under these conditions, Eq. 12.2 may be used with predetermined α values to simulate the basin discharges generated by various critical rainfall events (possible design storms), on the basis of which design discharges may be established (examples are presented in de Zeeuw, 1973).

The reaction factor α in this case applies to the basin. Its value may be estimated from the recession limb of a representative basin discharge hydrograph, similar to the method outlined in section 11.4.2. The most suitable hydrographs are those generated by storms having a total rainfall depth close to that of the design storm and which have a recession limb which is unaffected by further rainfall (see Figure 12.6; methods do however exist to derive α-values from more complex hydrographs, de Zeeuw, 1973).

Reaction factors for a medium sized agricultural basin in the Netherlands are typically of the order of 0.3 to 0.7 d^{-1}. For basins with a rather slow response, values may be as low as 0.05–0.10 d^{-1} while for basins with a rapid response they may rise to 5–10 d^{-1}.

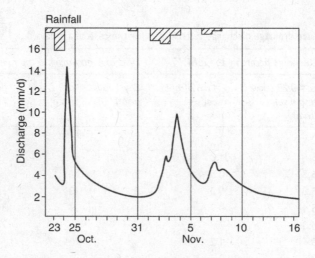

Figure 12.6 Recession curves

High values often indicate that a considerable part of the rain is discharged as shallow flow, particularly overland flow, in which case Eq. 12.2 is not strictly valid (although it seems to describe some types of shallow flow rather well). Reaction factors tend to increase as basins become smaller (approaching the values for field drainage mentioned in section 11.4.2).

12.4.2   Shallow drainage

Shallow drainage discharge from flat land mostly takes the form of interflow, although considerable overland flow is also to be expected when field drainage systems are especially designed to enhance this type of flow (bedding, row drainage, etc., see Chapter 9). The discharge characteristics of mole-drained land will generally also fit into this category.

As stated earlier, the hydrographs for shallow field drainage are much more peaked than those generated by subsurface drainage and the variation in inflow into the main system during the typical half to one day travel time through this system is too great to permit basin discharges to be established by simple averaging of the inflows from the field systems (as allowed for subsurface drainage). Instead, basin discharges should be determined by the following two step method:

1)   determination of the field drainage discharge from design rainfall
2)   conversion of these field discharges into basin discharges, taking into account the effects of differences in travel time but ignoring the effects of storage in the main system.

Field discharges

The determination of rainfall-induced shallow drainage discharge cannot be based on flow formulae as for subsurface drainage since shallow drainage flows, such as interflow and mole drainage flow, have not as yet been found amenable to physical-mathematical analysis. As a rough approximation it may be assumed that half of a 6 hrs rainfall will be discharged from the field during the rainfall period while the remainder will be discharged in the following 6-hour period. (the 1/2-1/2 rule, comparable to the 1/3-1/3-/1/3 rule used in Table 12.2.) In the example in Table 12.3, this rule is used for a 12 hour, 1 × 5 year design rainfall of 50 mm.

Table 12.3 Determination of the shallow field drainage discharge from 12 hrs, 1 × 5 year design rainfall

Rainfall period (mm)	Rainfall depth (mm)	Discharge from rainfall during periods		Total discharge (mm)
		0–6 hrs (mm)	6–12 hrs (mm)	
0–6 hr	32	1 (15mm storage)*		1
6–12 hr	18 a	16	9	25
12–18 hr			9	9
Total	50	17	18	35

* Estimated conservatively low as is prudent for design conditions; note, however, that the combination of a 1x5 year storm with minimal storage establishes a conditional probability with the probability of the combined event being less than the 1x5 year storm probability.

Basin discharges

Field discharges may be converted into basin discharge by dividing the basin into zones of equal time of travel (time-area principle, Shaw 1983) whereby a division into zones at 6-hour intervals will normally suffice. In the case of Figure 12.7 the basin is divided into three zones:

time of travel   0-6 hours: 40% of the basin
6-12 hours: 40% of the basin
>12 hours: 20% of the basin
100%

The discharge hydrograph for the outlet point may be composed from the field drainage hydrograph derived in Table 12.3 as illustrated in Table 12.4. In this example, the maximum discharge at the outlet occurs during the 12 to 18 hour period. Compared with the rainfall peak of 32 mm occurring during the 0–6 hour period, the peak is shifted by some 12 hours and is attenuated by some 18 mm.

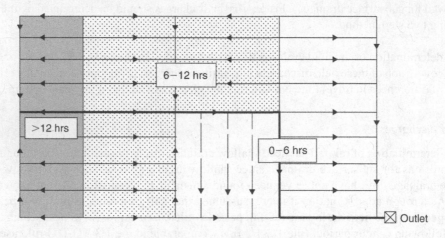

Figure 12.7 Agricultural basin divided into three travelling time zones

Table 12.4 Outlet discharge composition on the basis of field drainage discharge, taking into account times of travel in the basin

Time of travel hrs	Basin area (%)	Discharge during time interval, mm				
		0–6 hrs	6–12 hrs	12–18 hrs	18–24 hrs	24–30 hrs
Field drainage discharge in mm		1	25	9		
0–6	40	0.4	10.0	3.6		
6–12	40		0.4	10.0	3.6	1.8
12–18	20			0.2	5.0	
total mm/6 hrs		0.4	10.4	13.8	8.6	1.8
mm/24 hrs		1.6	41.6	55.2	34.4	7.2
l/s/ha		0.2	4.8	6.4	4.0	0.8

12.4.3   Further guidance for flat basins

The rainfall-discharge relationships described in the preceding sections may be used to convert a historical rainfall series, covering for example some 15 to 25 years, into discharges. The latter may then be statistically analysed as in Table 12.1 to arrive at a suitable design discharge. The rainfall data may also be statistically analysed beforehand to establish a design storm which may then be converted into discharge as in Table 12.2, 14.3 and 14.4 or by means of the linear reservoir model method described in section 12.4.1. All these calculations can be done by hand but may also readily be computer programmed (see Chapter 21).

In Table 12.2, a 1×1 year, 3-day design rainfall is used and the Q_{design} may be taken equal to the average discharge during a 2 to 3 day period, depending on how much canal storage can be permitted in the basin. The calculated drainage coefficient of about 10 mm/d is typical throughout NW Europe for areas with mainly subsurface drainage. Discharge capacities of subsurface drainage systems are limited (maximal when watertable at soil surface) and higher inflow into the main system usually indicates contributions of another type of field drainage.

The discharge from a mole drainage system often rises rapidly to a peak, which may be of the order of 20–30 mm/d, but this high discharge is only sustained for short periods of time. In general, the discharge hydrograph for a mole system generated by a critical 12 to 24 hour rainfall event is entirely contained within a 24 to 36 hour period. In the UK, the design capacity of the intercepting pipe drains (Figure 9.13) tends to be based upon 24 hours, 1 × 1 to 2 year rainfall, to be discharged within the same 24 hour period (applies to grassland; somewhat stricter criteria apply to arable and horticultural land). Some 20% of the rain is assumed to be retained in/on the land, so that only 80% of the total rainfall appears as discharge.

A 1×5 year, 12 hour storm as used in Table 12.3 is quite normal for shallow drainage although other storm durations and response times may also be used (e.g., use 6 hour storm periods for a well maintained flat bedding system and for clean mildly sloping furrows, while a 12 hour storm period would be more suitable for slower responding interflow systems). With shallow drainage, discharge can reach peaks several times higher than with subsurface drainage systems, but the peaks are of short duration (usually less than a couple of hours). Under normal freeboard conditions (section 13.1.6), the average discharges during 6–12 hour periods are more relevant for design. In the example in Table 12.4, the design discharge would normally be based on the average discharge during the 3x6 hour period of high discharge, which in this case is the 6–24 hrs period and averages ~5 l/s/ha.

Table 12.5 gives an idea of the order of magnitude and global variation of drainage coefficients for flat medium sized (10 000–25 000 ha) agricultural basins. The values given for the Netherlands apply to basins with predominantly subsurface drainage, while those for Germany, Canada and Croatia apply to a combination of subsurface and interflow drainage (for high values mostly the latter). The higher values for higher annual rainfall in Germany reflect the higher chance of rainfall occurring at a time of reduced storage (due to previous rainfall). The value of 7.0 l/sec/ha for Tanzania refers to composed interflow/overland. Drainage of rice fields has been described in detail in Chapter 19.

Table 12.5 Typical basin drainage coefficients from flat basins

	Annual rainfall (mm)	1 x 5 24–48 hrs rainfall (mm)	slope of the land (%)	drainage coefficient (l/sec/ha)
Netherlands (de Bilt, 1901–2017) (grassland/mixed cropping, rainfed)	799	30–45	0.05–0.20	1.0–1.3
Germany (mixed cropping, rainfed)	< 600	35–50	0.05–0.20	0.8
	600–1000	35–50	0.20–0.50	1.2
	>1000	35–50	0.05–0.20	1.0
		35–50	0.20–0.50	2.0
		35–50	0.05–0.20	2.0
		35–50	0.20–0.50	3.0
Canada (mixed cropping, rainfed)		40–50	0.05–0.20	1.3–1.4
			0.20–0.50	2.0–2.2
Croatia (mixed cropping, rainfed)		50–65	0.25–0.50	2.5
Sudan (cotton, furrow irrigation)		65–90	0.05–0.20	4.0
Japan (flooded rice)		80–115	0.05–0.20	5.0
Tanzania (sprinkler irrigated sugar cane)		145–165	0.20	7.0

12.5   Sloping basins

The highest discharges from sloping land are generated by storms lasting a few hours, which combine high rain volume with high intensity. Discharge occurs mostly as overland flow although the contribution of other rapidly responding types of shallow drainage may also be significant. High peak flows of short duration may be expected which can be very devastating if left uncontrolled. The methods described therefore focus on estimating these peak discharges.

Discharges in sloping basins are more dependent on the basins size and shape than for flat basins. The use of the drainage coefficient and the formula $Q = q \times A$ (Eq. 12.1) are therefore less relevant to sloping basins than to flat basins. Many methods have been developed to determine design discharge for sloping basins. However, most are highly empirical in nature and consequently are of limited local applicability. Two of the more soundly based methods are the *Rational Formula* and the *Curve Number* method, described below.

12.5.1   Rational formula

Discharge in a sloping basin starts as soon as the rain has replenished the surface retention/detention capacity. Assuming that this occurs at the same time throughout the basin, the order in which the discharges from various locations in the basin pass through the outlet (point B in Figure 12.8) will be determined by the time of travel. The longest time applies to water travelling from the hydraulically most remote point (point A). This time is called the *time of concentration* (T_c) and it is an important hydrological characteristic of the basin.

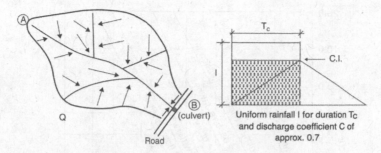

Figure 12.8 Concentration of discharge in a basin

The basic principle underlying the rational formula is that the highest discharge occurs in response to a storm with duration equal to the time of concentration, the reasoning being:

a) when the storm duration $<$ T_c, only part of the basin contributes to the total discharge at the outlet as the discharge from the area close to the outlet B will already have passed through B by the time the discharge from point A arrives

b) when the storm duration $>$ T_c, there is a time at which the entire basin contributes to the discharge through B but the discharge rates will normally be less than those generated by a storm of duration T_c since rainfall intensities decrease with increasing storm duration (see Table 5.2).

This reasoning only applies to rapidly responsive types of field drainage (overland flow, rapid interflow) where very little transformation occurs in the field drainage system (the field drainage hydrograph closely resembles the hyetograph). Under these conditions, the peak discharge may be estimated by the following formula[2]:

$$Q_p = \frac{C\,I\,A}{360}$$

Eq. 12.3

where:
Q_p = peak discharge (m³/s)
I = intensity of the design storm of duration T_c (mm/hr)
C = discharge coefficient (dimensionless)
A = basin area (ha)

The formula was originally developed to estimate peak discharges from small urban basins, generally having a large proportion of impervious area (for which C approaches 1.0). Its extension to agricultural areas is most appropriate for basins not exceeding 100–200 ha. For larger basins with T_c values of several hours, the assumed steady uniform intensity of the design storm is less realistic and considerable transformation of the discharge will occur in the main system.

Design storm: as explained, storms with duration T_c are adopted for design. Recurrence intervals are normally of the order of 10 to 25 years (this method typically being used to establish outlet capacities of small basins controlled by a culvert, spillway or other type of outlet structure, warranting rather high recurrence intervals).

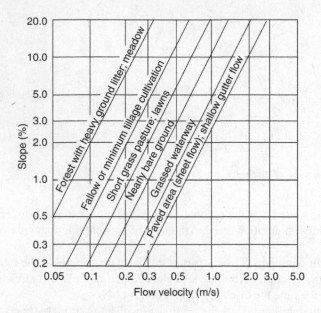

Figure 12.9 Overland flow velocities/or different land surfaces (Wanielista 1978)

Time of concentration (T_c): when observed values are not available one has to make do with estimates. Rough estimates may be made by dividing the travelling distance (L) by the flow velocity (v), according to $T_c = L/v$ whereby the flow path may for example be divided into an overland flow reach, an upper canal reach with predominantly small canals and moderate to fairly steep gradients, and a lower canal reach with large canals and rather flat gradients. The velocity of overland flow may be estimated using Figure 12.9 while the flow velocity in the drainage canals may be estimated using the Manning formula (Eq. 12.3). Several formulae have also been developed relating T_c to the relevant basin characteristics such as the basin area, basin shape, slope, soil conditions, etc. The most widely accepted formula is the one developed by Kirpich (1940), which applies to small agricultural basins (A < 50 ha):

$$T_c = \frac{L^{1.15}}{3080H^{0.38}} \text{(hours)}$$

Eq. 12.4

where:
T_c = time of concentration (hours);
L = maximum travelling distance in the basin (m) and
H = difference in elevation over the distance L (m).

Discharge coefficient (C): this coefficient indicates the proportion of the design rainfall that actually discharges rapidly from the basin and which contributes to the peak discharge (Figure 12.8). Its value is of course directly dependent on the infiltration and retention/detention characteristics of the land/basin, as indicated in Table 12.6 (retention/detention conditions are

Table 12.6 Guidelines for the determination of the discharge coefficient C in the Rational Formula

1. Agricultural land (SCS 1972)

	Infiltrability of the soil		
	High	*Medium*	*Low*
Arable land: Slope <5%	C = 0.30	C = 0.50	C = 0.60
5–10%	0.40	0.60	0.70
10–30%	0.50	0.70	0.80
Pasture: Slope < 5%	C = 0.10	C = 0.30	C = 0.40
5–10%	0,15	0.35	0.55
10.30%	0.20	0.40	0.60
Forest: Slope < 5%	C = 0.10	C = 0.30	C = 0.40
5–10%	0.25	0.35	0.50
10–30%	0.30	0.50	0.60

2. Urban Areas (ASCE 1969)

Residential areas: C = 0.30–0.50 for homes;
                            C = 0.5–0.75 for apartments

Industrial area: C = 0.50–0.80 for light industry;
                            C = 0.60–0.90 for heavy industry

Pavements: C = 0.7–0.95

represented by the slope of the land and by the land use). The values also vary somewhat with the rainfall intensity and the storm duration, all in all making C the weakest part of the formula. *Example*: determination of the Q_{design} for the small (96 ha) agricultural basin shown in Figure 12.10 with soil of medium infiltrability and (during the critical season) 40% arable and 60% pasture.

Time of concentration (T_c):

- reach A-B: overland flow, L_1 = 200 m, H_1 (fall) = 4 m (slope is 2%); T_c = 0.19 hr
  from Figure 12.9 select v = 0.3 m/s (pasture)
- reach B-C: channel flow, L_2 = 600 m, H_2= 0.2 m; use the Manning/ T_c = 0.74 hr
  Strickler formula with I = 0.2/600 = 0.0003, k_m = 20, b = 1m, d = 1m,
  side slope 1:1 (v = 0.22 m/s)
- reach C-D: channel flow, L_3 = 600 m, H_3 = 2 m; use the Manning/ T_c = 0.20 hr
  Strickler formula with I = 2/600 = 0.003, k_m = 20, b = 2m, d = 1m,
  side slope 1:2 (v = 0.84 m/s)

                                    Total from A to D T_c = 1.13 hr

Design storm: 1 × 10 yrs, T_c = 1.13 hr, P = 32 mm (over the T_c period),
                            I = 30 mm/hr

Discharge coefficient: C = (40% × 0.5) + (60% × 0.3) = 0.38, using the guidelines as in Table 12.6.

Rational formula (Eq. 12.3): Q= CIA/360 = 0.38 × 30 × 96/360 = 3.04 m³/s

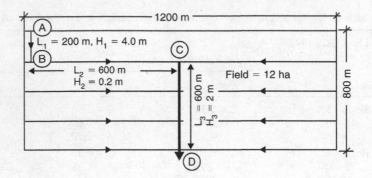

Figure 12.10 Example of small agricultural basin

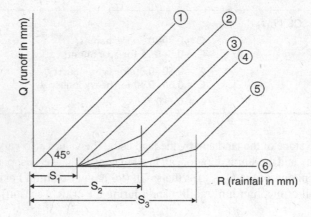

Figure 12.11 Relationships between rainfall, runoff and storage in a sloping basin

12.5.2 Curve number method

This method, developed by the US Soil Conservation Service, is applicable to larger basins than the rational formula (up to several thousand ha) and also has the advantage that the complete hydrograph, including the peak discharge, may be determined.

Underlying relationships

The discharge is estimated on the basis of the assessed storage capacity of the land/basin. When the land has no storage capacity, all the rain will be fully discharged ($S = 0$, $Q = R$, curve 1 in Figure 12.11). When the storage capacity is infinitely high, there is no runoff, not even after prolonged heavy rainfall ($S =$ infinite, $Q = 0$, see curve 6). When the storage capacity $S = S_1$, the runoff will start after this capacity has been filled while from this point on all the rain will be fully discharged (for $R < S_1$, $Q = 0$ and for $R > S_1$, $Q = R$, see curve 2).

When the storage capacity is not uniform but for example $S = S_1$ for half of the basin (e.g., high watertable area) while $S = S_2$ for the other half (low watertable area), the discharge will start at half rate (Q = 0.5 R) when R = S_1 but will not reach the full rate (Q = R) until R = S_2 (curve 3). When S = S_1 applies to only 20% of the basin and S = S_2 for the other 80%, the discharge will follow curve 4 (Q = 0.2 R for R < S_1 and Q = R for R > S_2). Curve 5 refers to a case with three different storage capacities.

CN graph and number

A large number of such curves can be composed for the widely varying storage conditions which may occur in a drainage basin. In the curve number method (CN method), these curves have been standardised. Each curve is characterised by a certain storage condition (S) and is identified by a CN-number whereby S and CN-number are related as shown in Figure 12.13. The shape of the curves is based on the relationships between Q, R and S as described previously which relationships were empirically found to be sufficiently well described by the given analytical formula.

The Natural Resources Conservation Service (NRCS) formerly known as the Soil Conservation Service (SCS) has prepared very detailed guidelines for selecting the appropriate CN curve to be used. In these guidelines the storage capacity of the basin is captured by the antecedent rainfall, the infiltration rate, the land slope, the surface relief, the land use and other relevant soil/land/basin features (for details see USDA/SCS 1985). Here only a simplified method is presented by which the curve selection is made on the basis of three land/basin features (Figure 12.12).

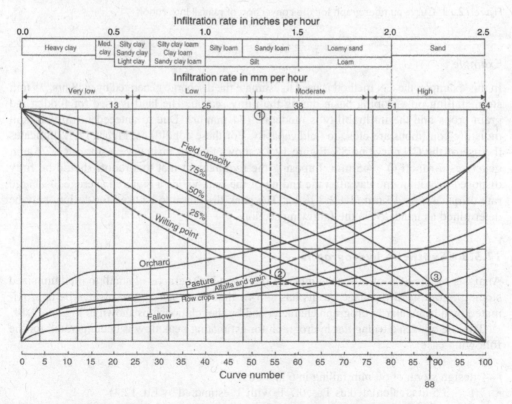

Figure 12.12 Simplified method for determining Curve Numbers (after USBR 1984)

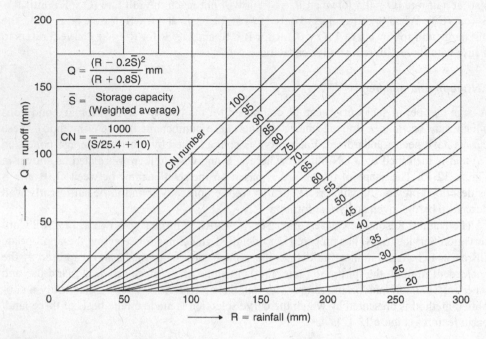

Figure 12.13 Curve number graph for the conversion of rainfall into runoff

Example

In this example, the CN method is used to estimate the discharge generated by a 48 hrs, 75 mm storm falling on a sloping basin during the rainy season. The land is used for fodder and grain crops and the infiltrability is moderate (34 mm/hr). Due to antecedent rainfall, soil moisture conditions are close to field capacity. For these conditions, Figure 12.12 indicates the use of the CN curve no 88. Figure 12.13 shows that for this curve a storm R = 75 mm generates a runoff Q = 45 mm. Imposing the requirement that this runoff should be fully disposed of within one day after the ending of the rain, yields a required average discharge rate of q = 45/3 = 15 mm/d = 1.8 l/s/ha. The peak discharge to be used for design, may be determined as indicated in the following section.

12.5.3 Synthetic hydrographs

Analysis by the SCS of a large number of observed hydrographs of small to medium sized sloping agricultural drainage basins has shown that these can almost always be approximated by a triangular hydrograph characterised by the relationships shown in Figure 12.14.

The use of such a triangular hydrograph for estimating peak flow is demonstrated for the following case:

- design storm of 60 mm falling in 6 hrs
- T_p = 3.5 hrs (calculated as T_p = 0.7 T_c with T_c estimated by Eq. 12.4)
- CN = 80.

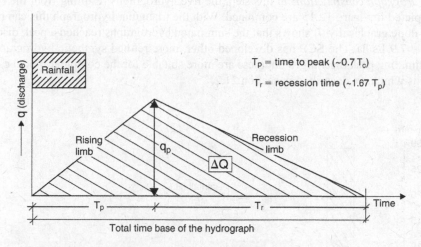

Figure 12.14 Characteristics of the SCS triangular hydrograph

Table 12.7 Runoff calculations

Design storm							
1. T (hrs)	0	1	2	3	4	5	6
2. R_t (mm)	0	5	13	42	50	55	60
Runoff							
3. Q_t (mm)	0	0	1	11	13	15	16
4. ΔQ (mm)	0	0	1	10	2	2	1
5. q_p (l/sec/ha)*	0	0	0.6	6.0	1.2	1.2	0.6

* according to Figure 12.14: $\Delta Q = \frac{1}{2}$ $(T_p+1.67T_p)$ x q_p = $1.34T_pq_p$ so for T_p = 3.5 hrs = 0.146 d $\rightarrow$ q_p = 5.1 x ΔQ (mm/d) and q_p = 0.6 x ΔQ (l/s/ha)

The procedure involves the following two steps:

(a) *Runoff calculations* (lines 1 to 5 in Table 12.7)

line 1: time (T) since beginning of the storm; the time interval (ΔT) should not greatly exceed 0.25 T_p, therefore in this case ΔT = 1.0 hours

line 2: design storm, cumulative in time (R_t), distributed as an S type storm (highest intensities during the mid-storm period, see WMO 1974)

line 3: R_t from line 2 converted into runoff (Q_t) using Figure 12.13 (CN = 80)

line 4: runoff (ΔQ) per time interval, calculated as $\Delta Q = Q_{t+1} - Q_t$

line 5: peak discharge (q_p) generated by each ΔQ.

(b) *Hydrograph construction:* in this step, the five hydrographs resulting from the rainfall depicted in Figure 12.15 are combined. With the triangular hydrograph this can readily be done graphically. It shows that the summated hydrographs reached a peak discharge $q_p = 7.9$ l/s/ha. The SCS has developed other, more refined synthetic hydrographs for estimating peak discharges but these are more suitable for the computer model calculations which are described in section 21.2.

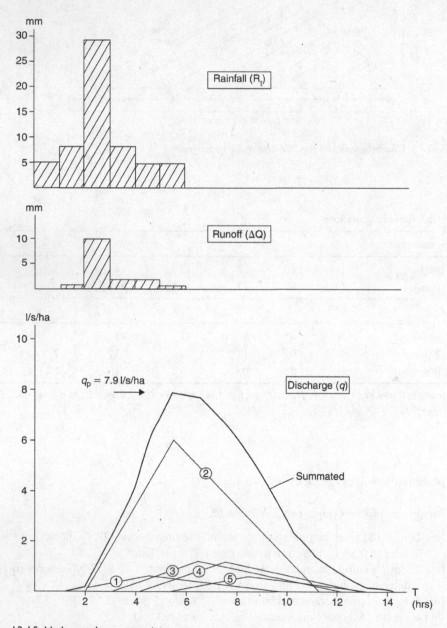

Figure 12.15 Hydrograph construction

12.6   Area reduction formulae

As described earlier the drainage coefficient (q) in the general formula Q = qA (Eq. 12.1 where Q = basin discharge and A basin area), should not in general be expected to be independent of the area A. This is due to the discharge transformation occurring in the main system, the effect of which increases as the size of the basin increases, generally resulting in lower values for q with increasing area A (Figure 12.1). It should also be considered that in large basins, the rainfall distribution is unlikely to be uniform (especially not for the short duration, convective type of storms used for sloping basins, which are typically of a highly scattered, localised nature). Also, with increasing times of concentration in large basins, the average rainfall intensity decreases. The combined effect of these last two factors is that both the average basin rainfall depth and intensity decrease with increasing basin area and a similar trend should apply to the drainage coefficients.

Flat basins

For flat basins the area reduction effects are often taken into account by using the empirical formula:

$$Q = q_0 A^\alpha \hspace{5cm} \text{Eq. 12.5}$$

where:
Q = discharge (m^3/s)
q_0 = drainage coefficient (m^3/s/km^2)
A = basin area (km^2; A > 1 km^2)
α = area reduction factor

This formula shows that the reduction is applied to the area A rather than to the drainage coefficient (not physically sensible but mathematically of course equally effective). The reduction depends of course on the nature of the drainage process. For shallow field drainage discharge, a large reduction may be applicable to include the effects of both the discharge transformation in the main system and the rainfall non-uniformity. When q_0 has been determined as in Table 12.4, most of the discharge transformation effects have already been taken into account in the time of travel but a further correction due to the rainfall non-uniformity might still be justified.

Area reduction formulae are widely used throughout the world. The nature of q_0 in these formulae is not always clear although they may generally be assumed to apply to field drainage discharge. Some guidance for selecting α has been presented in Table 12.8.

A well-known example of the Eq. 12.5 type of formula is the *Cypress Creek formula,* reading Q = q_0 A$^{5/6}$ with Q in cusec, A in sq. miles and q_0 in cusec per sq. miles. This formula

Table 12.8 Recommended area reduction factors (for 1–2 days, 1 x 5–10 yrs rainfall)

- moderate climates: coastal zones	α = 5/6 = 0.83
inland areas	α = 4/5 = 0.80
- (semi) humid tropics	α = 3/4 = 0.75
- (semi) arid tropics	α = 2/3 = 0.67

Table 12.9 Ratios between average basin rainfall and point rainfall for basins of different sizes

Basin area	USA	Northern India	West Africa
	6 hrs storms, 1x10 yr	24 hrs rainfall, 1 x 5 yrs	
100 ha	1.00	1.00	1.00
1 000 ha			0.87
5 000 ha	0.98	0.98	0.80
10 000 ha	0.94	0.96	0.75
25 000 ha	0.89		
50 000 ha	0.85	0.86	0.70

was derived in the USA for basins < 5 000 ha, slopes ⩽ 0.5% and for a mix of shallow/subsurface drainage discharge (at high discharges mostly the former). For normal farm crops, a 48 hours, 1 × 2 to 5 years design storm is used and in the case of high value crops a 24 hours, 1 × 10 years storm. The excess rainfall may be determined by means of the CN-graph (Figure 12.13).

Generally, area reduction factors are not applied in the Netherlands for subsurface drainage and the low intensity frontal type of rainfall. However, a reduction of 10% may be considered for areas exceeding 100 000 ha.

Sloping basins

The formula Q = qA and the underlying concept of the drainage coefficient are most relevant to flat basins. In sloping basins the influence of the area is taken into account partly implicitly in the different empirical relationships and factors (emphasising once again that these should only be used for the range of basin areas for which they were established) and partly explicitly in the time of concentration which takes into account the decrease of rainfall intensity with increasing storm duration. Basin size has thus partially been accounted for in the development of the formulae. The rainfall variability is only important for large basins. It can be accounted for either by using a known average basin rainfall based upon sufficient point measurements or by using ratios such as those in Table 12.9 between the point rainfall and the average basin rainfall.

12.7   Discharge reduction through storage

Flood storage involves the temporary impounding of drainage discharge in order to reduce downstream peak flows. This might provide a more economical solution to the prevention of flooding than the alternative of providing for a large disposal capacity in the canals and structures, especially when this is liable to involve pumping. Flood storage may be particularly suited to agricultural basins faced with increased urban development. The increased proportion of impervious surface and the installation of gutters and storm sewers in these areas increases the surface runoff while reducing the time of concentration, leading to high urban discharges into the main system (Figure 12.16).

Flood storage in drainage basins may take different forms. It may be localised in the form of a *retention reservoir* or be spread out over (part of) the basin in the form of *canal storage*.

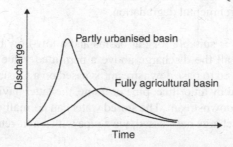

Figure 12.16 Characteristic hydrograph from an agricultural and a partly urbanised basin

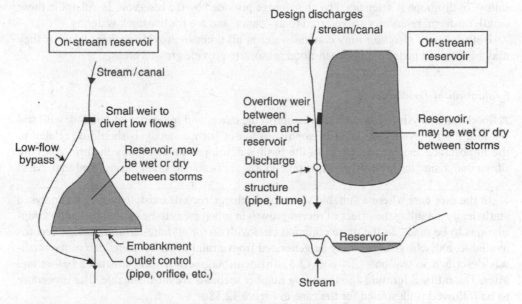

Figure 12.17 Types of retention reservoirs (modified from Hall and Hockin, 1980)

Canal storage may involve either allowing temporarily higher water levels in the canals or the provision of larger cross-sections than are needed for transport alone. Canal storage involves backed-up flow and is generally only applicable in flat basins (typically applied in polders to reduce pumping capacities).

12.7.1 Retention reservoirs

These reservoirs may be distinguished into the off-stream type and the on-stream type. The characteristic differences between these two types is illustrated in Figure 12.17, showing that the choice between them depends mostly on the prevailing local conditions. Low-flow-bypass systems are recommended for on-stream reservoirs in Australia for environmental benefits downstream (Darley 2017). In 2017/18 the South Australian state government allocated funds in the budget to install 500 low-flow-bypass systems. Low-flow bypass systems cannot be closed thus ensuring that low flows downstream of a reservoir are maintained when low flows generated upstream by runoff occurs. In the past these low flows were fully intercepted causing environmental degradation.

The off-stream reservoir is suitable for flat basins and a low-flow bypass can be included. It operates by diverting all the discharge above a permitted value into the reservoir. The discharge, which is allowed to pass downstream, is based on a sensible downstream channel capacity and is controlled by an orifice, pipe or flume, the latter having the advantage that it will allow debris to pass downstream. The stored water can normally drain back to the main-stream, once the high discharge has passed, via a pipe with a non-return flap valve at its end.

The on-stream reservoir is usually constructed in valley situations by building an embank-ment across the valley. The stream flow is controlled by the outlet from the reservoir and the inflow hydrograph is attenuated by the storage provided by the reservoir. In Australia these small, on-farm, reservoirs are referred to as "dams" and are for livestock watering.

Both types of reservoir may contain water at all times to provide shallow lakes or they may be allowed to dry out between flood seasons to provide grassed areas.

Evaluation of flood risks

A flood risk analysis is carried out to establish the extent and frequency of flooding with and without the reservoir. The level of sophistication of such an analysis should be related to the importance and consequences of the basin's development. Here only the principles are illustrated, using the relatively simple procedures applicable to small basins (upstream basin < 200 ha).

In the rare case where a suitable series of discharge records exist, they may be analysed statistically, enabling the effect of recent growth in urban areas to be established and extrap-olations to be made. In the more common case with no (or no long-term) discharge records available, a discharge series may be generated from a rainfall series, using one of the meth-ods described in sections 12.4 and 12.5, distinguishing between the situations before and after. The rational formula and the curve number methods are most suitable. The procedure to be followed is illustrated for the case in Figure 12.18.

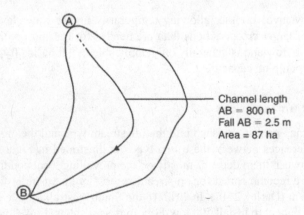

Channel length
AB = 800 m
Fall AB = 2.5 m
Area = 87 ha

Figure 12.18 Small rural basin

a) *Time of concentration* (T_c): rural basin: $T_c = \dfrac{L^{1.15}}{3080H^{0.38}} = 0.5\,hr$ (using Eq.12.4) development of the area for light industry with gutters, sewers, etc. is estimated to reduce the time of concentration by some 0.2 hours to $T_c = 0.3$ hours

b) *Runoff coefficients* (C-values, using Table 12.6): rural basin with pasture, low infiltrability and slope < 5% with estimated C = 0.4, assumed to rise to C = 0.7 at a return period > 10 years; for light industrial basin C = 0.8

c) *Rainfall intensity* (i): as given in Table 12.10

d) *Peak discharge* (Q): calculated with the Rational Formula (Eq. 12.3, results shown in Table 12.10).

The development of the area as an industrial estate leads to higher discharges at all return periods due to the increased runoff coefficients and the shortened time of concentration. The extent and damage of the flooding can be assessed provided the gradient and cross sectional areas of the disposal streams or canals are known as well as the topography and land use (see section 2.1).

Required reservoir volume

The required storage volume depends on the maximum permitted (safe) downstream discharge and upon the total discharge volume produced by the design storm. For a fixed return period of for example T = 20 years, the discharge volume varies directly with the storm durations. The storage provided should protect areas downstream against storms of specified return periods, for agricultural areas usually taken as 15 to 25 years and for residential/industrial areas as 35 to 100 years. The safe downstream discharge may either be taken as the downstream canal/stream capacity allowing for 0.3–0.4 m freeboard or be taken equal to the peak discharge occurring at a frequency of once or twice a year before development of the area.

The principles of the estimation of the required storage are illustrated in Figure 12.19. For the case of the off-stream reservoir, the maximum controlled downstream discharge is kept

Table 12.10 Calculation of peak discharges on the basis of the rational formula

Return period years	Rainfall intensity (mm/hr)		C-value		Peak discharge (m³/s)	
	$T_c = 0.5\,hr$	$T_c = 0.3\,hr$	Rural	Industrial	$T_c = 0.5\,hr$	$T_c = 0.3\,hr$
1	21	26	0.4	0.8	2.0	5.0
2	25	33	0.4	0.8	2.4	6.4
5	37	52	0.4	0.8	3.6	10.1
10	42	60	0.4	0.8	4.1	11.6
20	50	70	0.5	0.8	6.0	13.5
40	59	84	0.6	0.8	8.6	16.2
100	76	110	0.7	0.8	12.9	21.3

virtually constant (Figure 12.19a). Upstream discharges above this level are simply diverted into the reservoir. For the case of the on-stream reservoir in Figure 12.19b, the downstream discharge increases with the rising water level in the reservoir.

The required storage in the two cases is equal to the hatched areas of the hydrographs. These simple principles enable the identification, as a first step, of the approximate storage volume required and also the critical storm duration. This may suffice for small reservoirs (farm ponds, etc.) but more elaborate analyses are required for larger reservoirs, examples of which are presented below.

Storage calculations for an off-stream reservoir

The critical storage required for an off-stream reservoir may be calculated simply by evaluating the discharge from storms of variable duration as demonstrated below for the case of the light industrial estate of 87 ha. Using the curve number method, discharge volumes are calculated as in Table 12.11 for the selected return period of 20 years. The basin is assumed to be composed of 56.6 ha (65% of the basin) of industrial area with CN = 96 and 30.4

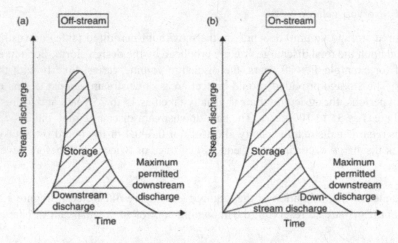

Figure 12.19 Principles of storage estimation for off-stream and on-stream retention reservoirs

Table 12.11 Estimation of discharge volume on the basis of the curve number method

Storm Duration (hrs)	Storm depth (mm)	Rainfall Intensity (mm/hr)	Discharge volumes (m³)		
			56.6 ha CN = 96	30.4 ha CN = 82	87 ha Total
0.1	15	150	4 004	76	4 080
0.2	19	95	5 875	290	6 165
0.5	25	50	8 858	840	9 698
1.0	30	30	11 439	1 450	12 889
1.5	34.5	23	13 816	2 090	15 906
2.0	38	19	15 684	2 651	18 335

ha (35%) of estate area with CN = 82. The maximum permitted downstream discharge is 2.0 m³/s, equal to the 1×1 year peak discharge before the development of the area.

Further analysis proceeds as in Figure 12.20. The discharge volumes for the various storm durations as calculated in Table 12.11 are plotted, forming a curved line. A tangent to this curve is drawn with a slope equal to the permitted downstream discharge of 2.0 m³/s. The origin of this line is set back by the time of concentration (T_c = 0.3 hrs) as the permitted downstream discharge will be reached within minutes of the commencement of the storm whereas the total volume of discharge from the same storm will only be accumulated over a period of at least T_c + storm duration. The required storage can now be read from the set-back vertical axis, in this case amounting to some 4 000 m³.

Storage calculations for an on-stream reservoir

The required storage depends on the storm duration and may be evaluated by analysing the consequences of storms of different durations (range from T_c to 5 T_c) at the selected return period (e.g., 20 years). The basic simplifying assumptions for an approximate, though often sufficiently accurate, procedure for estimating the required reservoir capacity are sketched out in Figure 12.21. The inflow hydrograph is approximated by a trapezium and the rising

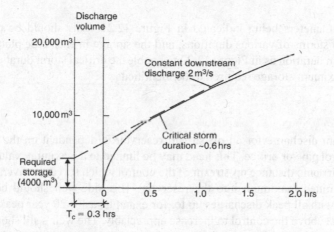

Figure 12.20 Maximum storage determination for off-stream reservoirs

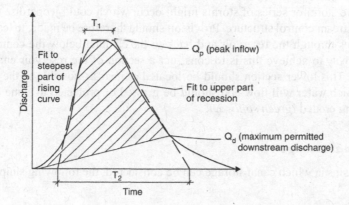

Figure 12.21 An approximate method to estimate reservoir capacity of on-stream reservoirs

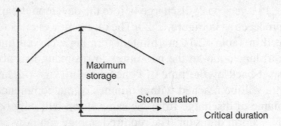

Figure 12.22 Optimisation of required reservoir storage

limb of the outflow hydrograph by a straight line (outflow is assumed to increase linearly in time) up to the permitted peak discharge (e.g., taken equal to the downstream canal/stream capacity with 0.3–0.4 m freeboard). The required storage (hatched area of the inflow hydrograph) may be approximated as:

$$S(\text{storage}) = Q_p \left(\frac{T_1 + T_2}{2} \right) - \frac{Q_d T_2}{2}$$ Eq. 12.6

The various parameters being indicated in Figure 12.21. This should be repeated using hydrographs of storms of various durations, and the storage in each case plotted against the respective storm duration as in Figure 12.22 to enable the critical storm duration and the corresponding maximum storage required to be identified.

Spillways

The downstream discharge for an off-stream reservoir is dependent on the head of water above the control pipe or orifice. This head may be limited to a maximum value by installing a long side-weir some distance up-stream of the control which serves to divert all discharges above the permitted maximum into the reservoir. This side-weir should be of sufficient length to cope with all peak discharges up to, for example, a 1 × 20 year peak without causing the head just above the control to increase appreciably. The weir's sill should actually be set so as to begin diverting flow before the maximum downstream discharge is attained, thus ensuring that the reservoir will function once or twice per annum. On rare occasions a particularly severe storm or series of storms might occur which could jeopardise the reservoir and the downstream control structure. Provision should therefore be made, to lead these very high discharges through the reservoir and back to the channel below the control structure. The simplest way to achieve this is to construct a section of the reservoir embankment at a lower level. This lower section should be located as near as possible to the downstream channel to which water will flow and should be grassed or protected in some other way to prevent it being eroded *(green spillway)*.

12.7.2   Canal storage

For the flat basins in which canal storage can be considered, the following simple water balance applies:

$$I-O = \Delta S \text{ (change in storage S)} = (A_w/A) \, \Delta d$$ Eq. 12.7

where:

A   = basin area

A_w = canal surface area (or more generally open water surface area)

Δd = permitted (temporary) rise of canal level in the design situation (rise above the HW-line as defined in section 13.1.2 or, in the case of a polder, rise above the polder level)

I   = inflow into the canal system (outflow from field drains)

O   = outflow from the canal system (outlet through sluices, pumps, etc.).

Limits are imposed on the area A_w (costly large canals) and Δd (damaging to the land use). A compromise should be sought between the open water area, permitted rise in open water level and the outlet capacity for the basin. One extreme is to provide a lot of open water area in the basin and permit considerable rise of water level, in which case the outlet capacity can be low; the other is to have a minimal A_w (canals only for water transport) and permit no rise in water level, in which case a large outlet capacity is required. The relevant considerations and calculation procedures are outlined below:

a) *Inflow:* the field drain discharge into the main system may be estimated as outlined in section 12.4.1 (subsurface drainage) and section 12.4.2 (shallow drainage). The influence of discharge transformation can generally be neglected when the water balance (Eq. 12.7) is applied to daily periods. In the example elaborated in Table 12.12, the design storm and drain discharge are the same as those in Table 12.2

b) *Pump outlet*: In Table 12.12 three different cases are considered for each of which the required maximum storage is calculated:

	Pump capacity	Pump start	Maximum storage
Case 1	7 mm/d	day 1	9 mm
2	7 mm/d	day 2	16 mm
3	10 mm/d	day 2	10 mm

Table 12.12 Calculation of canal storage requirements for the case of a pumped outlet

	Day 1	Day 2	Day 3	Day 4	Day 5	Day 6
- Design rainfall (mm)	20	10	8			
- Drain discharge (mm): day 1 rain	7	7	6			
day 2 rain	7	4	3	3		
day 3 rain	7	11	3	3	2	
daily inflow		18	12	6	2	
cumulative inflow		30	36	38		
Outflow by pumping (capacity = P)						
(1) P = 7 mm/d; start on day 1						
cumulative outflow	7	14	21	28	35	
to be stored	0	4	9	8	3	
(2) P = 7 mm/d; start on day 2						
cumulative outflow	0	7	14	21	28	35
to be stored	7	11	16	15	10	3
(3) P = 10 mm/d; start on day 2						
cumulative outflow	0	10	20	30	40	
to be stored	7	8	10	6	0	

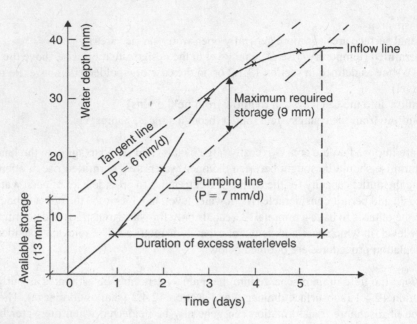

Figure 12.23 Graphical determination of storage requirements for the case of a pumped outlet

In this example, the maximum storage requirement is estimated for a known mode of pump operation. Alternatively, the procedure can be used to establish the required pumping capacity for a predetermined maximum storage.

Storage requirements may also be estimated by graphical means as shown in Figure 12.23. The inflow line corresponds to the cumulative drain inflow as calculated in Table 12.12. The pumping line is straight and has a slope equivalent to the pumping capacity (in this case P = 7 mm/d) whilst its point of origin coincides with the point in time that pumping commences (in this case the start of day 1). At any moment in time, the vertical separation between the pumping line and the inflow line corresponds to the storage requirements at that time. By adjusting the origin (start of pumping) and the slope (pump capacity) of the pumping line, a suitable combination of mode of pump operation, pump capacity and required storage may be selected, furthermore the duration of the period with extra high canal levels may be taken into account.

If the maximum storage is known, its value may be plotted as the ordinate at the start of pumping, and the required pumping capacity may then be determined from the slope of the tangent line to the inflow curve as 6 mm/d (Figure 12.23).

c) *Tidal sluice outlet:* in this case water is only able to discharge for part of the day (during the tide free period only, roughly lasting half a day, see Figure 13.14). The discharge through the sluice is directly related to the inner water level just upstream of the sluice, which is itself related to the current canal storage (S) in the basin. Storage and discharge normally begin to increase in parallel with the onset of rain, rising ultimately to peak values, after which they begin to decline. This is shown in Figure 12.24. The sluice line is shown as a broken line, being horizontal during tide locked periods and sloping during tide free

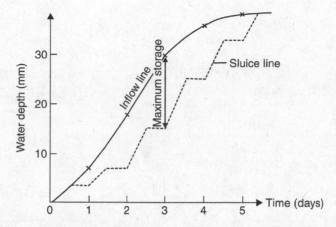

Figure 12.24 Graphical determination of storage requirements for a tidal outlet

periods. The slope of the sluice line (= rate of discharge through the sluice) is related to the current storage (slope steepens when storage increases the latter corresponding to the vertical separation between the sluice and the inflow line). This illustrates the principles of the storage requirement analysis, although actual determinations are best done by tabular calculations as in Table 12.12 with the slight difference that in this case half-day intervals should be used and also the current storage (S) should be determined (using Eq. 12.7).

Notes

1 The stage level (h) is directly related to the water depth 'd' (Figure 12.5). According to the Manning formula (Eq. 13.3, section 13.1), Q = function (d) for a canal when the gradient i = constant, so also Q = function (h). Rating curves for trapezoidal canals are typically of the type depicted in Figure 12.5 with Q increasing more than proportionally with increasing h.
2 The term rational formula was coined at the time of its development to distinguish this formula from the then widely used empirical formulae.

Chapter 13

Design of drainage canals, pumps and structures

The design of a drainage canal system may be divided into:

- selection of type and layout of the system (alignment of the canals, type and location of the structures, etc.)
- determination of the hydraulic dimensions of the different components of the system *(hydraulic design)*

The various considerations underlying the choice of the drainage infrastructure have been described in Chapter 10. In this chapter, the main aspects of the hydraulic design are outlined. In general, this part of the design is straightforward, based on well-established hydraulic engineering principles and practices. For details on the structural design of the drainage works reference is made to the listed handbooks (NRCS handbooks).

13.1 Drainage canals

The hydraulic design of drainage canals may be based in almost all cases on *steady uniform flow* (discharge is constant in time and conditions do not change along the length of the canal section considered). Under these circumstances, the following formulae apply:

$$Q = vA \tag{13.1}$$

$$Q = CAR^{1/2}i^{1/2} \qquad \text{(Chézy 1769) Eq. 13.2}$$

$$Q = \frac{1}{n}AR^{2/3}i^{1/2} \qquad \text{(Manning 1889) Eq. 13.3}$$

$$Q = k_mAR^{2/3}i^{1/2} \qquad \text{(Strickler 1923) Eq. 13.4}$$

where:
v = flow velocity (m/s)
Q = discharge rate (m³/s)
R = hydraulic radius (m)
i = hydraulic gradient (m/m)
A = wet cross-section (m²)
C = Chézy roughness coefficient (m$^{1/2}$/s)
n = Manning roughness coefficient (s/m$^{1/3}$)
k_m = Strickler roughness coefficient (m$^{1/3}$/s)

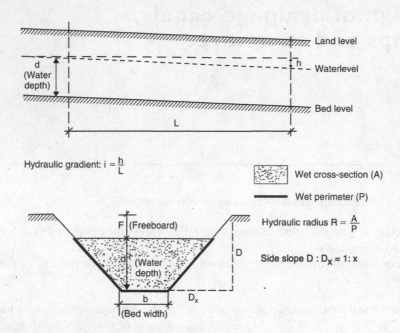

Figure 13.1 Main parameters in the Manning/Strickler formulae for canal design

Each of these formulae for uniform flow was developed independently. The meaning of the different parameters is illustrated in Figure 13.1. The given three roughness coefficients are closely related:

$$k_m = 1/n \qquad\qquad\qquad\qquad\qquad\qquad\qquad\qquad \text{Eq. 13.5}$$

$$C = k_m R^{1/6} \qquad\qquad\qquad\qquad\qquad\qquad\qquad\qquad \text{Eq. 13.6}$$

The Manning formula is widely used in English speaking countries whereas the Strickler formula is more common on the European continent. In this book both formulae will be used interchangeably.

Calculations: given the discharge (Q) to be carried by the canal and having established the hydraulic gradient (i) and some characteristics of the canal (n, x and b/d ratio), the main dimensions of the required canal can be calculated with the Manning/Strickler formula as in the following example (see also the flow chart in Figure 13.2).

$Q = 3.5$ m³/s, $n = 0.033$ ($k_m = 1/n = 30$), $i = 0.02\% = 0.0002$ m/m, $x = 1.5$; b/d = 3

$A = d(b + dx) = d(3d + 1.5d) = 4.5d^2$

$P = b + 2\sqrt{d^2 + (dx)^2} = b + 2d\sqrt{1 + x^2}$

$R = \dfrac{A}{P} = \dfrac{d(b + dx)}{b + 2d\sqrt{1 + x^2}} = \dfrac{b + dx}{b/d + 2\sqrt{1 + x^2}} \rightarrow R = \dfrac{3d + 1.5d}{3 + 2\sqrt{3.25}} = 0.68d$

$Q = k_m AR^{2/3}i^{1/2} \rightarrow 3.5 = 30(4.5\,d^2)(0.68\,d)^{2/3}(0.0002)^{1/2} \rightarrow d = 1.40$m; $b = 4.3$m (rounded)

Check on the velocity: $v = Q/A = 3.5/4.5d^2 = 3.5/8.82 = 0.40$ m/s → OK!

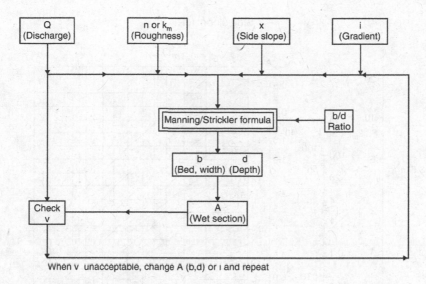

Figure 13.2 Flow chart for canal design

These calculations may best be done with a small electronic calculator or computer (see Chapter 21). The Manning/Strickler formula, however, may also be conveniently solved using nomographs, an example of which is presented in Figure 13.3. The determination of the different input parameters of the formula is described in the following sections.

Nomograph (Figure 13.3): connect Q = 3.5 m³/s on the Q-scale with i = 0.2 m/km on the i-scale (for $k_m = 1/n = 40$). Extend this line until it connects with the b-d graph and enter this graph horizontally. On this horizontal AR²/³ is constant and of the desired value, so all b, d combinations on this line fulfil the Manning/Strickler equation. The combination of d = 1.40 m and b = 4.15 m is selected in this case as it gives a b/d ratio closest to 3, which is the desired ratio for a canal of this depth (section 13.1.3). The wet cross-section of the canal may be determined by following the direction of the dashed lines from the b-d point to the A-scale (A is constant along these dashed lines). This leads in this case to A = 8.8 m². The flow velocity may then be readily calculated as v = Q/A = 3.5/8.8 = 0.40 m/s. When the velocity exceeds the permissible limits as given in Table 13.1, the hydraulic gradient should be reduced so as to arrive at a velocity within the permitted range. The nomograph in Figure 13.3 is for canals with side slope 1:1½. Similar graphs are available for other side slopes (three types of graphs with side slopes 1:1, 1:1½ and 1:2 will cover most requirements, ILRI 1964).

13.1.1   Discharge rate

The discharge rate (Q) is normally taken equal to the design discharge (Q_{design}) when designing new canals. The determination of design discharges was extensively described in Chapter 12. These methods establish the range, but the final selection and judgement should be left to the designer. Somewhat lower or higher values than those calculated might be taken depending on how the costs, risks and damage are evaluated in each particular case. For example, design discharges for structures are often taken some 25 to 50% higher than those for canals, the argument being that cross-sections in structures are generally more

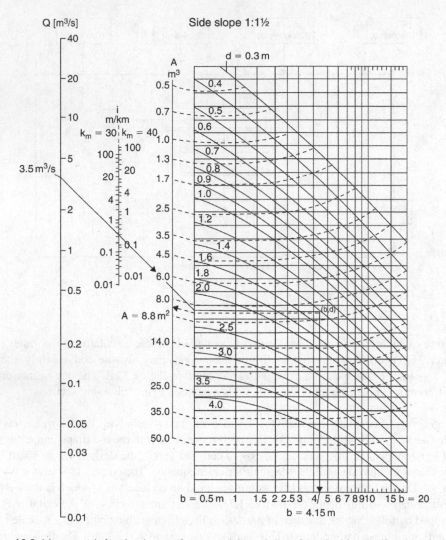

Figure 13.3 Nomograph for the design of trapezoidal canals based on the Manning/Strickler formula

constrained than in the canals and under-design may lead to unacceptably high flow velocities and backing-up of the flow. Also, it is generally more costly to increase the discharge capacity of a structure than that of a canal once the system is constructed and the adopted design discharge proves to be too low.

The determination of the discharges normally starts at the most upstream end of the system with the discharges being combined at the junctions as the calculations move downstream. At the junctions, the discharges may simply be added but this ignores the earlier described discharge transformation (section 12.1) and area reduction effects (section 12.6) and may lead to over-estimation. These effects can be determined by applying routing techniques as described in section 21.2. For initial design, when the system dimensions are not yet known, the following rules of thumb may be applied:

a) *the 20–40 rule:* under this rule (method 1), the combined discharge Q is calculated as $Q = Q_1 + Q_2$ as long as the sizes of contributing areas A_1 and A_2 are each only 40–60% of the total area A. However, when either A_1 or A_2 is < 20% of A, Q is determined on the basis of an undivided A using e.g., an area reduction formula (method 2). For all intermediate cases (A_1 or A_2 between 20–40% of A), Q is calculated both by method 1 and method 2 and the final value is determined as $Q = Q_2 + \beta$ $(Q_1 - Q_2)$ where ($\beta = \%/20$) with % being the actual % of A_1 or A_2 within the 20–40% range

b) *area reduction formula:* over-estimation of discharges at junctions for large basins can also be avoided by repeated application of an area reduction formula (section 12.6, Eq. 12.5) for the increasingly larger basins (Q_1 for A_1, Q_2 for A_2, Q_3 for A_3, etc. for increasingly larger basins $A_1 < A_2 < A_3$, etc.).

The above rules apply to major junctions (contributing area > 1 000–2 000 ha). For minor junctions, simple addition of discharge will suffice.

13.1.2 Hydraulic gradient and water levels

The hydraulic gradient should be neither too small nor too large. When the gradient is small, a large cross-section (A) is required to carry the flow (high excavation costs). Values should generally not be less than 5–10 cm/km (i = 0.00005–0.00010 m/m) but for large drains in very flat basins lower values may have to be used when the total available head between fields and outlet is small and will partly be expended in unavoidable head losses in the structures in the canal system (culverts, bridges, etc.).

High values for i, on the other hand, lead to high flow velocities in the canals with the attendant danger of scouring of beds and erosion of banks. Permissible flow velocities are described in section 13.1.3. Drop structures may be used to dissipate surplus head where canals run down steep slopes.

In practice, hydraulic grade lines often closely follow the slope of the land in the direction of discharge. Under these conditions, in the case of steady uniform flow, the land slope, the hydraulic grade line and the bed slope are all parallel. The velocity *head* ($v^2/2g$ with g = 9.81 m/sec^2) in drainage canals is normally negligible so that the energy grade line and the water surface line may be assumed to coincide. In these circumstances, the water depth in the canal section under consideration will be constant and the water surface will be at a constant depth below the land surface (this situation is depicted in Figure 13.1).

Water levels

The water level in drainage canals must meet certain freeboard requirements (section 13.1.6) and in the case of lower order canals serving as collectors, the water level must in principle remain 5–10 cm below the field drainage base to permit unimpeded outflow from the field drains. For deep pipe drains, temporary impedance in outflow is generally not harmful and the hydraulic grade line at the design discharge can be allowed to run above the outlet levels of such drains[1]. Collectors and main canals are in fact commonly designed on the premise that there should be unimpeded outflow from the field drains for the much more frequently occurring event of Q = 0.5 Q_{design}. Under normal conditions a discharge of this magnitude may on average be expected to occur several times per year whereas Q_{design} would occur only once per year (Box 13.1).

Box 13.1　Design discharges in the Netherlands

In the Netherlands the following discharges are identified:

design discharge Q_{design} to be expected about once per year
normal discharge $Q_n = 0.5\ Q_{design}$ to be expected several times per year
extreme discharge $Q_e = 2\ Q_{design}$ to be expected about once per 25 years

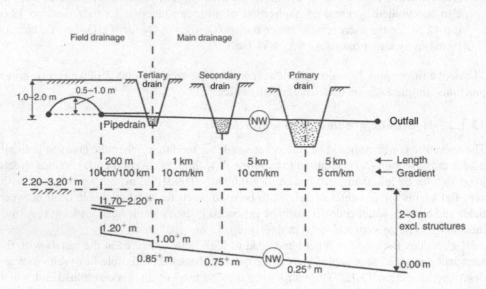

Figure 13.4 Hydraulic grade lines and losses for a subsurface field drainage system

The hydraulic grade line at $Q = Q_{design}$ is designated as the *HW-line* (high water level) while the grade line at $Q = 0.5Q_{design}$ is designated as the *NW-line* (normal water level). The NW-line may be determined with the Manning/Strickler formula assuming a flow equal to $0.5Q_{design}$ passing through the canals designed to carry Q_{design}. For many drainage canals, water depths roughly vary with $Q^{1/2}$, so a canal which fills to a depth 'd' while carrying flow Q, will fill to 0.7d when the flow is reduced to 0.5Q. This rule holds best for medium sized trapezoidal canals with 1:1 side slope and may be applied to approximately locate the NW-line in a canal, knowing the HW-line.

The total headloss from field to outlet depends on the length of the canal system and on the number and types of structures (Figure 13.4). The slope of the land may provide part of the head but in most cases increasing the installation/bed depth of the drains in the down-stream direction must also create part of the head.

Longitudinal sections, showing the slope and elevations of the land along the canal alignment, are useful in establishing grade lines. Various water level requirements (*control points*) to be met at various places can be marked on the section (field drainage bases, freeboard requirements, levels of existing structures, etc.). Longitudinal sections are almost always indispensable in fixing grade lines of canals running through sloping, irregular land (section 13.2.4).

13.1.3 Permissible flow velocities

For the flows normally encountered in drainage canals, the tractive force exerted by the flowing water on the bed material is proportional to the square of the average flow velocity. When the tractive force exceeds those forces retaining the bed material, the constituent particles begin to move with the water causing the bed to erode. The movement of non-cohesive (single grain) bed material is resisted by friction which increases with the weight of the particle. Safe, non-erosive flow velocity limits may be established on the basis of the particle size distribution of the bed material. For the cohesive materials encountered in most drainage canals, there is no straightforward relationship between susceptibility to erosion and particle size of the bed material and limits are based mostly upon experience. Permissible flow velocities for drainage canals as used in the Netherlands are given in Table 13.1.

The erosion hazard is always most severe in newly constructed canals. Higher velocities can generally be permitted once the canals have matured and the bed has become vegetated. Higher velocities can also be permitted when the canals are large and deep since in these canals the actual velocities along the bed are considerably lower than the average velocity. Erosion hazards are increased by curves in the canal (permissible flow velocities to be reduced by 25–40% when protection is not to be provided).

13.1.4 Cross-section

Steep side slopes are desirable in that they save on excavation as well as on land loss. However, they can only be used in cohesive, well-aggregated soils or where bank protection measures are adopted. Unprotected side slopes in unstable soils should generally not be steeper than 1:2 (V:H) to avoid serious sloughing/caving in. Some general guidelines for design are presented in Table 13.1.

Side slopes generally become more stable once they have become vegetated, but design should mostly be based on the initial period after excavation. Factors other than soil type

Table 13.1 Limitations on flow velocity and on side slope in drainage canals
(adapted from ILRI 1964)

Soil type	Maximum permissible mean flow velocity (in m/s)	Maximum permissible side slope
Fine sand	0.15–0.30	1:2 to 3
Coarse sand	0.20–0.50	1:1½ to 3
Loam	0.30–0.60	1:1½ to 2
Heavy clay	0.60–0.80	1:1 to 2

Note: highest velocities and steepest side slopes apply to well vegetated canals.

should also be considered. Seepage inflow into the canal for example promotes caving in, and gentle side slopes (or even toe protection) are to be recommended in situations where this occurs. It is also sound practice to adopt gentler side slopes with increasing canal depth and for canals with widely fluctuating water levels. Special conditions, e.g., a road running along the canal or the provision of drinking access for cattle, may also dictate gentler side slopes.

Canals with a semi-circular wet cross-section are hydraulically the most efficient (minimum excavation per unit wet cross-section). Other considerations (construction, maintenance, etc.), however, also exert an influence so that, all in all, canals with a trapezoidal cross-section have been found to be most suitable for drainage canals. As a compromise between hydraulic efficiency and other considerations the following ratios between bed width (b) and water depth (d) should be approximately adhered to:

- small canals (d < 0.75 m)   b/d = 1 (clay) – 2 (sand)
- medium canals (d = 0.75–1.5 m) b/d = 2–3
- large canals (d > 1.5 m)   b/d > 3–4

For reasons of construction and maintenance, canals should not be made too small. Depending on the method of construction, minimum sections should have bed widths of 0.50–1.00 m, side slopes of 1:¾, 1:1 or 1:1½ and water depths of 0.30–0.50 m. Minimum sections typically apply to ditches bordering fields and road ditches, carrying flows up to 0.5 m^3/s. These ditches are often constructed without grade (no fall in bed line).

13.1.5   Roughness coefficient

The roughness coefficients n and k_m should be understood as the (inverse) constant of proportionality between the average flow velocity (v) and the combination $R^{2/3}i^{1/2}$. As Eq. 13.6 shows, it not only depends on the bed roughness but also on the shape of the canal (expressed by the parameter R). The bed roughness, however, is the most important influencing factor so that the roughness coefficients depends primarily on the type of bed material and bed vegetation. Minor influencing factors are the canal alignment (number and smoothness of curves) and the size of the canal. Values vary from as low as n=0.012–0.015 (k_m=70–80) for good sized concrete lined canals to as high as n=0.10 (k_m=10) for small heavily overgrown ditches.

The roughness coefficient used for design is of considerable importance as canal capacities and excavation costs depend directly on the value taken. Its estimation can however seldom be done objectively as there is usually little assurance as to how well the canals will be maintained. Most designs are based on the assumption of fair maintenance, to which the general guidelines of Table 13.2 apply. Medium sized, earthen drainage canals are commonly designed with n=0.025–0.033 (k_m=1/n=30–40, the values used in Figure 13.3).

13.1.6   Freeboard

Freeboard is meant to provide a margin of safety in case the actual discharge exceeds the designed discharge capacity of the canal (Figure 13.5). It follows that less freeboard needs to be provided when the design discharges are estimated conservatively.

Table 13.2 Guidelines for the selection of roughness values for the design of drainage canals (adapted from ILRI 1964)

Canal description	k_m-value (= 1/n)	n-value
(a) Small canals (water depth <0.75 m)		
sandy soil.	20	0.050
clayey soil	15	0.065
(b) Medium canals (water depth 0.75–1.50 m)		
sandy soil	30	0.035
clayey soil	20	0.050
(c) Large canals (water depth > 150 m).	40–50	0.020.0.025
(d) Concrete lined canals	60–80	0.0125–0.017

Note: these guidelines apply to canals with their wet cross-section well covered by a weedy or grassy vegetation, kept reasonably short by fair maintenance (the kind of conditions typically assumed for design). Lower k_m values should be taken when poor maintenance is anticipated and when the canal system is composed of short lengths with variable cross-sections and sharp curves/junctions.

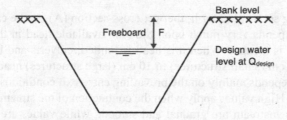

Figure 13.5 Freeboard in drainage canals

The freeboard required is often established on the basis of experience, typical guidelines being:

- small canals (Q < 1–2 m³/s) freeboard F = 30 cm
- large canals (Q > 5 m³/s)   freeboard F = 50 cm

A more rational approach towards defining freeboard requirements is to specify the *bankfull capacity* of the canal. In situations where the Q_{design} is taken for example as the 1 × 5 year event, the bankfull capacity could be based on the 1 × 25 year frequency. Alternatively, freeboard may be chosen to provide adequate over-capacity for the canal to carry 2 × Q_{design} without over-topping of the banks. For a Q_{design} based on a 1x 5 year frequency, Q = 2 × Q_{design} may well represent the 1 × 25 year event. The water depth would be roughly 1.4 × the water depth of the canal when carrying Q_{design} (so the freeboard for a canal with a water depth of 125 cm would be some 50 cm).

The concept of freeboard is most applicable to drainage where the actual discharge may be a multiple of the Q_{design}. Subsurface discharge is quite steady and flooding risks are low. Water levels in canals receiving subsurface drainage are in the first instance determined by the field drainage base level, and are in fact often deeper than required simply on the basis of freeboard considerations (see section 13.1.2).

13.2   Structures in drainage canals

Common structures in drainage canals are culverts and bridges. Drop structures are used where gradients are too steep. Weirs may be installed to retain water in the canals during dry periods. At the outfall points, sluices, gates or pumps may be encountered.

13.2.1   Culverts and bridges

The flow through culverts and bridges may in most drainage design work be calculated with the simple formula:

$$Q = \mu A \sqrt{2gh}$$
Eq. 13.7

where:
Q = discharge rate (m³/s)
A = wet cross-section (m²)
g = gravitational acceleration (= 9.81 m/s²)
h = headloss across the culvert (m)
μ = coefficient

Given Q and having set a value for h, the wet cross-section (A) can be calculated. The permitted head loss depends very much upon the total available head in the canal section in which the structure is located. When this head is limited, culverts and bridges are usually designed for only 5 cm (small structures) to 10 cm (large structures) head loss.

The value of μ depends mainly on the prevailing entry/exit conditions for the water into/out of the structure. High values apply when the contraction of the streamlines upstream and their divergence downstream are gradual and smooth, while values are lower when these transitions are accompanied by considerable turbulence in the water. High friction encountered in the flow through the structures also depresses the μ-value (high wall roughness, long barrel, etc.). Under normal, favourable, flow conditions the following values may be used:

- short culverts (length < 10 m)   μ = 0.8
- long culverts (length 20–30 m)   μ = 0.7

These values apply to culverts operating under *inlet control* as is usually assumed for design (barrel only partly filled). Different values apply when the barrel is completely filled (*outlet control*). Common values used for bridges are μ = 0.8–0.9. For bridges with a wet cross-section almost equal to that of the canal, hardly any headloss needs to be accounted for. Further details can be found in the hydraulic handbooks.

Culvert barrels usually consists of pipes when the discharge is small (Q < 0.5 m³/s). For larger discharges, more pipes may be installed but when the discharge exceeds 1–2 m³/s, box type barrels are generally preferred. The invert of the barrel is usually laid flush with the canal bed whilst the top should normally have a minimum 10 cm clearance, especially when the water carries debris (Figure 13.6). Non-reinforced pipe culverts should be buried at least 50 cm deep when they have to support normal wheel loads due to on-farm or rural traffic.

In the design, the outflow velocities should always be checked and if found to be too high, either anti-scouring measures should be provided, or the cross-section should be enlarged (the latter is usually the cheapest solution). Outflow velocities into unprotected earthen canals should be no more than 1½-2 times the maximum velocities given in Table 13.1.

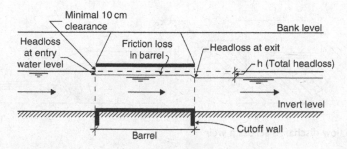

Figure 13.6 Longitudinal section of a typical drainage culvert

Corrugated steel pipe is commonly used in the USA for drainage system culverts. They range in sizes from 300 mm through 2 500 mm in diameter with a typical maximum length of 10 m. The corrugation is either sinusoidal or spiral (Figure 8.5). The pipes are generally not perforated, and they also come in arc-shaped sections that are bolted together on site. Prefabricated concrete culvert sections in all kinds of shapes are commonly used in the Netherlands. Connections between sections are of the bell and spigot type with special O-rings to seal the connection. For more important structures special post tension reinforcement bars are applied.

13.2.2 Weirs

Weirs are installed in drainage canals for different purposes. They may serve as drop structures to dissipate excess head in the canal system. More commonly weirs are used to prevent water levels in certain canal sections from falling too low during periods of low discharge, thus preventing over-drainage of the adjacent land or even promoting lateral inflow of water from the canals into the land (applications described in section 10.2.1 and in Chapter 20). They are also used in the design of low-flow bypass structures with small reservoirs (section 12.7.1).

The free flow discharge over a rectangular weir as depicted in Figure 13.7 may be calculated as:

$$Q = \frac{2}{3}\mu bh_1\sqrt{\frac{2}{3}gh_1} = 1.7\mu bh_1^{3/2} \qquad \text{Eq. 13.8}$$

where:
Q = discharge rate(m^3/s)
g = gravitational acceleration (9.81 m/sec^2)
b = crest width (m)
h_1 = upstream water level above crest level (m)
μ = coefficient

Eq. 13.8 actually applies to broad crested weirs only (in which the streamlines over the crest are essentially horizontal) but may also be used for sharp crested weirs (in which the streamlines over the crest are quite curved) by using suitable values for the coefficient. For the semi-sharp crested weirs normally encountered in drainage canals, μ-values are of the order of 1.0–1.1. The higher values apply when the width of the weir is almost the same as that of the approaching canal and when the water training edges of the structure are smooth and well rounded. When these edges are rather sharp and the weir is of the so-called *contracted* type (i.e., the weir crest << canal width as is the case in Figure 13.7), the lower values of μ should be used.

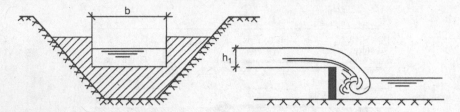

Figure 13.7 Free flow discharge across a weir

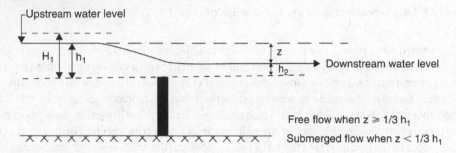

Figure 13.8 Conditions for free and submerged flow over a weir

Submerged discharge occurs when the flow over the weir is affected by a high down-stream water level (the weir is 'drowned', see Figure 13.8). For this type of discharge the following formula applies:

$$Q = \mu b h_2 \sqrt{2g(h_1 - h_2)} \approx \mu b h_2 \sqrt{2gz} \qquad\qquad \text{Eq. 13.9}$$

Actually, the energy head $H = h + v^2/2g$ should be used rather than simply the water depths (h_1 and h_2). However, provided the flow velocity remains low (as is usually the case, seldom exceeding 0.5–1.0 m/s) there are no significant errors made when using water depths in Eqs. 13.8 and 13.9 instead of energy heads.

Moveable crest weirs are most suitable when the main purpose is to retain water during the dry season. Crest levels can be lowered during the wet season to permit unimpeded drainage flow. Conversely, during the dry season the weir crest may be raised. The simplest form of a moveable crest weir is the stoplog weir. To save on costs (construction and operation), moveable crest weirs may be installed at the most critical locations, supplemented with fixed crest weirs at various other locations (Figure 20.7).

13.2.3   Backwater curves

Structures in a canal system will often create some obstruction to the discharge. The canal levels will adjust to this by allocating additional head to the flow through the structure to maintain flow throughout the system. This extra head will create a backwater curve as in Figure 13.9, which merges asymptotically with the normal hydraulic grade line (i.e., the grade line which would establish in the absence of structures).

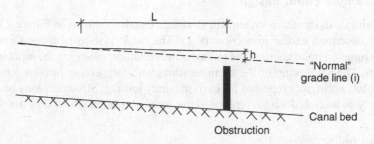

Figure 13.9 Backwater curve upstream of a structure

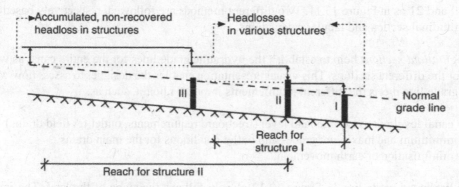

Figure 13.10 Overlapping backwater curves

The reach of the backwater curve (= distance L in Figure 13.9) may be approximated as:

$$L = \frac{2h}{i}$$ Eq. 13.10

where:
L = reach of the backwater curve (m)
i = hydraulic gradient/bed slope (m/m)
h = head loss over the structure (m)

Example calculation:
$$\left.\begin{array}{l} h = 0.10 \text{ m} \\ i = 0.02\% = 0.0002 \text{ m/m} \end{array}\right\} L = \frac{2 \times 0.10}{0.0002} = 1000 \text{ m.}$$

When the nearest upstream structure is outside the reach of the backwater curve, the head-loss h will be fully recovered in the canal section between the two structures. When the backwater curves overlap as in Figure 13.10 the headloss is not fully recovered in the canal sections and so part of the total available head in the system must be specifically allocated to the flow through the structures. A rule of thumb states that head does not need to be allocated to structures located at the head of canals and to small structures designed for h < 5 cm when there is no adjacent structure in the immediate upstream vicinity (say within 500 m). For large structures, the available head in a canal section between two such structures should be equally allocated to the flow through the structure located at the downstream end of the section and the canal flow in the section.

13.2.4   Example canal design

A suitable, simple designation system for drainage canals is shown in Figure 13.11 (refers to the basin described earlier in section 10.1). The canal system is divided into sections, bounded by junction points, structures or any other condition changing the input used for the canal design. The numbering of the sections starts with the section furthest removed from the outlet point; a principle repeated for each tributary joining. Structures may be designated independently or be coded with reference to the canal section in which they are situated.

Canal design practices

Procedures commonly used are illustrated for the canal line composed of the sections 1, 3, 5, 7, 19 and 21 as in Figure 13.11. Two different methods are followed, respectively based on longitudinal section and tabular calculations.

Longitudinal sections help to establish the hydraulic grade lines for the entire canal as well as for the different sections. This visual presentation enables the designer to assess how well the grade line meets the different requirements imposed upon it, such as:

*   canal level relative to the land level (freeboard requirements, outlet for field drains)
*   minimum and maximum grades and outlet conditions for the main drains
*   minimisation of earth movement.

The projected grade line in Figure 13.12 closely follows the slope of the land. The inner water level at the outlet is 6.60 m + which is 10 cm above the maximum outer level needed for gravity outlet.

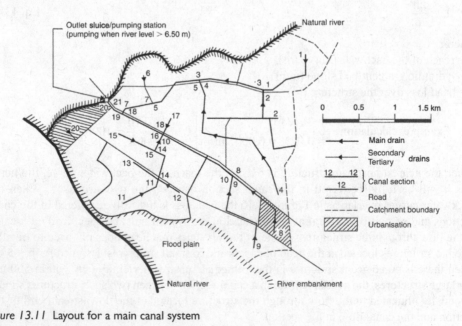

Figure 13.11 Layout for a main canal system

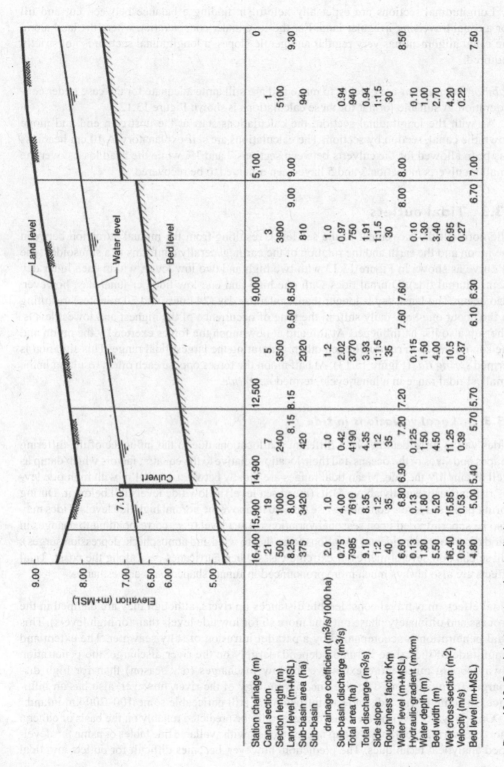

Station chainage (m)	16,400	15,900	14,900	12,500	9,000	5,100	0
Canal section	21	19	7	5	3	1	
Section length (m)	500	1000	2400	3500	3900	5100	
Land level (m+MSL)	8.25	8.00	8.00	8.15	8.50	9.00	9.30
Sub-basin area (ha)	375	3420	420	2020	810	940	
Sub-basin drainage coefficient (m³/s/1000 ha)	2.0	1.0	1.0	1.0	1.2	1.0	
Sub-basin discharge (m³/s)	0.75	4.00	0.42	2.02	0.97	0.94	
Total area (ha)	7985	7610	4190	3770	1750	940	
Total discharge (m³/s)	9.10	8.35	4.35	3.93	1.91	0.94	
Side slope	1:2	1:2	1:2	1:1.5	1:1.5	1:1.5	
Roughness factor K_m	40	35	35	35	30	30	
Water level (m+MSL)	6.60	6.80	6.90	7.20	7.60	8.00	8.50
Hydraulic gradient (m/km)	0.13	0.13	0.125	0.115	0.10	0.10	
Water depth (m)	1.80	1.80	1.50	1.50	1.30	1.00	
Bed width (m)	5.50	5.20	4.50	4.00	3.40	2.70	
Wet cross-section (m²)	16.40	15.85	11.25	10.5	6.95	4.20	
Velocity (m/s)	0.55	0.53	0.39	0.37	0.27	0.22	
Bed level (m+MSL)	4.80	5.00	5.40	5.70	6.30	7.00	7.50

Figure 13.12 Canal design using a longitudinal section

Longitudinal sections are especially helpful in finding a balance between cut and fill for a canal traversing irregular land. For the case under consideration where the land along the canal alignment has very regular and gentle slopes, a longitudinal section is not strictly required.

Tabular calculations are quicker to make and are still quite adequate for the case under consideration. A suitable format for these calculations is shown Figure 13.12.

As with the longitudinal section, the calculations start at the upstream end and move down the canal, section by section. The calculations are self-explanatory. A 10 cm headloss has been allowed for the culverts between sections 7 and 19, while the head losses over the smaller culverts in section 3 and 5 have been assumed to be recovered.

13.3   Tidal outlets

The normal tidal movement of the sea level, resulting from the mutual attraction between the moon and the earth and the rotation of the earth, generally conforms to a sinusoidal type of curve as shown in Figure 13.13 with two high and two low levels within each lunar day (semi-diurnal tide). Diurnal tides with one high and one low tide per lunar day, however, also occur. The lunar day is longer than a calendar day (24 hours and 50 minutes), resulting in the about one-hour daily shift in the time of occurrence of the highest and lowest levels. The sun also has an influence. At full and at new moon the forces exerted by the moon and the sun on the earth reinforce each other, generating the largest tidal range. This situation is termed *spring tide* (Figure 13.13). At half moon the forces oppose each other, resulting in the smallest tidal range in a lunar cycle, termed *neap tide*.

13.3.1   Local variations in tide

Tides vary a great deal between different sea locations due to the influence of the differing shapes and sizes of the oceans and their/location relative to the equator, factors which damp as well as amplify the tide. Mean tidal ranges are mostly between 1 and 4 m with high tide levels rising nearly equally above MSL (mean sea level) as low tide levels fall below it. During spring tides, sea levels commonly rise 0.5–1.0 m above the normal high tide levels. Tides may also be superimposed upon seasonal variations in sea level (e.g., corresponding to changes in wind direction) or incidental variations (e.g., due to wind and atmospheric depression surges). These can add several metres in extreme cases when hurricanes pass along the coast. Tidal effects are also always much more pronounced in funnel shaped bays and estuaries.

Tidal effects may travel considerable distances up rivers, although they are damped in the process and ultimately phase out (the more so for low tide levels than for high levels). This tidal penetration is accompanied by a parallel intrusion of salty seawater. The extent and magnitude of the saline intrusion depends largely on the river discharge, the penetration always being much more extensive for low discharges (dry season) than for high discharges (rainy season). The size and morphology of the river, however, also has an influence. In some rivers, tide and salt intrusion are still noticeable some 100–200 km inland.

Critical water levels at the coast can generally be predicted reliably on the basis of either a short measuring programme set up in conjunction with available tide tables or using the developed analytical techniques. The prediction, however, becomes difficult for outlets into tidal

Table 13.3 Canal design in tabular form

Section no in Figure 9.12	Chainage (m)	Area (ha)	Drainage coef. (m³/s/1 000 ha)	Discharge (m3/s)	Total discharge (m3/s)	Side slopes	k_m (or 1/n)	Grade (m/km)	Water depth (m)	Bed width (m)	Wet cross-section (m2)	Velocity (m/s)	Land level	Water level	Bed level
		Contributing sub-basin			Hydraulic calculations								Levels m + MSL*		
1	5100	940	1.0	0.94	0.94	1.5	30	0.10	1.00	2.70	4.20	0.22	9.30	8.50	7.50
													9.00	8.00	7.00
3	3900	810	1.2	0.97	1.91	1.5	30	0.10	1.30	3.40	6.96	0.27	9.00	8.00	6.70
													8.50	7.60	6.30
5	3500	2020	1.0	2.02	3.93	2	35	0.115	1.50	4.00	10.50	0.37	8.50	7.60	6.10
													8.15	7.20	5.70
7	2400	420		0.42	4.35	2	35	0.125	1.50	4.50	11.25	0.39	8.15	7.20	5.70
	Culvert												8.00	6.90	5.40
19	1000	3420	2.0	4.00	8.35	2	40	0.13	1.00	5.20	15.85	0.53	8.00	6.80	5.00
													8.15	6.67	4.87
21	500	375		0.75	9.10	2	40	0.13	1.00	5.50	16.90	0.55	8.15	6.67	4.87
													8.25	6.60	4.80

* Upper lines give levels at the upstream end and lower lines at the downstream end of the section

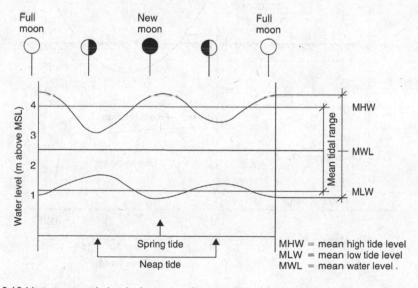

Figure 13.13 Variation in tide levels during one lunar cycle

rivers. Methods are available to calculate water levels at different distances up the river under various discharge conditions, the reliability of which depends on the accuracy of the input for the formula adopted. Estimates based on locally recorded values are always preferable.

13.3.2 Discharge through a sluice

Figure 13.14 illustrates the principles of a gravity outlet in a tidal situation. Discharge takes place during the tide free period which lasts from A to D. The gate opens when the inner level is slightly above the outer level (point A in Figure 13.14), a 10 cm head difference being sufficient for outlet into the sea when the gates are well maintained (in freshwater a 5 cm head difference should suffice). Closure occurs at point E when the outer level has risen a few cm above the inner level. During the period from A to B, the outflow is restricted by the still high outer levels and the same applies to the last part of the tide free period from C to D when the outer level has once again risen. In between there is the period of free outflow with the outer levels being so far below the inner levels as to have no influence on the discharge through the sluice (period lasting from B to C).

The discharge through a sluice may be calculated using the broad crested weir formula presented earlier as Eq. 13.9 (for notation see Figure 13.8):

$$Q = \mu b h_2 \sqrt{2g(h_1 - h_2)}$$

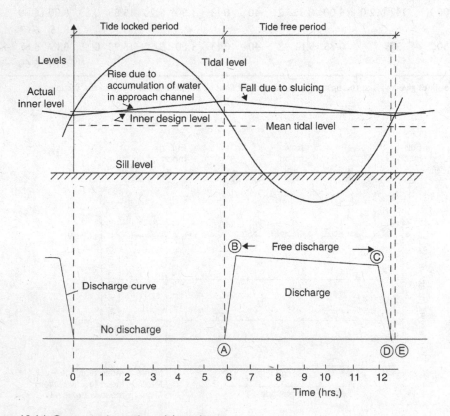

Figure 13.14 Gravity outlet under tidal conditions

where in this case b = the width of the sluice and the value of μ would be of the order of 0.9–1.1. Free flow occurs when $h_2 < 2/3h_1$. Inserting $h_2 = 2/3h_1$ in the above formula yields:

$$Q = 2/3\mu bh_1\sqrt{2/3gh_1} \qquad \text{which is similar to Eq. 13.8}$$

For discharge into the sea, the water depth h_2 should be increased by 3.5% to account for the higher density of sea water (also explains closure in Figure 13.14 at point D rather than at point E). Paradoxically, neap tides often correspond to the most critical condition for design since the tidal range is then narrow, thus ensuring that the gates only open for a relatively short period of time, or in extreme circumstances, not at all. This can lead to water accumulating upstream and causing flooding, a situation that will continue until the inner levels exceed a later falling tidal level.

13.3.3 Example of calculations

The variations of the inner water level (h_1) during a sluicing cycle may be determined by water balance calculations, treating the lower part of the main drainage system behind the sluice as a reservoir (storage of backed-up discharge in the canals, see section 12.7.2). Given the reservoir area and the inflow and outflow rates, the variation in h_1 may be determined as the variation of the reservoir level. The inflow may be taken equal to the design discharge of the basin draining through the sluice and is assumed to remain constant throughout the sluicing cycle. The outflow, which is zero during the tide locked period, varies once the sluice gates have opened depending especially on the value of h_1. This implies repetitive calculations in which due account is taken of the relationships between h_1, reservoir storage and outflow. However, outflow calculations are often sufficiently accurate when the inner water level is assumed to fall linearly as in Figure 13.15 (and also in the calculations in Table 13.4).

The example applies to a case in which the design discharge is 10 m³/s (area = 8 000 ha, drainage coefficient = 1.25 l/s/ha). The reservoir area is 200 ha (2½% of the basin area). During the 11 hours tide locked period, the inner water level rises 0.20 m [(11x10x3 600)/ (200x10⁴)] above its normal level. Sufficient sluice capacity should be installed in order to return the reservoir to the normal level during a single discharge period. The calculations proceed as in Table 13.4 using average levels during hourly periods (this is generally satisfactory for the limits of accuracy required). The discharge through the sluice is calculated for each hourly period using either Eq. 13.9 (submerged flow) or Eq. 13.8 (free flow). The total outflow during the 13 hours and 50 minutes discharge period amounts to 135 420xb m³. Equating

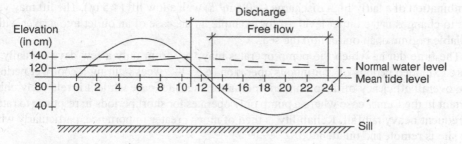

Figure 13.15 Water levels at gravity outlet during a tidal cycle

Table 13.4 Calculation of discharge through a tidal sluice

Period (hours^{minutes})	h_1 *(m)	h_2* (m)	Type of discharge	μ	Outflow (m³)
11–12	1.39	1.18	submerged	1.1	2.63 × b × 3600
12–13	1.37	1.00	submerged	1.1	2.96 × b × 3600
13–14	1.36	0.82	free	1.1	3.00 × b × 3600
14–15	1.34		free	1.1	2.92 × b × 3600
15–16	1.33		free	1.1	2.88 × b × 3600
16–17	1.31		free	1.1	2.82 × b × 3600
17–18	1.30		free	1.1	2.78 × b × 3600
18–19	1.28		free	1.1	2.72 × b × 3600
19–20	1.27		free	1.1	2.70 × b × 3600
20–21	1.25		free	1.1	2.63 × b × 3600
21–22	1.24		free	1.1	2.58 × b × 3600
22–23	1.22		free	1.1	2.52 × b × 3600
23–24	1.21	0.64	free	1.1	2.50 × b × 3600
24–24^{50}	1.20	0.87	submerged	1.1	2.43 × b × 3600
24^{50} → negligible				Total	135 420 × b

* Values read from Figure 13.15 and, in the case of h_2, increased by 3.5%

this outflow to the total discharge per sluice cycle of $(24 + 50/60) \times 10 \times 3\,600$ m^{32}, it follows that a sluice with a width b = 6.60 m is required. In this case the level of the sill was fixed in advance although it could of course also be a design variable. Different combinations of sill-level and sluice width are possible, each satisfying the required discharge capacity, enabling a suitable choice to be made. Such calculations are best done by computer (see section 21.1).

13.4   Pumps

Pumped drainage is required in low-lying areas from which water is unable to drain by gravity. These typically are flat with poor natural outlets, often with the drainage base level lying below the outlet level (as in the case of polders, see section 10.2). Figure 13.16 shows some typical pumping stations of the Netherlands.

The duty of a pump refers to the work to be performed as determined by the quantity of water to be discharged and the difference between the inlet and delivery levels (the lift, see Figure 13.17 and Figure 13.19). A common feature of the duty of a land drainage pump is the combination of a fairly high discharge (> 0.5 m³/s) with a low lift (< 5 m). The lift may vary due to changes in the outlet level as, for example, in the case of an outlet into a river with a variable regime or an outlet into the sea.

The time during which the pump operates may vary from only a few days annually, to more extended periods of continuous operation (e.g., 1–2 weeks during a cool wet period). The overall efficiency of the pump is important in the latter case while it is relatively unimportant in the former case where a pump only operates for short periods in response to rather infrequent heavy rainfall. Reliability is then of much greater importance, particularly when the site is remote and unattended.

The pump, drive-unit and control works may be sited in various ways to suit local conditions. A fairly standard arrangement is shown in Figure 13.17. The pump itself is normally

(a) small automatic float controlled pump for local waterlevel control, downstream view
(b) small automatic float controlled pump for local waterlevel control, upstream view
(c) upstream view of Mr. G.J.H. Kuijkgemaal (pumping station) on the Linge river of the Betuwe Polder District in the Netherlands, discharging in the river Neder Rijn; note trash or weed rack and cleaning arrangements
(d) downstream view of (c); note water inlet option in centre and two hardwood flap-gate outlets
(e) Archimedean screw pumping station at Kinderdijk, the Netherlands (all photos W.F. Vlotman)

Figure 13.16 Pumping stations

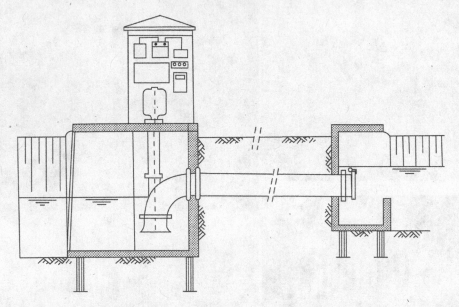

Figure 13.17 Main characteristics and layout of a low head land drainage pumping station (courtesy Bosman, the Netherlands)

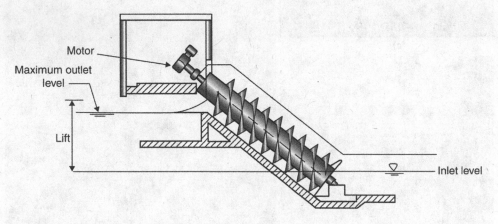

Figure 13.18 Archimedean screw pump

placed in, or close to a sump, in which the drainage water from the area collects by gravity flow. Pumps normally operate in cycles, starting and stopping automatically in response to pre-set water levels in the sump (or in the approach canal).

13.4.1   Types of pumps

Several types of pumps are used for land drainage. The Archimedean screw *pump* consists of an inclined spindle fitted with a surrounding, spirally wound blade, which rotates within a fixed semi-circular casing. Water trapped between the blades is lifted from the lower to the higher level (Figure 13.18). Rotation speeds normally vary between 20 and 120 rpm.

The much more widely used rotodynamic pumps consist of an impeller which rotates within a totally enclosing casing. There are three principal types: *radial flow*, *mixed flow* and *axial flow*, the terms describing the direction of flow through the pump. These three types of pumps have different shaped impellers and casings, each type being suited to a particular range of duties. Radial flow pumps derive most of the pressure head from the impeller energy while with the axial flow pumps this head is mostly built up by the lifting action of the vanes. Radial flow pumps are also commonly referred to as centrifugal pumps and axial pumps as propeller pumps.

13.4.2   Pumping head and characteristics

Pumps raise water from a lower to a higher elevation over a vertical distance termed the *lift (static head*, see Figure 13.19). During pumping additional head is expended as frictional losses in the suction line leading the water into the pump and in the pressure line leading water away from the pump, and as a loss of kinetic energy when the water discharges into the delivery bay. The pump has to be capable of imparting a total head *(dynamic head)* to the water which is equivalent to the sum of the lift plus these losses:

$$H = H_s + h_{fs} + H_d + h_{fd} + \frac{v_d^2}{2g}$$

Eq. 13.11

where:
H = total dynamic head (m)
H_s = suction head (m)
H_d = delivery head (m)
h_{fs}, h_{fd} = frictional losses in the suction and pressure lines (m)
v_d = flow velocity at the outlet of the pressure line (m/s) and
g = gravitational acceleration (m/sec²)

The performance of a pump may be evaluated from *the pump characteristics* which show the relationships between the discharge and respectively the head, the power used and the efficiency (= power expended to the water as a percentage of the total power used by the pump). Typical examples of the characteristics of the Archimedean and the three types of rotodynamic pumps are presented in Figure 13.20.

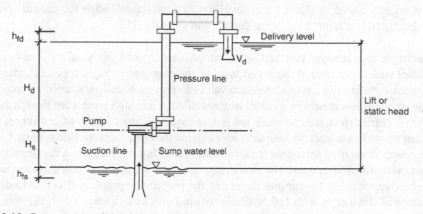

Figure 13.19 Components of the dynamic head

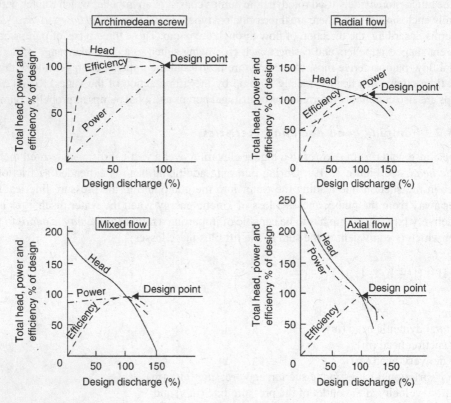

Figure 13.20 Typical pump characteristics of the major types of pumps

Archimedean pump

The head and efficiency characteristics of the Archimedean screw pump are quite different from those of the rotodynamic pumps because the depth of submergence of the screw impeller changes with the water level at the inlet. The discharge increases from zero, when the screw is wholly above the inlet water level, up to a maximum when the central spindle is fully submerged (the latter case being depicted in Figure 13.18).

The discharge then remains constant with further submergence of the screw. The maximum outlet level should not exceed the invert level of the casing at its highest point (higher levels would require the installation of non-return valves to prevent backflow when the pump stops). With the fixed casings generally used all water is lifted to this high level, even though the outer level, for example a river, may be much lower. Consequently, variations in lift are almost wholly dependent upon the variations of the inlet level so that the head characteristic is fairly flat.

The power consumed increases steadily with increasing discharge to a maximum, which coincides with the design duty. The drive unit, designed for this maximum, is thus unlikely ever to be overloaded, a favourable feature of the pump. The pump is also able to deliver a wide range of discharges with rather similar overall efficiency (range 60–75%) thus giving the pump a certain operational flexibility.

Rotodynamic pumps

The characteristics of these types of pumps indicate that the head developed by the pump decreases as the discharge increases, the maximum head coinciding with zero discharge (i.e., pumping against a closed delivery valve). The efficiency increases with increasing discharge up to a maximum (usually of the order of 75–85%), which ideally should coincide with the design point.

The power characteristics of the three types of rotodynamic pump are quite different. In the case of a radial flow pump the power consumed increases steadily with the discharge (the pump is so called self-regulating), whereas in contrast the peak power demand of an axial flow pump occurs at zero discharge. This condition exists when the pump is started and a necessary precaution adopted for this type of pump, to prevent overloading of the motor, is the avoidance of the use of any valve in the pipeline which might inadvertently be left in a permanently closed position (i.e., gate valves). Some provision must, however, be included to prevent reverse flow siphoning back through the pump once it has stopped. This may take the form of either a flap valve placed at the end of the pressure line or, for large installations, a siphon-breaker placed at the highest point in the pressure line.

13.4.3 Pump selection

Pump selection should be based on a careful analysis of the prevailing Q (discharge) and H (head) combinations under which the pump is to operate, in relation to the pump characteristics. A pump may in fact be regulated by means of valves to develop any combination of discharge and head from its unique characteristic, although there is only one combination, which coincides with the peak efficiency. This is obviously an important combination since the corresponding discharge and head represent the optimal pump duty.

In the selection of pumps future discharge requirements should be considered as well as such practical points as facilities for maintenance. Long-term costs, not just the initial capital outlay, should be considered bearing in mind that an initially cheap pump may in the long-term prove to be more expensive due to rapid wear and high costs of maintenance.

The design discharges as determined in Chapter 12 apply to the design of canals and related structures (including outlet sluices). As pointed out above, pump selection should not be based on a single Q value, but rather the expected range and the frequency of occurrence should also be considered. Where the expected discharge is fairly uniform as, for example, is usually the case with subsurface drainage, the discharge used for the canal system, would also apply to pump design. Accordingly, a pump should be selected which has its peak efficiency at this discharge (and the corresponding, required lift). When the discharge is composed of subsurface drainage and shallow drainage water, peak discharges will be high but of short duration. The pump should of course be able to cope with these high discharges (unless storage is provided, see section 12.7), although it need not operate at its highest efficiency. Instead, the peak efficiency of operation should normally be based on a lower, more frequently occurring discharge.

Number of units

The number of pump units to be installed in a station should also be based on a discharge analysis. More units allow more flexibility of operation although one full capacity pump is usually cheaper to install and to operate than two half capacity pumps. If the discharge generally ranges between Q and 2Q (= maximum), two pumps, each with a capacity of Q, would be suitable. If standby capacity is required, three units with a capacity of Q each could be used, thus permitting one unit to be non-operational, e.g., for overhaul, while preserving the required maximum capacity.

Type of pump

The Archimedean type of pump is particularly suitable for situations where the outlet level is approximately constant, and the converse follows that it is unsuitable in situations where this level varies considerably. It can raise discharges ranging from 0.05 m³/s up to 6 m³/s through lifts of 2–10 m but is most commonly used in small to medium sized pumping stations with lifts in the range 2–4 m. It is sturdy and is able to raise weed and debris (a weed screen, or trash rack, is not needed), which makes it suitable for remote unattended stations.

A rotodynamic pump should, other things being equal, be selected primarily on the basis of its suitability for the design duty. Radial flow pumps should preferably be used in situations with low Q and high H, while axial flow pumps are to be preferred in situations with high Q and low H. For the typical duty of a land drainage pump, consisting of high discharge (Q > 0.5 m³/s) and a low head (H < 5 m), the axial flow pump is well suited. This type of pump is also attractive on the basis of its capital costs, for the pump itself is compact and the small propeller shaped impeller operates at high speed, lending itself to the use of relatively cheap electric motors. This high speed, however, increases the risk of cavitation (section 13.4.4). These and other considerations have led to selection charts such as in Figure 13.21. Particular conditions may prevail justifying deviation from the general guidelines laid down in this chart.

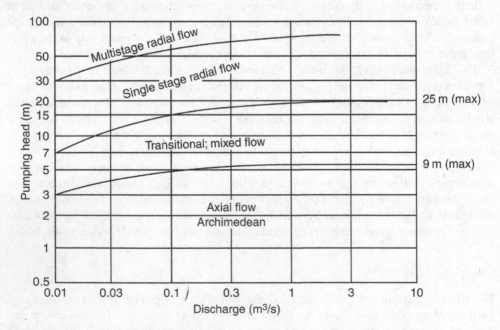

Figure 13.21 Selection chart for pumps

13.4.4 Sump and intake design

The design of the sump depends on the type of installation and the number of pumps. Generally, the final 100–200 m of the main canal should be straight and have no inflow from the side, to make the approach flow uniform and undisturbed. The sump and the positioning of the inlet pipe to a rotodynamic pump should be such as to prevent the formation of air entraining vortices (which reduce the pumps efficiency) and to exclude weed and debris (which might clog the impellers). The height of the bell mouth inlet above the floor and the minimum depth of the water above this inlet are also critical factors, significantly affecting the pumping efficiency.

Rotodynamic pumps may be situated either above or below the water level in the sump. In the former situation, the water pressure in that section of the suction pipeline above the inlet water level is sub-atmospheric. This partial vacuum situation may lead to separation of the water column (so that the pump fails to draw water) or to boiling of the water causing inefficient operation and cavitation (i.e., vapour bubbles collapsing with violent force near the tips of the impellers, causing damage to the blades, noise and vibration). This imposes a limitation on the maximum height between the pump and the lowest inlet water level in the sump, a limitation, which varies with the type of pump. The speed of operation of the pump also has an influence and axial flow pumps which operate at a higher speed than radial flow pumps, are more susceptible to cavitation and therefore have a lower suction capability than the latter (respectively < 1.5 m compared to » 7.5 m). Axial flow pumps are often installed permanently submerged which has the added advantage that the pump is primed at all times (pump must be full of water to start).

13.4.5 Power and cost calculations

The power requirements (E) of a pump may be calculated as:

$$E = \frac{9.8Q \times \rho \times H \times e_1 \times e_2}{\eta_t \times \eta_p}$$

Eq. 13.12

where:
E = power consumption (kW)
Q = discharge (m³/s)
H = total dynamic head (m)
ρ = density of water (kg/l) [normally 1.0 but higher for salty/dirty water]
e_1 = site correction factor [1.0 at MSL and 15°C, but increasing by 1% for each 100 m rise in elevation and by another 1% for each 5°C rise in temperature]
e_2 = safe load factor [takes into account that the engine should not continuously run fully loaded, normally e_2 = 1.3–1.4 meaning engine 70–75% loaded]
η_t = transmission efficiency [normally of the order of 0.90 to 0.95]
η_p = pump efficiency [for axial and mixed flow pumps η_p = 0.65 for H~1.0 m, increasing to η_p = 0.80 for H = 2.5–3.0 m; for radial flow pumps η_p = 0.60 [for H = 1.0 m increasing to η_p = 0.80–0.85 for H > 4.0 m].

Annual cost calculations are based on:

a) *Capital costs*: depend on interest and operational lifetime (ranging from 10 000–50 000 running hours or 4–25 years; on average take 10 years depreciation period)
b) *Energy* costs: for electric power: E × N × P_1

for fuel: E × N × P_2 × U

where:

E = power consumption as calculated by Eq. 13.12 (e_2=1.0)

N = running hours per annum

P_1 = price per kWh

P_2 = price per litre fuel

U = fuel consumption per kWh (for gasoline 0.50–0.80 l/kWh, diesel 0.30–0.65 l/kWh, natural gas 0.65 m^3/kWh)

c) *Annual maintenance and repairs*: as a rough approximation taken as 20–40% of the annual energy costs (20% for gasoline engines, 40% for diesel engines); alternatively estimated as 5% of the capital costs;

d) *Other costs* like personnel for operation, insurance, etc.

For rough and ready estimates: E = 13 × Q × H (kW); energy costs = E × N × P, as in b) above; total annual pump costs = 2 × energy costs.

13.4.6   Example cost calculations of an electrically driven pump

A pump station is required to drain 650 ha of flat arable land in the west of England. The pump station is to be capable of evacuating the total rainfall occurring over a 5 day consecutive period and at a return period of 10 years which is 60 mm. The average winter excess rainfall of 190 mm occurs on average over a 145 day time period. The highest average monthly excess rainfall of 93 mm occurs in December. The water has to be raised to an embanked river which is approximately 4 m above the normal level in the channel leading to the pump station.

Discharge requirements

Design peak discharge: $Q = \dfrac{60}{1000 \times 5} \times 650 \times 10000 \times \dfrac{1}{3600 \times 24} = 0.90 \, m^3/s$

Average flow in winter: $Q = \dfrac{190}{1000 \times 145} \times 650 \times 10000 \times \dfrac{1}{3600 \times 24} = 0.10 \, m^3/s$

High flow in December: $Q = \dfrac{93}{1000 \times 31} \times 650 \times 10000 \times \dfrac{1}{3600 \times 24} = 0.225 \, m^3/s$

The numbers show that the pump station capacity (0.90 m^3/s for 650 ha =12 mm/d) is much greater than the normal discharge (3 mm/d in December, 1.3 mm/d over the whole winter). Analysis of the rainfall records indicates that the rainfall rate over durations shorter than five days but at the same frequency of occurrence i.e., 10 years, exceeds the actual design rate, e.g., the 1-day rainfall event at T_p = 10 years is 30 mm. As the pump station is unable to evacuate more than 12 mm/d, some localised flooding may occur.

In this example the range of variation of the discharge is such that two similar pumps will be chosen although two pumps with capacities of 0.3 m^3/s and 0.6 m^3/s could also be considered. The selection chart, Figure 13.21, indicates that the axial flow pump will be suitable for the required duties.

Power requirements and cost estimates

According to Eq. 13.12, section 13.4.5:

$$E = \frac{9.8Q \times \rho \times H \times e_1 \times e_2}{\eta_t \times \eta_p} = \frac{9.8 \times 0.45 \times 1 \times 4.4 \times 1 \times 1}{0.9 \times 0.8} = 27 \, kW \text{ per pump}$$

Annual running time = $\frac{190 \, mm}{6 \, mm} \cong 32$ days or 760 hours; annual running costs based upon 1 000 hours operation at 27 kW; =>27 × 1000 × 0.06 = £ 1 620/annum (electricity £ 0.06 / kWh; early nineteen eighties; replace by current costs). The comparative running cost of diesel assuming this to be used at 0.4 litres/kWh and at a cost of £ 0.25 /litre is: £ 2 700/ annum. In general diesel units are considered more economic for the larger pump stations (> 150 kW, or > 3–4 m³/s) although much depends on the availability and costs of electrical power supply.

Notes

1 High water levels in the collector ditches raise the field drainage base. In the case of pipe drains, the water will rise above the pipe up into the trench. The pipes will continue to discharge, although at a reduced rate, as long as the watertable in the land is above the water level in the collector ditch.
2 The formula takes into account that the length of the lunar day exceeds that of the calendar day by some 50 mins.

Part V

Salinity control

Chapter 14

Soil salinity

All soils contain salts as these remain when rocks weather into soil. During the weathering process, most excess salts are carried downwards *(leached)* with the percolating (rain) water. They may precipitate at lower depths or continue to be transported in solution, ending up, ultimately in the sea or in a terminal lake.

Under normal conditions, the salinity in the upper soil layers is quite moderate and the occurrence of high soil salinity *(salty soils)* is the exception rather than the rule. When the high soil salinity is directly related to the soil's parent material and its formation, it is referred to as *primary* or *residual salinity*. The salinity of marine soils is a special case of residual salinity. Marine deposits may remain saline from past geological periods up to the present time in situations where there is very little leaching (arid climates, poor drainage). The most common cause of high soil salinity in agricultural land, however, is *salinisation*, i.e., the accumulation of salts in the upper layers of the soil. Frequently, salinisation involves a reversal of the leaching process i.e., the return of the leached salt (therefore often termed *secondary salinity*). A great deal of contemporary salinisation is caused by man's activities, especially by irrigation development (Chapter 15), but may originate from different circumstances as well (Box 14.1). Atmospheric fall-out may be a significant source of salt in coastal land and near deserts (annual salt loads of up to 100–200 kg per ha have been reported).

The occurrence of high soil salinity, although not restricted to hot, dry climates, is much more prevalent under these conditions than in the temperate, humid climates. In temperate climates there is usually sufficient excess water percolating downwards through the soil to maintain low salt levels in the upper layers. Soil and drainage conditions, however, also have a great influence for they largely determine the physical possibilities for leaching and removal of the excess salts from the land.

14.1 Forms of occurrence and distribution of salts in the soil

Salts occur in the soil in one of the following three forms:

- dissolved salts: salt ions dissolved in the soil water (the soil solution)
- adsorbed salts: salt cations adsorbed on the negatively charged surfaces of the soil particles (adsorption complex)
- precipitated salts.

Box 14.1   Dryland salinity in the Murray-Darling Basin, Australia

Salt is a natural part of the Australian landscape. Over geological time the Murray-Darling Basin has been a natural salt trap. The clearing of native vegetation and its replacement with annual crops and pastures, irrigated agriculture, town gardens and lawns, has unleashed a hydrological disequilibrium that brings this vast salt store to the land surface and increases its seepage to river systems (MDBC, 2001). Two indices reflect the state of matters in the basin: the salinity on land and in the rivers and the position of the (ground) watertable with respect to the land surface.

Using the 1995 rates at which groundwater levels are rising, most of the irrigation areas in the southern region of the Basin will have watertables within two metres of the surface by the year 2020. Where the sub-soils contain large volumes of salt, salinised water begins to affect vegetation when watertables rise to within two metres of the surface. At this stage roots can extract water from the watertable but will leave the salts in the root zone. Water and salts can than reach the surface through capillary action, where water evaporates and leaves the salts on the land surface. This happens both in irrigated regions and in dryland areas. Rising groundwatertables, and sub-soils with a high salt content, together create dryland salinity. In 1995 approximately 500 000 hectares of agricultural (non-irrigated) land were affected. The underlying cause of dryland salinity is the increased volume of rainwater being added to the watertable in the uplands of the Basin. This has been caused by the large-scale clearing of trees and the introduction of agricultural practices such as shallow rooted crops and pasture, which take up less rainfall than the native vegetation cover. Dryland salinity also affects urban areas by corroding foundations of buildings etc. In addition to areas affected by dryland salinity, approximately 333 000 hectares of irrigated, agricultural land are affected by salinity.

As shown in Figure 14.1, dynamic exchange equilibrium exists between the cations in the soil solution and those adsorbed on the complex, and also between dissolved and precipitated salts.

The salt composition in the soil is generally a reflection of the salt composition at its source of origin (parent-rock material, groundwater, sea water, etc.) but various dilution-concentration, leaching-precipitation and desalinisation-re-salinisation cycles may have resulted in considerable changes in the original composition.

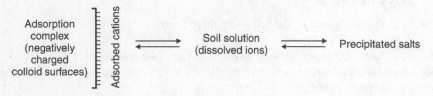

Figure 14.1 Forms of occurrence of salts in the soil

14.1.1   The soil solution

The main salt ions found in the soil solution of a salty soil are (see also Table 14.3):

- cations: Na^+ Ca^{++} Mg^{++} (K^+)
- anions: Cl^- SO_4^{--} HCO_3^- (CO_3^-) (NO_3^-)

A characteristic feature of many salty soils is the high concentration of Na. This contrasts with normal (non-salty) soils in which < 5% of the soluble cations normally consist of Na, while Ca or Ca+Mg are by far the most dominant cations in the soil solution. In many salty soils the concentration of Na exceeds that of Ca and Mg. The concentration of K is usually only slightly higher in salty soils when compared to non-salty soils, remaining low in both cases.

On the anion side, there is often a predominance of Cl and SO_4 in salty soils. At normal pH values (pH 6–8), carbonates are only present in the form of bicarbonates (HCO_3). At higher pH values HCO_3 changes into CO_3 although up to pH ~ 8.5 the quantity of CO_3 in most soil solutions remains quite insignificant.

The relatively high salt concentrations found in the soil solutions of salty soils frequently exceed the solubility limits of the poorly soluble alkaline earth carbonates ($CaCO_3$ and $MgCO_3$) and also that of gypsum ($CaSO_4.2H_2O$). In these circumstances the salts precipitate, withdrawing Ca, Mg and SO_4 ions from solution[1]. Sodium salts and chlorides are generally highly soluble hence the preponderance of Na and Cl in many salty soils. However, this high solubility ensures that their concentration also falls rapidly upon leaching of the soil. When the less soluble salts are leached, their concentrations in the soil solution fall more slowly as previously precipitated salts come into solution and replace the leached salts. Therefore, in a soil exposed to cyclic salinisation-desalinisation, the presence of Na and Cl in the soil solutions is often of the same order or even less than that of Ca, Mg and SO_4.

14.1.2   Adsorbed cations

Soils have the capacity to adsorb cations. This capacity is vested in the negatively charged surfaces of the soil colloids which attract cations from the soil solution. It is quantified by the so-called *cation exchange capacity* of the soil (CEC, expressed in meq/100 g dry soil).

A dynamic exchange equilibrium exists between the adsorbed cations and the cations in the soil solution. As the concentration of cation 'b' in the soil solution increases relative to cation 'a', the cation composition on the adsorption complex will adjust itself by exchanging adsorbed 'a' cations for 'b' cations from the soil solution until a new equilibrium becomes established (Figure 14.2).

This cation exchange process is complicated by the fact that divalent cations are adsorbed more readily than monovalent cations, whilst in addition there is also a difference in adsorption preference between cations having the same valence (see Box 14.2). Roughly, the order of adsorption preference is:

$$Ca^{++} > Mg^{++} > H^+ > K^+ > Na^+$$

Therefore, as long as Ca is the dominant cation in the soil solution, as is generally the case in non-salty soils, little Na will be adsorbed since it is very difficult for Na to displace

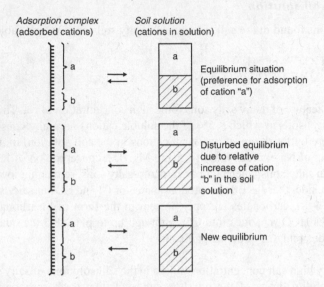

Figure 14.2 Principles of the cation exchange process

Ca on the complex. This is fortunate because a high Ca occupancy on the complex is much more favourable than a high Na occupancy (Section 14.2.3). A significant Na occupancy is only to be expected when more than half of the cations in the soil solution consist of Na (Figure 14.3).

Box 14.2   Selected atomic weights and their valences

Salt concentrations are normally expressed in mg/l (or in ppm, its numerical equivalent). They may also be expressed in meq/l. The equivalent weight of an element or compound is the weight (in g) of that element or compound that will combine with or displace one g of hydrogen. Numerically it is determined by dividing the atomic weight of the element or the molecular weight of the compound by the relevant valence. The equivalent weight of, for example, $CaCO_3$ is: Ca (40) + C (12) + 3 O (3 × 16) = 100 (molecular weight), divided by the valence (in this case the valence of Ca or CO_3, both 2), giving an equivalent weight 50 g (meq weight = 50 mg).

	Weight	Valence		Weight	Valence
Aluminium (Al)	27	2 or 3	Magnesium (Mg)	24	2
Boron (B)	11	3	Nitrogen (N)	14	3 or 5
Calcium (Ca)	40	2	Oxygen (O)	16	2
Carbon (C)	12	4	Potassium (K)	39	1
Chlorine (Cl)	35	1	Sodium (Na)	23	1
Hydrogen (H)	1	1	Sulphur (S)	32	2, 4 or 6
Iron (Fe)	56	2 or 3			

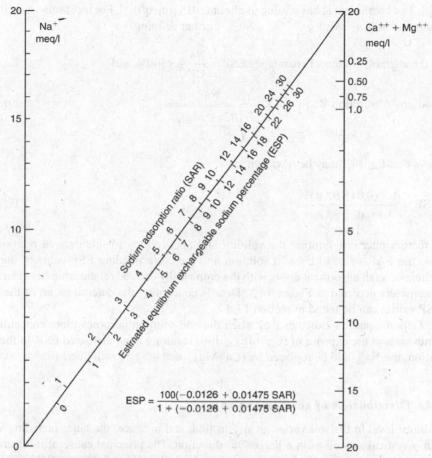

Use of the graph: plot the Na-value on the vertical left-hand scale and the (Ca + Mg)-value on the vertical right-hand scale; connect these two points and read the SARvalue at the point where the connecting line crosses the diagonal scale. The corresponding ESP-value is read on the opposite diagonal scale. For example: Na = 12 meq/l, Ca = 3 meq/l, Mg=2 meq/l, read SAR = 7.5 and ESP = 9. The shown correlation between ESP and SAR actually holds only for the SAR$_e$ values in the extract from a saturated paste (see Figure 14.6)

Figure 14.3 Experimental SAR-ESP relationships (USDA1954)

14.1.3 Equilibrium relationships

The equilibrium relationships between the adsorbed cations and those in the soil solution are described by the *Gapon equation* (Tanji 1990). For salty soils containing mainly Ca, Mg and Na cations, this equation reads:

$$\frac{Na^+_{ads}}{(Ca+Mg)^{++}_{ads}} = K \frac{Na^+_{sol}}{\sqrt{\dfrac{(Ca+Mg)^{++}_{sol}}{2}}}$$

Eq. 14.1

The equation shown in the figure reads:

$$ESP = \frac{100(-0.0126 + 0.01475\ SAR)}{1 + (-0.0126 + 0.01475\ SAR)}$$

with the adsorbed (ads) cations expressed in meq/100 g dry soil and the soluble (sol) cations in meq/l. The coefficient K has a value of about 0.015 [(meq/l)$^{-\frac{1}{2}}$]. For most salty soils it may be assumed that $(Ca + Mg)_{ads} = CEC - Na_{ads}$. Further defining:

$$\text{Exchangeable Sodium Percentage}: ESP = \frac{Na^+_{ads}}{CEC} \times 100\%, \text{and} \qquad \text{Eq. 14.2}$$

$$\text{Sodium Absorption Ratio}: SAR = \frac{Na^+_{sol}}{\sqrt{\dfrac{(Ca+Mg)^{++}_{sol}}{2}}} \qquad \text{Eq. 14.3}$$

it follows that Eq. 14.1 may be rewritten as:

$$ESP \approx \frac{100(0.015\ SAR)}{1 + 0.015\ SAR} \qquad \text{Eq. 14.4}$$

Many factors interfere, limiting the validity and applicability of this derived relationship between the SAR-value of the soil solution and the corresponding ESP-value of the soil. Nevertheless, a fair agreement exists with the empirically derived relationship between these two parameters depicted in Figure 14.3. Details concerning the determination of the SAR and ESP values can be found in section 14.4.2.

The Gapon equation indicates that when the soil solution becomes more concentrated, Na_{ads} increases at the expense of $(Ca+Mg)_{ads}$, thus resulting in an increased ESP. In the case of dilution, the Na_{ads} will be replaced by $(Ca+Mg)_{ads}$ and the ESP will decrease.

14.1.4   Distribution of salts in the soil

The salinity level in the soil varies greatly in time and in space, the latter including variations in a vertical as well as in a horizontal direction. The principal cause of this variation is the movement of salts with the movement of the soil water and because field soil water regimes are highly dynamic, the salt distribution in soil is also highly dynamic. The upward and downward flow of water in the profile frequently changes the vertical distribution of salts. Following rain (or irrigation) the salt content in the upper layers may be low while the reverse situation is likely to be found at the end of a dry period. At greater depths where moisture conditions change less, the salt content is more constant with only slight seasonal variations.

An example of horizontal variation in soil salinity is presented in Figure 14.4, which depicts the salt distribution in a cross-section through two ridges in furrow irrigated land, some days after it has been watered. The highest salinity is found in the top of the ridges which are most exposed to evaporation, towards which the water (and thus the salts) from the surrounding soil are drawn. Variation in soil salinity is the rule rather the exception. Changes in soil hydrological regime, soil variations, root distribution, micro relief all contribute to this variation. Any soil salinity assessment should take this into account. The related sampling problems are described in section 14.4.1.

14.2   Agricultural impacts; diagnosis and assessment

The problems caused by soil salinity may be classified into three types with each type being related to a particular aspect of soil salinity:

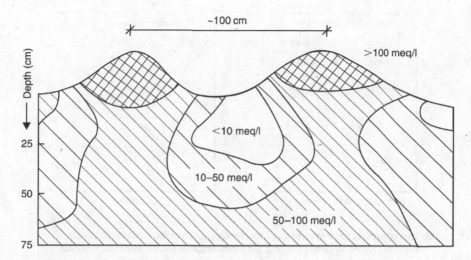

Figure 14.4 Examples of the distribution of salts in the soil

(a) *Osmotic problems:* high total salt concentration of the soil solution which raises the osmotic pressure of the soil water and makes it more difficult for the plant roots to extract water from the soil

(b) *Toxicity problems:* high concentration in the soil solution of some specific ions or an imbalance between two or more ions that harms plant growth

(c) *Dispersion problems:* relatively high occupancy of the soil's exchange complex by Na which allows for easy dispersion of the colloids in the soil and poor soil structure.

A fourth problem i.e., the chemical aggressiveness of some salts with respect to some construction materials is described briefly in section 14.2.4

14.2.1   Osmotic problems

Salts dissolved in water exert binding forces on the water, these being termed osmotic forces. A plant extracting water from a salty soil must overcome these forces, in addition to the normal soil moisture retention forces (capillary forces etc., see section 6.2). Not only must the plant apply greater suction forces (or otherwise accept lower uptake rates) but also, a larger proportion of the soil moisture will be held by strong forces and thus be less readily available or even unavailable to the plants. At high soil salinity, crops show early signs of moisture stress, which is one of the principal causes of the poor growth of crops in salty soils.

Measurement

The osmotic forces increase in direct proportion to the total salt concentration of the soil solution for the range of salt mixtures normally found in salty soils. The total salt concentration (TDS = total dissolved salts) may thus be used to indirectly assess the osmotic forces and problems to be expected. The TDS should preferably be measured under the conditions under which the plant takes up most of its water from the soil i.e., in the moisture range from field capacity (FC) to wilting point (WP). Since it is difficult to extract a soil solution sample from the soil in this moisture range, the salt concentration is determined at a standardised higher moisture content, the *saturation point* (SP), see Box 14.3.

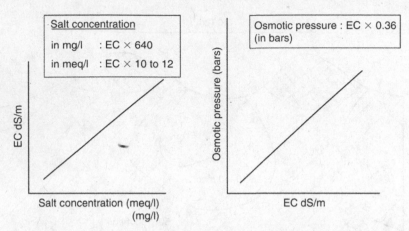

NB. The given relationships are valid for the range of salt mixtures and concentrations normally found in salty soils and are most accurate for the saturation extract.

Figure 14.5 Linear relationships between total soluble salt concentration, electrical conductivity and osmotic pressure

Soil salinity is commonly expressed by the *electrical conductivity* (EC-value) of the soil solution. For the salt mixtures normally found in salty soils, the EC-value and the TDS are linearly related and since the osmotic forces increase linearly with the salt concentration, a linear relationship also exists between the osmotic forces in the soil solution and its EC-value (USDA 1954, see Figure 14.4). The SI unit of electrical conductivity is the S/m (Siemens per meter). Soil and water salinity are mostly expressed in dS/m (deci-Siemens per meter) as this smaller unit gives values which are numerically equal to the values expressed in the formerly used mmhos/cm units. The EC measurement is temperature sensitive and is therefore standardised to 25°C.

The EC-value of the soil solution at saturation point is designated as the EC_e-value. It has proved to be a suitable diagnostic parameter for the osmotic problems caused by high soil salinity. The procedure and main features of the determination of the EC_e-value of the soil are detailed in Figure 14.6.

The agricultural interpretation of EC_e-values

Good correlations between the EC_e-values and yields have been established for many crops. The following general criteria apply (Schofield scale, USDA 1954):

EC_e = 0–2 dS/m : negligible effect on all crops
  2–4 dS/m : slight effect on sensitive crops
  4–8 dS/m : significant effect on many common crops
  8–16 dS/m : only salt tolerant crops can be grown
  >16 dS/m : only highly resistant plants can survive.

Tolerance tables compiled by the *US Salinity Laboratory* (Table 14.1) give for various crops the EC_e-value at which yield reductions of respectively 0, 10, 25 and 50% are to be expected. This table shows that the crops vary in their salt tolerance. Beans, for example, are highly sensitive with yield reductions to be expected for EC_e < 1.0 dS/m while barley is able to tolerate EC_e=8 dS/m without yields being affected. Tolerance may also vary between different varieties of the same plant. Conversion from dS/m to other units is shown in Box 14.4.

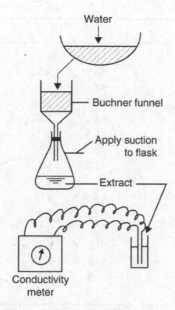

(1) Make a saturated paste by adding water to the soil while stirring, until the saturation point is reached, i.e, the point at which the soil surface glistens although there is still no free water.

(2) Extract a soil solution sample (approx 10 cc) from the paste using suction equipment (this extract is referred as the saturation extract).

(3) Measure the EC-value of the saturation extract using a conductivity meter.

Figure 14.6 Outline of the EC_e measurement

For most crops the yield response to soil salinity can quite adequately be described by a broken-linear model as depicted in Figure 14.7. By fitting the yield reduction data of Table 14.1 to this model, these data have been translated into values for the broken-linear model parameters i.e., the parameters T = threshold value and α = percentage yield reduction (Tanji 1990, FAO 2002).

Box 14.3   Soil moisture and salinity relationships at the saturation point

The saturation point (SP) refers to a puddled soil and should not be confused with the moisture content of a saturated undisturbed soil. Sandy soils generally hold less water at SP than in saturated undisturbed conditions, while the opposite generally applies to clay soils. For medium and medium-fine textured soils it roughly holds that (θ = moisture content in %w [by weight]):

$$\left. \begin{array}{l} \theta \text{ at SP} \approx 2 \times \theta \text{ at FC} \\ \theta \text{ at SP} \approx 4 \times \theta \text{ at WP} \end{array} \right\} \text{ so } \theta \text{ at SP} \approx 3 \times \text{ average } \theta \text{ in the available moisture (AM) range}$$

As the salt concentration of the soil solution is inversely proportional to the soil moisture content, it follows that for medium textured soils, the soil solution in the available moisture (AM) range is on average 3 × as concentrated than at Saturation Point, SP, meaning that the EC-value faced by the plants is *about* 3 × EC_e. For sandy soils, a somewhat higher concentration ratio applies; for clays it would normally be slightly lower. The inverse proportionality between salt concentration and soil moisture content of course applies only when all salts remain dissolved. In the presence of large quantities of only slightly soluble salts (lime, gypsum), somewhat lower concentration ratios apply due to precipitation at high concentrations.

Table 14.1 Yield reduction at different EC$_e$ (dS/m) for various crops (FAO 1994)

	Yield reduction: 0%	10% EC$_e$ value	25%	50%
FIELD CROPS				
Barley (Hordeum vulgare)	<8	10	13	18
Cotton (Gossypium hirsutum)	<7.5	9.5	13	17
Sugarbeet (Beta vulgaris)	<7	8.5	11	15
Sorghum (Sorghum bicolor)	<4.7	7.5	8.5	10
Wheat (Triticum aestivum)	<6	7.5	9.5	13
Safflower (Carthamus tinctorius)	<5.5	6	7.5	10
Soybean (Glycine max)	<5	5.5	6.5	7.5
Cowpea (Vigna sinensis)	<5.0	5.5	7	9
Groundnut (Arachis hypogaea)	<3	3.5	4	5
Rice (paddy) (Oryza sativa)	<3	4	5	7
Sesbania (Sesbania macrocarpa)	<2.5	3.5	6	9.5
Sugar cane (Saccharum officinarum)	<2	3.5	6	10
Corn (Zea mays)	<1.5	2.5	4	6
Flax (Linum usitatissimum)	<1.5	2.5	4	6
Broadbean (Vicia faba)	<1.5	2.5	4	7
Beans (Phaseolus vulgaris)	<1.0	1.5	2.5	3.5
FRUIT CROPS				
Date palm (Phoenix dactylifera)	<4	7	11	18
Fig (Ficus carica) Olive (Olea-europea) Pomegranate (Punica granatum)	<2.5	4	5.5	8.5
Grapefruit (Citrus paradisi)	<2	2.5	3.5	5
Orange (Citrus sinensis)	<1.5	2.5	3	5
Lemon (Citrus limonea)	<1.5	2.5	3.5	5
Apple (Pyrus malus) Pear (Pyrus communis)	<1.5	2.5	3.5	5
Walnut (Juglans regia)	<1.5	2.5	3.5	5
Peach (Prunus persica)	<1.5	2	3	4
Apricot (Pyrus armeniaca)	<1.5	2	2.5	3.5
Grape (Vitis spp.)	<1.5	2.5	4	6.5
Almond (Prunus amygdalus)	<1.5	2	3	4
Plum (Prunus domestica)	<1.5	2	3	4.5
Blackberry (Rubus spp.)	<1.5	2	2.5	4
Boysenberry (Rubus spp.)	<1.5	2	2.5	4
Avocado (Persea americana)	<1.5	2	2.5	3.5
Raspberry (Rubus idaecus)	<1.0	1.5	2	3
Strawberry (Fragaria spp.)	<1.0	1.5	2	2.5
VEGETABLE CROPS				
Squash/Courgette (Cucurbitapepo melopepo)	<4.5	6	7.5	10
Beets (Beta vulgaris)	<4	5	7	9.5
Broccoli (Brassica italica)	<3	4	5.5	8
Tomato (Lycopersicon esculentum)	<2.5	3.5	5	7.5
Cucumber (Cucumis sativus)	<2.5	3.5	4.5	6.5
Cantaloupe (Cucumis melo)	<2	3.5	5.5	9
Spinach (Spinacia oleracea)	<2	3.5	5.5	8.5
Celery (Apium graveolens)	<2	3.5	6.0	10

Yield reduction:	0%	10%	25%	50%
		EC_e value		

VEGETABLE CROPS

	0%	10%	25%	50%
Cabbage (Brassica oleracea capitata)	<2	3	4.5	7
Potato (Solanum tuberosum)	<1.5	2.5	4	6
Sweet corn (Zea mays)	<1.5	2.5	4	6
Sweet potato (Ipomea batatas)	<1.5	2.5	4	6
Pepper (Capsicum frutescens)	<1.5	2	3.5	5
Lettuce (Lactuca sativa)	<1.5	2	3	5
Turnip (Brassica rapa)	<1	2	3.5	6.5
Radish (Raphanus salivas)	<1	2	3	5
Onion (Alliutn cepa)	<1	2	3	4.5
Carrot (Daucus carota)	<1	1.5	3	4.5
Beans (Phaseolus vulgaris)	<1	1.5	2.5	3.5

FORAGE CROPS

	0%	10%	25%	50%
Wheat grass (Agropyron elongatum)	<7.5	10	13	19
Wheat grass (Agropyron cristalam)	<7.5	9	11	15
Bermuda grass (Cynodon dactylon)	<7	8.5	11	15
Barley (hay) (Hordeum vulgare)	<6	7.5	9.5	13
Perennial rye grass (Lolium perenne)	<5.5	7	9	12
Trefoil, birdsfoot narrow leaf (L. corniculatus tenuifolius)	<5	6	7.5	10
Harding grass (Phalaris tuberosa)	<4.5	6	8	12
Tall fescue (Festuca elatior)	<4	5.5	8	12
Crested Wheat grass (Agropyron desertorum)	<3.5	6	10	16
Vetch (Vicia sativa)	<3	4	5.5	7.5
Sudan grass (Sorghum sudanense)	<3	5	8.5	14
Sesbania (Destania exaltata)	<2.5	4	6	9.5
Wildrye, beardless (Elymus triticoides)	<2.5	4.5	7	11
Trefoil, big (Lotus uliginosis)	<2.5	3	3.5	5
Alfalfa (Medicago sativa)	<2	3.5	5.5	9
Lovegrass (Eragrostis spp.)	<2	3	5	8
Corn (forage) (Zea mays)	<2	3	5	8.5
Clover, berseem (Trifolium alexandrinum)	<1.5	3	6	10
Orchard grass (Dactylis glomerata)	<1.5	3	5.5	9.5
Meadow foxtail (Alopecurus pratensis)	<1.5	2.5	4	6.5
Clover, alsike, ladino, red, strawberry (Trifolium spp.)	<1.5	2.5	3.5	5.5

Note: these data should be applied with caution, particularly keeping in mind that:

a) the data are based on results obtained in small plots, artificially salinised by irrigation water

b) the EC_e values indicated are average values over the active rootzone, although the salinity is normally somewhat higher in the upper than in the lower rootzone

c) the data apply to the growth period from seedling stage to early maturity. Generally, crops are slightly more sensitive during germination. Part of the salinity problems commonly experienced during germination are due to the fact that seeds are placed where soil salinity is highest (e.g., top of ridges) rather than due to a higher sensitivity to salinity during germination

d) the data best applies to medium to medium-fine textured soils. In light textured soils the soil solution between FC and WP is more concentrated than in the medium textured reference soils (see Box 14.3) and so slightly lower EC_e values should be adopted when crops are to be grown in light soils. In heavy clay soils the reverse argument applies and so slightly higher values may be adopted

e) the data apply when highly soluble salts dominate the soil solution. Somewhat higher EC_e values may be adopted when only slightly soluble salts are present in large quantities (> 20–30 meq/l) or when the $CO_3^{2-}+HCO_3^-$ concentration of the soil solution is high

f) salt concentrations around the roots will generally be higher, especially at high evapotranspiration rates (ET). Therefore, it is prudent to take somewhat lower EC_e values when the evapotranspiration is high.

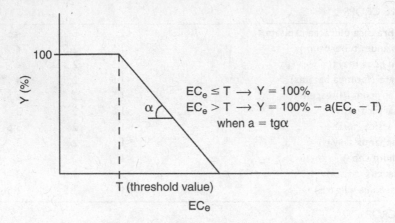

Figure 14.7 Broken-linear soil salinity response model (see also Figure 1.7)

Salinity conditions vary spatially within the rootzone as well as in time. Plant response reflects the integration of these variable conditions experienced by the plant. As the root activity and water uptake generally decrease with depth, conditions in the upper rootzone may normally be expected to carry more weight than conditions in the lower rootzone. Which rootzone salinity parameter best relates to crop response is still the subject of further research.

Box 14.4   Average conversion factors of EC_e to different units (after van Hoorn and van Alphen, 1994)

Different countries use different unit to express salinity unit. The following are conversion factors that can be used:

dS/m	meq/l	mg/l	mg/meq	(meq/l)/(dS/m)
1	10	640	64	10
10	120	7000	58.3	12

NB in Australia 1 EC= 1 μS/cm or 100 EC = 1 dS/m
See also Table 14.1, Figure 14.5 and Table 15.1.

14.2.2   Toxicity problems

Whereas osmotic problems are caused by the high total salt concentration of the soil solution, the cause of the toxicity problems appears to be either a high concentration of a particular cation or anion (see also Table 1.5 for indicator values of selected water quality parameters), or an unfavourable salt composition in the soil solution resulting in an excess or unbalanced uptake by the plants.

Both soil and foliar analysis are used in the diagnosis of toxicity problems. These tests are often most conclusive if used in combination. Diagnostic parameters and criteria have been established for some forms of toxicity, notably toxicity resulting from excess uptake, but the salinity conditions underlying other forms of toxicity remain obscure. The occurrence of toxicity is often linked with a high salt concentration in the soil solution, thus occurring concurrently with osmotic effects. Part of the toxicity problems, therefore, is assessed jointly, although usually not explicitly, with the osmotic problems by the EC_e parameter, and the tolerance tables (Table 14.1) reflecting a combination of toxicity and osmotic problems (although dominantly the latter).

Most serious toxicity problems are related to excess uptake of sodium, chloride and boron. Excess uptake leads to accumulation, especially in the leaves (concentration effect due to transpiration). Typical symptoms are leaf burn, scorch, dead outer leaf edges, especially occurring in the older leaves where the accumulation first reaches critical levels. With time the symptoms may progress between the veins towards the leaf centre. Fruit trees and other woody perennials are particularly sensitive to these problems while most annual crops are rather tolerant.

Sodium: sodium toxicity may be expected when the Na concentration of the soil solution is high. The Ca concentration, however, is also important as the impact often reduces in the presence of soluble Ca in the soil. This is the reason why the SAR and ESP values are relevant for diagnosing Na-toxicity since these parameters reflect the Na/Ca ratio of the soil solution. Sensitive crops like deciduous fruits, nuts, citrus, avocado and beans are unable to tolerate SAR/ESP values in excess of 10–20. Most common crops are more tolerant but will suffer from the poor soil structure when ESP > 20% (section 14.2.3). Analysis of leaves from standing crops can help in the diagnosis of the problem. The Na content of leaves from affected and unaffected plants may be compared. Sensitive crops usually show symptoms when the Na content of the leaf tissue is in excess of 0.25–0.50% (dry weight basis).

Chlorides: excess uptake of Cl affects sensitive crops like various kinds of berries, many fruit trees, some orange and grape species. Damage may be expected when the Cl concentration in the saturation extract exceeds 10 meq/l or when the leaves contain > 0.5–1.0% Cl (dry weight basis).

Boron and other trace elements: plants require very little boron but are sensitive to excess uptake (especially various citrus, grapefruit, many fruit trees, different nuts, beans, many vegetables). Above normal boron concentration in the soil solution is seldom due to boron released in-situ from the soil minerals but is more likely to have originated from some external source, principally irrigation water. Few surface streams have significant concentrations of boron. It is, however, prevalent in well water and springs from geothermal areas or near earthquake faults. As a result, boron toxicity is typically a local/regional problem. The problems can usually be assessed adequately by determining the boron content of the irrigation water (section 15.5.3).

Selenium and molybdenum problems also are mostly related to specific geological conditions. Concentrations in the soil seldom reach toxic levels but problems may occur when these trace elements accumulate and become concentrated. A well-known case is the selenium poisoning of waterfowl in the Kesterson reservoir in the Central Valley of California, with the selenium carried into the reservoir by the drainage water from the irrigated land in the Valley.

14.2.3 Dispersion problems

The adsorbed cations on the surfaces of the soil colloids (*adsorption complex*) form the so called *diffuse double layer* (DDL). Its thickness largely determines whether the soil colloids are in a state of coagulation or in a state of dispersion. When the layer is compressed the soil colloids are able to come close enough together to coagulate. When the layer expands, the attraction forces become too weak to hold the colloids together, and as a result they disperse. The thickness of the DDL depends mainly on the following two factors:

1) *The salt concentration of the soil solution*: a high salt concentration in the soil moisture compresses the DDL while it expands when the salt concentration decreases. Thus, soils are more liable to dispersion when the salt concentration of the soil solution is low than when it is high
2) *The composition of the adsorbed cations*: when the adsorbed cations are mainly divalent cations (especially Ca), the DDL is compressed. It expands when there is a high proportion of monovalent cations (especially Na) on the adsorption complex. In the latter case, soils easily disperse.

It has been found that at normal salt concentrations, the state of dispersion closely correlates with the relative proportion of Na on the adsorption complex as expressed by the ESP-value (section 14.1.3). Dispersion problems generally increase in line with the increasing ESP-values. A state of easy dispersion, as exists at high ESP-values, is likely to result in poor physical soil conditions such as:

- *poor infiltration*, conceivably due to blockage of pores by dispersed colloids; downward movement by water of dispersed material may lead to the formation of an incipient clay pan, limiting root development and of drainage
- *unfavourable soil consistency*, hard when dry and plastic-sticky when moist; such soils are difficult to work
- *low resistance to slaking*, easily leading to the formation of surface crusts; these crusts hamper the infiltration of water into the soil and the mechanical strength of the crust is likely to hinder proper germination (irregular, patchy stands)
- *waterlogging*, resulting from the general deterioration of the drainage characteristics of the soil associated with the above effects; care must however be taken in relating poor drainage in the soil to high ESP since, although high ESP frequently causes poor drainage, inherently poor drainage characteristics of the soil may also lead to high soil salinity, including high ESP-values.

In most soils, no problems are experienced as long as ESP-values < 15%. However, as pointed out, the salt concentration of the soil solution also has an influence (Figure 14.8). This is why soils having an ESP > 15% may not disperse as long as the salt concentration

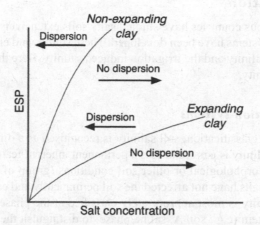

Figure 14.8 Dispersion of soil as influenced by the salt concentration, ESP and type of clay mineral (adapted from McNeal 1974)

remains high. Only when this high salt concentration decreases, for example, due to leaching, will dispersion problems arise. Of course, high salt concentrations in the soil are otherwise harmful (osmotic problems) and the limit of ESP = 15% applies to acceptably low concentrations ($EC_e \sim 1.0$ dS/m).

The limit of ESP = 15% is a very general limit. Soils and crops vary in their sensitivity to dispersion problems. In general, it can be stated that of two soils having the same ESP value, the soil with the highest percentage of colloids (especially when this involves clay minerals of the expanding lattice type) will experience the severest problems. Rather than using the general limit of ESP = 15%, separate limits may be used for different textural classes (e.g., a limit of ESP = 10–20% for fine textured soils and ESP = 20–30% for coarse textured soils). As the ESP and the SAR values are closely related (Jurinak and Suarez 1990), the more easily determined SAR is often used instead of ESP as the diagnostic parameter (see section 15.5.2).

14.2.4 Corrosion problems

Certain salts may attack and corrode construction materials such as steel and concrete, requiring special protective measures when these are present in the soil. Concrete is affected by sulphates with Mg type sulphates (prevalent in sea water and as a result in coastal areas and in marine soils) being more aggressive than Ca type sulphates. Porous concrete is more sensitive to attack by salts than dense concrete. When the sulphate content of the soil is high, dense concrete made with sulphate-resistant cement should be used (generally when $SO_4 >$ 1000 mg/l in a 2:1 water extract of the soil).

Steel is also corroded by sulphur in the water, especially by H_2S. Steel parts of tubewell pumps in particular are badly affected. The aggressiveness of the water is enhanced by low pH-values, presence of CO_2 in the water and by low redox potentials of the water. Protective paintings may help but, in severe cases, heavy-duty steel or stainless steel should be used, rather than normal mild steel.

14.3   Classification

Soil scientists in various countries have studied salty soils extensively and a number of different classification systems have been developed. These studies and classifications involved both the natural soil salinity and the irrigation induced salinity. Here the focus will be on the latter type of soil salinity.

14.3.1   Classification systems

In the taxonomic soil classification, soil salinity is recognised as a differentiating soil characteristic when the salinity is a more or less a permanent inherent feature of the soil and has resulted in specific morphological or other soil conditions (group of Halomorphic soils or Halosols). When the salts have not affected the soil permanently and can be readily removed by leaching, soil salinity is most appropriately classified at the phase level in a systematic soil classification system (e.g., soil A, saline phase, to distinguish the soil in question from the standard A soil which is non-saline).

The system developed by the *US Salinity Laboratory* (Riverside, California) is most commonly used for the classification of salty soils (see Table 14.2). The system is suitable for phase level differentiation in a systematic soil mapping as well as for single value mapping. The diagnostic parameters in this system are the EC_e-*value* (indicative of the osmotic problems) and the *ESP-value* (indicative of the dispersion problems). The limits used conform to those described in sections 14.2.1 and 14.2.3.

Typical analytical results of the above distinguished four different types of salty soils are presented in Table 14.3 to Table 14.5. The pH-values of salty soils often fit the following pattern:

saline soils pH < 8.5

saline sodic soils pH ~ 8.5 } determined in the saturation extract

sodic soils pH $= 8.5 - 10$

The pH should, however, not be used as a classification criterion as many other factors may interfere, viz in the presence of lime the pH may well exceed 8.5 even while the ESP is low, while in gypsiferous soils the pH seldom exceeds 8.2 regardless of the value of the ESP.

An older Russian classification which is still widely used in Russia, Eastern Europe, Central Asia and China distinguishes between *Solonchak* soils (close to the saline soils of

Table 14.2 The US Salinity Laboratory classification system for salt affected soils (USDA 1954)

	$EC_e \leq 4\ dS/m$	$EC_e > 4\ dS/m$
ESP ≤ 15 (SAR ≤ 13)	non-saline non-sodic soils	saline soils[2]
ESP > 15 (SAR > 13)	sodic soils[1]	saline sodic soils

1 Instead of *sodic* the term *alkali* is also used (older term).
2 This rather specific and limited definition of the term *saline soils* precludes its general use for soils suffering from salinity; such soils are therefore referred to as *salty soils*.

Table 14.3 Typical analytical data for different types of salt affected soils for the soils indicated in Table 14.2 (USDA 1954, 1/3)

Exchange Complex in meq/100g soil

Soil no	SP%	CEC*	Na^+	K^+	Ca^{++}	Mg^{++}	H^+
Non-saline, non-sodic soils							
1	36	20	< 1	2	11	6	1
2	32	29	3	<1			0
3	40	17	< 1	<1			0
Saline soils							
4	52	14	2	<1			0
5	47	17	2	<1			0
6	40	19	2	3			0
Sodic soils							
7	59	33	6	1	11	10	0
8	61	34	8	1			4
9	39	22	10	7			0
Saline-sodic soils							
10	62	36	9	1	10	12	4
11	60	40	11	1			0
12	36	26	17	2			0

* In principle: CEC = Na + K + Ca + Mg + H; Not all adsorbed cations are recorded in the table.

Table 14.4 Typical analytical data for different types of salt affected soils for the soils indicated in Table 14.2 (USDA 1954, 2/3)

Saturation extract in meq/l

Soil no	Ca^{++}	Mg^{++}	Na^+	K^+	Total	CO_3^{--}	HCO_3^-	SO_4^-	Cl^-	Total
Non-saline, non-sodic soils										
1	2.7	2.3	1.2	0.9	7.1	0.0	2.6	2.1	0.9	5.6
2	3.3	1.9	12.2	0.7	18.2	0.0	6.1	4.3	4.9	5.4
3	2.8	1.7	5.2	0.2	9.9	0.0	6.6	2.7	0.4	9.7
Saline soils										
4	31.5	37.2	102.0	0.2	170.9	0.0	4.5	90.0	78.0	172.6
5	37.0	4.0	79.0	0.4	150.4	0.0	7.2	62.2	47.0	148.4
6	28.4	22.8	53.0	1.1	105.3	0.0	5.2	74.0	29.0	108.2
Sodic soils										
7	1.1	1.4	15.6	4.0	18.5	0.0	6.5	8.5	2.9	17.9
8	1.4	1.0	21.5	0.3	24.2	0.0	3.3	3.8	16.7	23.8
9	1.1	0.3	29.2	4.1	34.7	8.4	18.7	4.6	7.5	39.2
Saline-sodic soils										
10	6.7	9.9	79.5	0.5	96.6	0.0	2.4	20.1	72.0	94.5
11	32.4	38.3	145.0	0.5	216.2	0.0	3.3	105.0	105.0	213.3
12	0.6	0.9	58.5	1.6	61.6	5.0	19.9	21.5	16.3	62.7

Table 14.5 Typical analytical data for different types of salt affected soils for the soils indicated in Table 14.2 (USDA 1954, 3/3)

Salinity and sodicity indicators

Soil no	EC_e dS/cm	ESP %	SAR_e	pH	Gypsum meq/ 100 g soil	Alkaline earth carbonates**
Non-saline, non-sodic soils						
1	0.6	2.0	0.8	6.4	0.0	–
2	1.7	10.0	8.0	7.8	0.0	+
3	0.8	3.0	3.5	7.9	0.0	+
Saline soils						
4	13.9	13.0	17.0	7.9	7.1	+
5	12.0	8.0	13.0	8.0	0.0	+
6	8.8	10.0	11.0	8.0	0.0	+
Sodic soils						
7	1.7	18.0	14.0	8.3	0.0	+
8	2.5	24.0	20.0	7.3	0.0	–
9	3.2	46.0	35.0	9.6	0.0	+
Saline-sodic soils						
10	9.2	26.0	28.0	7.3	0.0	–
11	16.7	26.0	24.0	7.8	42.2	+
12	5.6	63.0	68.0	9.3	0.0	+

** Present (+) or absent (–)

Table 14.2 but also including some saline sodic soils) and *Solonetz* soils (characterised by having a pronounced columnar or prismatic structure in the B-horizon). Such a horizon, more or less pronounced, develops eventually in sodic soils but its development takes time (20–100 years). So, a Solonetz is almost always a sodic soil, but not all sodic soils are Solonetz soils.

14.3.2   Field appearance

Many salty soils have a normal field appearance. The salt content must usually be quite high before salinity becomes observable in the field. Soil salinity problems cannot therefore be properly assessed in the field but should also be based on (laboratory) measurements. Field appearances of soils and vegetation are only able to give indications of soil salinity (usually of severe cases only). Typical symptoms in the soil are:

a)   efflorescence phenomena: powdery, crystalline salt deposits on exposed surfaces from which water evaporates (leaving the salts behind): high spots (e.g., top of ridges), side slopes of ditches, walls of soil pits, etc.
b)   damp, oily looking soil surface, slick spots (due to hygroscopic salts, especially $CaCl_2$)
c)   mycelia in the soil profile: salts precipitated in the fine pores, forming a pattern of thin white veins (usually carbonates)

d) crystals, clustered or scattered, usually encountered at specific depths (especially true with gypsum crystals)

e) crusts: concentration of crystalline salt at the soil surface or any other evaporating plane in the profile. leading to the formation of a cemented layer

f) dark film on the soil surface, left by evaporating soil moisture containing dispersed organic matter (especially in the presence of Na_2CO_3).

The features a), b), c), d) and e) mainly indicate a high salt content in the soil (high EC-value, saline soils) although the ESP may be high as well. Feature f) indicates sodicity (high ESP and especially high pH). Poor soil physical conditions (hard consistency when dry, plastic/sticky when wet; surface crusting; poor aeration conditions; low hydraulic conductivity due to movement and blockage of pores by dispersed particles) are also commonly associated with high sodicity, although it must be remembered that poor physical conditions in the soil can be the result as well as the cause of the salinity problems.

Typical salinity symptoms in vegetation are: stunted growth, patchy stands, dull/dark/bluish green colours and signs of toxicity as described in section 14.2.2. The salinity must be quite high for the signs to appear in crops. At low levels the crop growth is more likely to be rather uniformly depressed all over the field, without showing special signs. Such effects are easily overlooked, only yields will be reduced. In uncultivated land severe salinity is often apparent from (salt) indicator plants.

14.4 Conventional mapping and sampling

The merit of using the US Salinity Laboratory system of classification for mapping is its simplicity and the fact that the two classification parameters used, are diagnostic of the two main salinity problems. The established limits used for EC_e and ESP, however, are very general. In detailed soil mappings the specific tolerances of the soil and the crops to be grown may be used to adapt the classification criteria for the specific conditions for which it is used. The general limits are, however, appropriate for a preliminary assessment of the salinity conditions in an area.

Soil salinity is a highly dynamic soil characteristic, a fact that should be appreciated in salinity mapping. The time of the year, the crop, the irrigation schedule, soil management etc., all have an influence and mapping results may differ as conditions vary. As conventional salinity mapping is as a rule largely based on analysis of soil samples, considerable effort should be made to collect representative samples.

14.4.1 Sampling

Random sampling is appropriate when all samples form one population. Locations are selected at random over the entire area. Normally, grid locations are sufficiently randomised to constitute a valid sampling strategy. Stratified sampling is to be preferred to random sampling when there are indications that the salinity variation in the area is not entirely random but is influenced by identifiable factors. Differences in geomorphologic features and natural drainage conditions often provide a good basis for the design of a stratified sampling plan. In alluvial plains, the more elevated and naturally better drained levee soils are far less prone to salinisation than the lower lying, poorer drained basin soils and the sampling density may be varied accordingly (e.g., double or triple density in the basin areas). Variations in watertable regime/depths would generally also be a good basis for a stratified sampling design. Seepage areas may also be singled out for intensive sampling. Variation in sampling density may also be based on land use, irrigation regime and other relevant features and conditions which conceivably might have influenced the land salinisation.

Except for research, it is seldom feasible or necessary to collect the large number of samples required for a proper statistical analysis of the results. For project purposes, it is usually sufficient to establish general patterns and the main underlying causal factors, which can then be used to inter or extrapolate the findings. A stratified sampling, skilfully carried out, can often provide this information with an acceptably limited number of samples.

Composite sampling (average sample composed by mixing samples from a number of sites) is as a rule of little use in salinity investigations. The salinity patterns in a soil profile and in an area are often fairly unique and considerable insight can be gained into the underlying salinisation processes from a careful study of these patterns. Composite sampling largely obscures these patterns. It may only have value in cases of randomly occurring micro-salinity variations (may e.g., apply to topsoil sampling).

Sampling can best be done at standard depths, unless there are special reasons to assume a particular vertical salt distribution. As a rule, salinity variations decrease with depth and therefore the sampling intensity should be highest in the upper profile zone. Suitable standard sampling depths are: 10 cm, 30 cm, 60 cm, 100 cm, and 150 cm. Salt crusts should be sampled separately. The required sampling intensity of the deeper layers (e.g., 250/300 cm) will soon become evident during the course of the survey. This also applies to the sampling of groundwater for salinity evaluation.

14.4.2   Laboratory analysis

The burden of collection and laboratory analysis of the statistically required large number of samples can often be reduced by adopting some of the following short cuts. Samples do not need to be larger than 500 g while in arid climates air-drying (in the sun) is usually quite acceptable (periodic checking with oven-dried samples is still recommended). Laboratory EC_e determinations can often be minimised by establishing convenient conversion ratios with the much easier to determine EC_1, EC_2 or EC_5 values. Similarly, also the requirements for the highly complicated ESP determination can be greatly reduced by relying on SAR values and by establishing conversion ratios and relationships with easier to determine diagnostic parameters.

Relationships between EC_e and EC_1, EC_2, EC_5

Extraction of a soil moisture sample at SP requires vacuum suction equipment. At higher soil/water ratios there is enough free water to obtain such a sample by simple filtration. In large scale soil salinity survey work, therefore, EC values are often measured at lower soil: water ratios e.g.,

- EC_1 value: measured in a 1:1 soil:water extract (100 g dry soil in 100 g water)
- EC_2 value: measured in a 1:2 soil:water extract (100 g dry soil in 200 g water)
- EC_5 value: measured in a 1:5 soil:water extract (100 g dry soil in 500 g water)

These EC_x values are useful for a relative assessment of the soil salinity, e.g., the salinity distribution within an area. No set of standards exists as for EC_e-values to evaluate crop response, but conversions may be made to EC_e-values, and the interpretation of the EC_x

value is then done on the basis of the available EC_e standards (as given in Table 14.1). Appropriate conversion factors can be derived either by correlation or by calculation.

Conversion by correlation: an EC_x/EC_e conversion factor is derived by determining the EC_e-value as well as the EC_x-value on a selected number of samples, usually at the start of the survey. On all other samples, only the EC_x value is determined and the established conversion ratio is used to estimate the corresponding EC_e-value: $EC_e = (EC_x/EC_e) \times EC_x$. When the salt composition in an area is uniform, this method can be quite satisfactory. In the example shown in Figure 14.9, the EC_e/EC_2 ratio decreases with increasing soil salinity. This is quite common in the presence of poorly soluble salts. Compared to the EC_e value, the EC_2 value is measured in a more diluted extract so proportionally more of the poorly soluble salts are dissolved. When only highly soluble salts are present, a constant ratio is to be expected.

Conversion by calculation: assuming an inverse linear relationship between the EC-value and the soil moisture content (θ), the following relationship holds:

$EC_e = (\theta_x/\theta_{sp}) EC_x$. Given the moisture content of a puddled soil at saturation point (SP), conversion factors can be calculated as shown in the following example:
$EC_1 = 4$ dS/m $\theta_1 = 100$ %w; $\theta_{SP} = 40$ %w so $EC_e = (100/40) \times 4 = 10$ dS/m.

Roughly the following conversion factors apply:

	Coarse	Medium	Fine
θ_{SP} (w%)	25–35	35–50	50–70
Conversion factors $EC_1{\rightarrow}EC_e$	3–4	2–3	1½–2
Conversion factors $EC_5{\rightarrow}EC_e$	14–20	10–14	7–10

Since the assumed inverse linear relationship between EC-values and θ clearly does not hold in the presence of poorly soluble salts, conversion by calculation should be applied carefully.

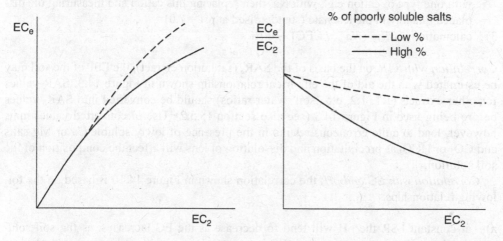

Figure 14.9 Typical relationships between the EC-values of the soil solution for different extraction ratios (see also Vlotman 2000b and van Hoorn and van Alphen 1994)

For example, in a lime containing, heavy textured soil the following conversion factors were determined by correlation:

- at low salinity levels, $EC_e/EC_5 = 3.5$
- at high salinity levels, $EC_e/EC_5 = 2.6$

which compare to the factors 7–10 calculated above. Calculated factors can easily overestimate the soil salinity by a factor of 2 to 3. Chances of overestimation decrease:

a)   the closer the moisture content of a puddled soil at the measurement is to SP
b)   the lower the salt content of the soil
c)   the lower the percentage of poorly soluble salts (e.g., $CaCO_3$, $MgCO_3$, gypsum, see section 14.1.1.

Determination of the ESP-value of the soil

Different methods can be followed to determine the ESP-value of a soil. Measured values may be expected to be more precise than the results of correlation methods. The accuracy of the SAR correlation method is often good enough for routine work. The determination of the ESP by correlation with EC and pH is less accurate and should only be used to obtain rough, indicative values.

Laboratory measurement: this measurement involves the following three steps:

1)   measurement of Na^+_{ads} (in meq/100 g soil), usually done by replacing Na and all other cations on the complex by NH_4^+, followed by measuring the displaced Na^+ in the leachate. All this is standardised at SP soil moisture content. At a more diluted soil/water ratio, Na^+_{ads} will be less due to replacement by Ca^{++} and Mg^{++} (see section 14.1.3)
2)   measurement of the CEC (in meq/100 g soil), usually done by saturating the complex with one type of cation e.g., with Na, then replacing this cation and measuring the displaced quantity in the leachate (standardised at pH = 7.0)
3)   calculation: $ESP = (Na^+_{ads} / CEC) \times 100\%$.

Correlation with SAR: on the basis of the SAR_e (saturation extract), the ESP of the soil may be estimated with the aid of the empirical relationship shown in Figure 14.3. SAR-values for other extracts (1:1, 1:2, etc., soil: water ratios) should be converted into SAR_e values before being used in Figure 14.3 (see also section 15.5.2). Use of converted values may, however, lead to quite erroneous results in the presence of lowly soluble Ca or Mg salts and CO_2 or HCO_3 as precipitation and dissolution of ions will affect the composition of the soil solution.

Correlation with EC and pH: the correlation shown in Figure 14.10 is based on the following relationships:

a)   at constant ESP, the pH will tend to decrease as the EC increases: as the soil solution becomes more concentrated (EC increases), H ions on the adsorption complex are exchanged for base type cations, thus lowering the pH of the soil solution

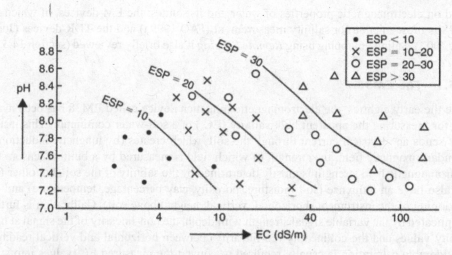

Figure 14.10 Relationships between the pH, the ESP and the EC frequently found in salty soils (ILRI 1963)

b) at constant EC, the pH increases with the ESP: a high ESP corresponds to a high Na concentration in the soil solution which leads to hydrolysis:

$$2Na^+ + CO_3^{--} + 2H_2O \, 2Na^+ + 2OH^- + H_2CO_3 \text{ or}$$
$$Na^+ + H_2O \, H^+ + Na^+ + OH^-$$

c) The ESP increases with the EC and vice versa: this relationship is not universal. Although it is inconceivable that the ESP will decrease with increasing EC, the ESP may well remain fairly constant.

Figure 14.10 enables the ESP to be estimated on the basis of two simple measurements viz. the pH and the EC (determined in a saturation extract, in a 1:1 or 1:2 or 1:5 soil-water extract, or whichever extract the correlation graph is based on).

14.5   New salinity measurement and mapping techniques

Conventional soil salinity mapping based on sampling requires considerable field and laboratory work and related logistics and organisation. For these reasons, more than cost considerations, salinity mapping for projects and for monitoring is seldom done up to the desired standards. Large scale salinity projects in Pakistan are commonly planned and designed on the basis of soil salinity sampling densities of only one point per 300–400 ha which is clearly far below even the most lenient statistical requirements. Throughout the world, important investment, management and policy decisions are taken on the basis of such sub-standard information and databases. Improved soil salinity measurement and mapping methods are urgently needed but major breakthroughs have not yet been achieved. Nevertheless, there are two relatively new methods that have reached maturity since the 80's that require less laboratory and organisational inputs and that can be automated using GPS and GIS for data collection and analysis. Neither of these methods has achieved the breakthrough referred to above but advances in both methods are such that they may be imminent. They are both based on electromagnetic properties of water and its solutes: the EM devices, of which the EM38 is most suitable for salinity measurement (FAO 1999a) and the TDR devices (Jones et al. 2002). Salinity mapping using *Remote Sensing* is also briefly reviewed (section 14.5.3).

14.5.1 The EM38

Since the early eighties, the electromagnetic induction device called EM38 has been available for measuring the apparent bulk salinity (EC_a) of a soil-water continuum. This instrument sends an electrical current through the soil, which creates (by magnetic induction) a secondary magnetic field, the strength of which field is measured by a built-in sensor. The electromagnetic field strength is mostly determined by the salinity of the soil, but other factors also have an influence (soil moisture, porosity, clay percentage, temperature) and the positioning of the instrument (horizontal, vertical, height above soil). Calibration is further complicated by the variable signal strength with depth, the non-linearity of the signal at high salinity values and the collinearity (co-linearity) between horizontal and vertical readings. Considerable calibration is usually required to convert the measured EC_a values into standard EC_e values of the soil.

Best calibration procedures for the EM38 have been developed by the US Salinity Laboratory (FAO 1999). As part of this procedure, a simplification of the traditional saturated paste method of salinity determination is proposed. The best use of the EM38 might be to work with the measured EC_a values, and develop separate EC_a tolerance criteria, similar to those existing for the EC_e.

Although not yet fully operational, indications are that the EM38 has scope to become a viable alternative for rapid low-cost salinity mapping. Automated soil salinity measurement and assessment methods with electromagnetic induction methods for precision agriculture are under development. The EM38 can be used for pre-investigations for planning and design of irrigation and drainage systems, salinity monitoring, assessment of mitigating measures and for localising of any other salinity feature that can be measured with the instrument (FAO 1999a, Vlotman 2000a). When these electromagnetic induction techniques are combined with precision agriculture, autonomous swarmfarm robots (GPS equipped, Figure 22.1) can provide a rapid and up-to-date record of salinity levels (Vlotman 2017).

14.5.2 Time Domain Reflectometry

Time Domain Reflectometry (TDR) is an accurate method for determining porous water content and electrical conductivity at the same time by the same instrument. It was first reported by Topp et al. (1980) and has since then undergone major improvements which combined with developments in electronics and computer analysis techniques make the method now practical for field application (Jones et al. 2002). Water content θ and apparent soil electrical conductivity (EC_{aTDR}, which is different from the EC_a determined with the EM38) are measured by different but comparable principles. Not all commercially available TDR devices are equipped with both measurement methods.

Water content is inferred from the dielectric permittivity of the medium[2], whereas electrical conductivity is inferred from TDR signal attenuation. Empirical and dielectric mixing methods are used to relate water content to measured dielectric permittivity. Clay and organic matter bind substantial amounts of water, such that measured bulk dielectric constant is reduced and the relationship with total water content can be calibrated. Soil texture, temperature and probe configuration, and coating on the probe rods affect the measurement. Depending on probe length TDR is generally suitable for salinity determination in low to moderate saline soils ($EC_a < 2$ dS/m, or $EC_w < 6$ dS/m) but this can be modified to measure highly saline soils.

The TDR method is at this stage less suitable for large area surveys but can provide intensive time-series measurements at multiple locations of both water content and salinity from the same location.

14.5.3   Remote Sensing methods

Remote Sensing (RS) involves the observation of earth features from a sensor loaded satellite platform. The most widely used sensor is the radiometer, which measures the electromagnetic radiation reflected/emitted by the concerned earth feature. Radiometers differ in the measured bandwidth. Soil salinity is best observed in the visible band (between 0.4–0.7 μm, from blue to red) and the infrared band (between 0.7–15 μm). Some platforms are also fitted with radar equipment. Widely used platforms are the SPOT-4 and the Landsat-7 which both have spatial resolutions of some 10–30 m and a return time of 1–2 times per month. For Landsat-7 an image series of some 30 years is now available while for SPOT satellites this is about 20 years. The most advanced ASTER[3] radiometer on board of NASA Terra spacecraft has been operating since 1999.

The use of RS for soil salinity mapping is still very much in an experimental state. Some indicative partial relationships have been established (e.g., between carbonate salts and signals in the red band and between chlorides and signals in the blue band). The same applies to some soil salinity associated soil surface features (darker colours and hard consistency associated with sodicity). The most clearly established relationship is, however, for salt crusts which are of course a feature of severe soil salinity and a feature which generally can also readily be observed by conventional aerial photography. RS can also map differences between crop appearances (colour, stand, uniformity, etc.) but these may have different causes and therefore are not directly diagnostic for soil salinity.

As unique, one-to-one comprehensive soil salinity signals have not yet been established, the best that can often be done is to develop tools (usually in the form of flow charts or tables), which interpret observed RS images/signals in terms of soil salinity occurrence. This of course can only be applied when the image/signal – soil salinity relationship is fairly unique and interferences /interactions by outside factors can readily be excluded. The development of these tools generally requires considerable fieldwork while the validity is often of rather limited local extent.

RS images, although not diagnostic enough to replace conventional salinity mapping, can greatly facilitate such mapping. Low cost RS images (often < 1 US$ per km² at 2018 price levels) can help to delineate salinisation prone landscape units (including already affected units) which mapping can then be used to design a cost effective stratified sampling program (see also section 14.4.1). Such delineation would generally be based on such landscape features as: broad textural soil classes (heavy vs. light textures), topographic differences (low vs. high), crop appearance (colour, stand and other features also mentioned above), water management regimes (noting over- and under-irrigation, drainage conditions, etc.). These features can often be recognized on the RS images with limited, judiciously conducted fieldwork especially when use can be made of available soil and topographic maps and information on water management practices.

See also section 3.2 for more broad applications of Remote Sensing. Many earth observation sensors have been designed, built and launched with primary objectives of either terrestrial or ocean remote sensing applications (CEOS 2018). Often the data from these sensors are also used for freshwater, estuarine and coastal water quality observations, bathymetric and bentic mapping.

Notes

1 The solubility of $CaCO_3$ (lime) at pH > 7.5 is < 1 meq/l whereby 1 meq weight = 50 mg. Its solubility sharply increases at pH < 7.5, especially in the presence of CO_2 produced e.g., by the plant roots [$CaCO_3 + CO_2 + H_2O \rightarrow Ca(HCO_3)_2$ which is highly soluble]. The solubility of $MgCO_3$ is slightly higher than that of $CaCO_3$. Gypsum ($CaSO_4.2H_2O$) has a solubility of 20–40 meq/l = 1 500–2 500 mg/l.
2 Dielectric constant is the ratio of the capacitance formed by two plates with a material between them to the capacitance of the same plates with air as the dielectric.
3 ASTER – Advanced Space Borne Thermal Emission and Reflection Radiometer (http://asterweb.jpl. nasa.gov/).

Chapter 15

Irrigation induced salinisation

Salinisation and sodification processes described in this Chapter are those occurring in irrigated land in the arid zone. A distinction is made between direct salinisation by the applied irrigation water and capillary salinisation from the underlying groundwater. This distinction is mostly conceptual as in practice these two processes often occur simultaneously. Both processes are largely controlled by drainage and irrigation. The type of irrigation system and the efficiency of the water delivery system determines the need for drainage (Table 2.3). From the efficiencies it may be apparent that higher efficiency may result in lower drainage need, i.e., less rise of the watertable and subsequent risk of capillary salinisation.

Although higher irrigation efficiency is desirable it sometimes causes also unexpected side effects such as salts collecting on the outside of the wetted drip area. These salts will be washed into the soil profile during rainfall. When rain is expected irrigators in California, and presumably elsewhere, start irrigating to prevent infiltrated rain washing salts into the rootzone; this assumes that the canopy of the crop prevents rain reaching the drip zone.

15.1 Salinisation by the applied irrigation water

All irrigation water contains some salts. Most irrigation water percolates through the soil towards the groundwater and onwards towards the rivers, collecting salts on its way. The use of groundwater for irrigation poses especial problems because this water in particular may contain a considerable salt load. This is especially true of arid climates where, due to low rainfall and high evaporation, groundwater is not refreshed so frequently as in humid climates and salts tend to become more concentrated. Rivers often have a higher salt content during the low flow season than during the flood season, while salt conditions may also vary along the course of a river. In Australia high salt concentrations in rivers also occur during high flow when dryland salts are washed into the river.

Each irrigation application imports a certain amount of salt into the rootzone. As the water is extracted by evapotranspiration, the salts remain behind in the rootzone/evaporation zone where they will accumulate unless an equivalent amount of salt is removed as is introduced by the irrigation water *(salt balance concept)*. As the salt uptake by crops is small, the salt removal depends almost entirely upon leaching by deep percolation i.e., the washing out of salt by water percolating through the soil to depths below the rootzone.

Salinisation by the applied irrigation water may be expected when the leaching is constrained by poor drainage or by a lack of leaching water. The latter may occur due to under-irrigation (with the applied water only barely meeting the evapotranspiration, leaving no excess for deep percolation). It may also occur due to high-efficiency sprinkler or drip irrigation. The salinity of the applied water is an important factor. In most cases, however, this type of salinisation is only temporary with most in-seasonal accumulation being "removed" or "controlled" or "reduced" by pre-irrigation before the next season. Winter rains (as e.g., occurs in the Mediterranean region) are also highly effective in controlling over-season salt accumulations.

15.2 Salinisation from the groundwater (capillary salinisation)

Evaporation of (saline) groundwater is by far the most common cause of soil salinisation. The groundwater may evaporate directly from the watertable when the latter occurs within the evaporation zone, or it may be drawn from deeper layers as the evaporation itself creates a gradient for upward capillary flow from a deeper watertable into the evaporation zone (Figure 15.1). Uptake of soil water by the roots also contributes to the salt accumulation as this process also helps to create gradients for upward flow and leaves salts behind (plants roots take up much more water than salts).

Prior to the introduction of irrigation, the groundwater recharge is usually quite small, only being fed by the minimal rainfall of the arid climatic setting of a typical irrigation scheme. The natural groundwater discharge is generally easily able to cope with such a low recharge, even with deep watertables (low heads). The deep percolation component of the newly introduced irrigation can easily amount to 20–30% of the applied irrigation water and recharge can easily rise to a multiple of the original recharge. In response, watertables will rise up to a level where the combination of the increased head, resulting in higher discharges, plus upward capillary flow is able to balance the increased recharge. There are many irrigation projects where watertables have risen from 20–30 m depth to 1–2 m depth below the soil surface over a period of 10–15 years following the start of the project (see Figure 16.1).

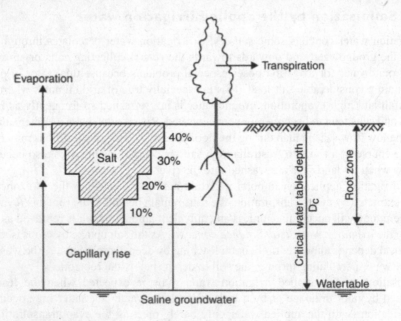

Figure 15.1 Capillary salinisation

15.2.1   Critical watertable depth

Upward capillary flow from a watertable can reach to great heights but the rate of flow generally decreases with increasing depth to the watertable (Figure 15.2a). The rate of upward salt movement, being proportional to the rate of upward flow, also decreases. The watertable depth at which the upward capillary flow becomes too small for any significant upward salt movement is called the *critical watertable depth* (D_c). Its value depends on:

a) *soil type:* soils with a high proportion of uniform small size pores have large D_c values (fine sandy loam, silty loam). Coarse sand has a small D_c value whereas well-structured medium to fine soils have intermediate values

b) *salt concentration of the groundwater:* upward salt movement is the product of flow rate x salt concentration. Therefore, required D_c values should increase with increasing groundwater salinity. In general, very little capillary salinisation will occur provided the salt concentration in the upper groundwater layer remains < 1000 mg/l (EC < 1.5 dS/m).

In situations where the regional groundwater inflow or outflow in the area is insignificant, the critical watertable depth (D_c) may be estimated as the depth to which the watertable falls towards the end of a long dry period. Rough estimates may also be derived from Figure 15.2 by assuming a critical upward flow of e.g., 0.5 mm/d for high groundwater salinity and 1.0 mm/d for low groundwater salinity and then reading the corresponding watertable depths. The values would apply to uniform profiles. Stratification will generally reduce the high values but may well increase the low values. It should also be duly noted that under the dynamic soil moisture regimes of irrigated land, the significance of the critical watertable depth concept is generally limited to the off-season fallow period (see also next section).

15.2.2   Factors influencing capillary salinisation

Capillary salinisation from ground water is influenced by a number of factors described in this section.

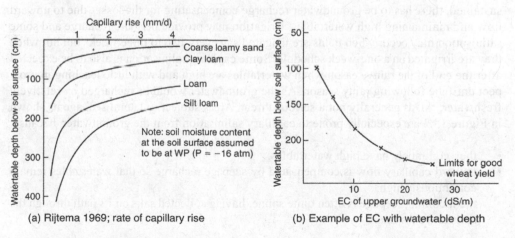

(a) Rijtema 1969; rate of capillary rise   (b) Example of EC with watertable depth

Figure 15.2 Rates of capillary rise to the soil surface from stationary watertables at different depths, Lower Indus Basin, Pakistan

Evaporation rate

A high evaporation rate leads to low moisture content in the upper soil layers and to correspondingly strong suction forces. The latter enhance upward capillary flow while this flow is restricted by the low hydraulic conductivity of the dried-out soil. The net result is that, after a short initial period of high soil moisture depletion, the dryness of the upper soil layers reduces the moisture losses from the layers below. As the upper zone become drier and thicker, losses will eventually become insignificantly small. Such a dry, protective layer is termed a *mulch layer*. The process described is usually referred to as self-mulching, indicating that the mulch establishes itself automatically as the soil dries out due to evaporation. This distinguishes it from the artificial establishment of a mulch layer by tillage or by spreading loose dry (organic) material on the soil surface.

When the evaporation rate is low, poorly developed mulch layers are formed and evaporation losses from the soil can continue (although at a rather low rate) for a long period, drawing water from great depths. This explains why capillary salinisation is often less hazardous under conditions of high evaporation than under moderate evaporation. In the former case the initial salinisation is strong but of short duration while in the latter case the salinisation rate is smaller but continues over a longer period, and the salt accumulation is often higher.

Land use

In bare soil, the evaporation zone is shallow (approx. 15–20 cm). When there is vegetation, the water loss will occur over a much deeper zone, about equal in depth to the main rootzone. Since crops are grown either under irrigation or in the rainy season, providing some leaching, it is mostly on fallowland during the dry season that capillary salinisation is a problem (enhanced by weed growth which negates the beneficial effects of the self-mulching).

Ground water recharge

As groundwater moves upwards by capillary flow, the watertable will fall to, or even beyond the depth D_c at which point capillary salinisation will virtually cease. For salinisation to be sustained, there has to be groundwater recharge compensating for the losses due to upward flow and maintaining high watertables. Irrigation may provide periodic recharge and some salinisation may occur when fields are irrigated say every two to three weeks but not when they are irrigated on a one-week schedule. Some capillary salinisation is also to be expected after the end of the rainy season when watertables are high and with little leaching in prospect until the following rainy season. As the groundwater would be recharged by relatively freshwater, this is generally not a serious threat. Areas subject to lateral seepage as shown in Figure 15.3 are especially prone to capillary salinisation from the groundwater, because:

- they commonly have high watertables;
- upward capillary flow is compensated by seepage recharge so that watertables remain continuously high;
- the incoming seepage is often quite saline, having collected salts on its path through the soils and substrata.

The seepage rate and the salt content of the seepage water vary throughout the year, in line with variations at the source. This is especially evident in cases like those depicted in Figures Figure 15.3b and Figure 15.3c. In the case of Figure 15.3a the seepage will continue,

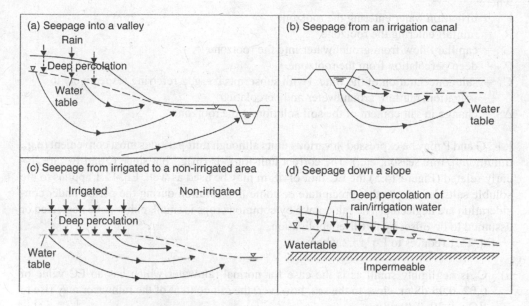

Figure 15.3 Some typical seepage situations

although at a diminishing rate, long after the supply of the source (rain) has stopped, or until the reservoir of water held in the soil at the source has been depleted.

15.3 Sodification

Sodification involves the replacement of other cations on the adsorption complex by sodium[1]. Significant replacement only occurs when Na becomes the dominant soluble cation in the soil solution (high SAR-value, see section 14.1). This may occur when the salinising source is Na rich, when the soil solution becomes more concentrated or when for other reasons the salinisation processes favour the accumulation of Na. The presence of CO_2 and HCO_2 in the soil solution is especially important. These anions form salts with Ca, which are only slightly soluble while the corresponding Na salts are highly soluble. Their presence thus leads to a relative enrichment of the soil solution with Na as the soil solution becomes more concentrated as the Ca salts precipitate.

Sodic soils with a low salt concentration in the soil solution but a high Na occupancy on the complex, also occur. The high ESP of these soils is unlikely to have developed under conditions of low salinity. Rather, these soils have almost always developed from saline-sodic soils by a leaching process in which the concentration of the soil solution has decreased more rapidly than the Na occupancy on the complex.

15.4 Salt balance of irrigated land

The salt balance of the rootzone within irrigated land may be expressed as in Eq. 15.1 (Figure 15.4):

$$IC_i + RC_r + GC_g = PC_p + \Delta S$$

Eq. 15.1

where:

I   = irrigation water entering the rootzone
R   = rainfall entering the rootzone
G   = capillary flow from groundwater into the rootzone
P   = deep percolation from the rootzone
C   = salt concentration of the water (with subscripts i, r, g, p referring respectively to
      irrigation, rainfall, groundwater and percolation)
ΔS = change in salt content of the soil solution in the rootzone

I, R, G and P may be expressed in various units although mm/period is most convenient (e.g.,
mm/month, mm/season, etc.). The correct unit for C is (m)g/l but since C and EC are linearly related (Figure 14.5) the EC-unit of dS/m may be used as well. Eq. 15.1 applies to the
soluble salts only. Salts that precipitate or come into solution during the period under consideration are neglected. The salt uptake by common crops is small, either to be neglected or
assumed to be offset by fertiliser application.

Eq. 15.1 reduces to Eq. 15.2 when:

a) C_r is negligibly small as is the case for normal rainwater which has an EC-value of
    0.02–0.05 dS/m; close to the sea, however, the salt content of the rainwater may rise to
    EC = 0.20–0.30 dS/m
b) $C_g = C_p$ which is a reasonable assumption for annual averages but less valid for short
    periods
c) $\Delta S = 0$ i.e., the salt balance is in equilibrium (the salt content in the rootzone at the
    beginning of the period and at the end being equal).

$$I\, C_i = (P\text{-}G)C_p = LR\, C_p \rightarrow I\, EC_i = (P\text{-}G)EC_p = LR\, EC_p \qquad \text{Eq. 15.2}$$

where *LR = leaching requirement* which is the excess of P over G. When a water quantity
LR satisfying Eq. 15.2 is drained from the rootzone, as much salt is leached as is brought in
by the irrigation. Eq. 15.2 expressed in EC-units, may be rewritten as:

$$LR = (EC_i/EC_p)\, I \qquad \text{Eq. 15.3}$$

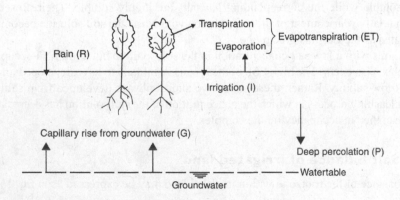

Figure 15.4 Water balance of irrigated land

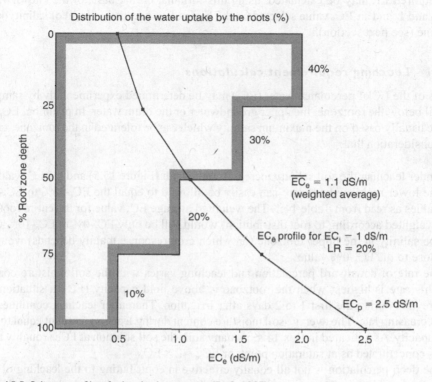

Figure 15.5 Salinity profile of a leached rootzone (FAO 1985)

Eq. 15.3 expresses LR as a fraction of I, i.e., the fraction of the infiltrated irrigation that must go into deep percolation in order to maintain a salt balance in the rootzone. The fractional factor EC_i/EC_p is called the *leaching fraction (LF)*.

In an equilibrium situation the following water balance also holds (Figure 15.4):

$$I = (ET - R) + (P - G)$$

Eq. 15.4

The difference (ET-R) is the rainfall deficit (= *net crop irrigation requirement*, designated as I_c) while the difference (P-G) represents the leaching requirement. Eq 15.4 may thus be rewritten as:

$$I = I_c + LR$$

Eq. 15.5

where I = the *total irrigation requirement* (= crop irrigation requirement + leaching requirement). Combining Eq. 16.3 and 16.5 gives:

$$LR = \frac{EC_i}{EC_p - EC_i} I_c$$

Eq. 15.6

The required LR may be calculated, using this formula, on the basis of the known values of EC_i and I_c and an EC_P value selected on the basis of an acceptable level of salinity in the rootzone (see next section).

15.4.1   Leaching requirement calculations

Values of the EC of percolated water (EC_p) may be determined experimentally by sampling the soil below the rootzone, the upper groundwater or the drain water. In planning, EC_p values are usually based on the maximum salinity which can be tolerated in the rootzone, taking into consideration that:

a)   under leaching the soil salinity increases with depth (Figure 15.5) and the EC_e values at the lower rootzone boundary can easily be allowed to equal the EC_e-25% to EC_e-50% values as read from Table 14.1. The weighted average EC_e value for the entire rootzone (weighted according to root distribution) would still be only EC_e-0% to EC_e-10% while the salinity in the upper rootzone (on which crop response mainly depends) would be close to the EC_e-0% value

b)   the rate of downward percolation and leaching varies with the soil moisture content. This rate is highest when the rootzone is above field capacity (FC), a situation that prevails during the first 1 to 2 days after irrigation. Thereafter leaching continues at a decreasing rate. The average soil moisture content during leaching is about equal to field capacity. As indicated in Box 14.4, in many soils the soil solution at FC is roughly twice as concentrated as at saturation point, or $EC_{fc} \sim 2 \times EC_e$

c)   the deep percolation is not all equally effective in contributing to the leaching of salts from the rootzone. The most effective leaching results from water which moves through the mass of the soil. The water that moves rapidly downwards through the larger pores, cracks, etc. (*preferential flow*) picks up very little salt and has only a minimal leaching effect. This may be expressed by introducing the *leaching efficiency factor* (f), being the mass flow proportion of the total deep percolation flow [the remainder (1-f) being the preferential flow]. Assuming the salinity of the mass flow to be equal to EC_{fc} and that of the preferential flow equal to EC_i, it follows that $EC_p = f.EC_{fc}+(1-f)EC_i$. The leaching efficiency depends on the soil texture/structure but also on the initial moisture content and on the mode of water application. Observed values range from $f = 0.4$–0.7 in Iraq to $f = 0.60$–0.95 in Tunisia.

On the basis of these considerations, values for EC_p in Eq. 15.6 are often taken as (Rhoades 1974):

$EC_p = 2EC_e$-25%: for mostly sensitive crops, when rather low leaching efficiencies are to be expected or when a high standard of salinity control is desirable

$EC_p = 2EC_e$-50%: for more tolerant crops, when high leaching efficiencies are to be expected or when somewhat lower salinity control is acceptable.

Example: $EC_i = 1.2$ dS/m, $EC_p = 12.0$ dS/m ($= 2 \times EC_e$-50% for the crop to be grown); $I_c = 6.0$ mm/d. In Eq. 15.6: LR = $[EC_i/(EC_p$-$EC_i)]$ I_c = $[1.2/(12.0$–$1.2)]$ 6.0 = 0.7 mm/d; I = I_c + LR = $6.0 + 0.7 = 6.7$ mm/d; LF = (EC_i/EC_p) = (LR/I) = (0.7/6.7) = 0.10 (or 10%)

In this example the salt balance in the rootzone is maintained when a minimum of 10% of the infiltrated irrigation goes into deep percolation. A deep percolation loss of this order or even higher is quite common under surface irrigation so that there is generally no need to over-irrigate to satisfy the leaching requirement.

Eq. 15.6 may straightforwardly be used to calculate the average leaching requirement during a rainless period. During the rainy season consideration also has to be given to the deep percolation from rainfall, which also contributes to the leaching of salts from the rootzone. The example in Box 15.1 illustrates both cases.

15.4.2 Regional salt balances

The principle of the salt balance can also be applied regionally:

(salt influx in the area) – (salt efflux from the area) = (change in salt storage in the area)

The salt influx and efflux can be calculated as: (water flux) x (salt concentration) x (time). Both the flux and the concentration are liable to vary during the year so that annual fluxes have to be determined by summating the uniform salt fluxes occurring during shorter periods of time. Various dimensional units may be used; m^3/s or mm/d for flux and $(m)g/l$, kg/m^3 or dS/m for salt concentration.

The water fluxes in an area, especially the groundwater fluxes, are quite difficult to quantify and one frequently has to rely on estimates. An example of a simple regional salt balance is illustrated in Figure 15.6. In this case there is one source of salt influx (irrigation water abstracted from the river), while all drainage water passes through one outlet point. The salt balance for this case reads: $\sum(Q_i \times C_i) - \sum(Q_d \times C_d) = \Delta S$. Such balance calculations can be used to check whether in the long-term salts are accumulating in irrigated basins.

Salt balance analyses of most irrigated basins are considerably more complicated than the simple and straightforward case depicted in Figure 15.6. In addition to salts imported by diverted river water, marine and fossil salts' may become mobilised by groundwater pumping and the provided drainage. Similarly, various salt disposal mechanisms and routes may be operative: disposal to rivers, disposal to sea, disposal to evaporation ponds, internal salt storage and re-use. Salt balances calculations at the basin level of course only show the aggregate result of a wide range of salinisation and desalinisation processes and as such may well be overly alarming as well as disguising. More detailed understanding, quantification, break-down and mapping of the underlying salt dynamics is generally required to arrive at a sound assessment of the need for remedial measures. For further details, reference is made to the salt balance studies conducted in five major arid zone irrigated basins by the International Water Management Institute (IWMI 2000).

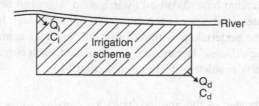

Figure 15.6 Salt balance for an irrigation scheme

Box 15.1   Examples of leaching requirement calculations

Data

Annual crop evapotranspiration	$ET = 1655$ mm
Total annual rainfall	$R = 695$ mm
Part of the rainfall going into deep percolation	$LR_r = 140$ mm
Part of annual irrigation going into deep percolation	LR_i = unknown
Capillary rise from groundwater	$G = 0$ mm
Irrigation water	$EC_i = 1.75$ dS/m
Saturation extract at 25% yield reduction	$EC_e = 3.5$ dS/m
Deep percolation water ($EC_p = 2EC_e - 25\%$)	$EC_p = 7.0$ dS/m

Calculations

Net rainfall utilised by crop $R_c = R - LR_r \rightarrow$ $R_c = 555$ mm
Net crop irrigation requirement $I_c = ET - Rc \rightarrow$ $I_c = 1\,100$ mm
Salt influx by irrigation: $(I_c + LR_i)EC_i = (1\,100 + LR_i)1.75 = 1925 + 1.75\,LR_i$ units
Salt efflux by deep percolation: $LR\;EC_p = (LR_i + LR_r)EC_p = (LR_i + 140)7 = 7LR_i + 980$ units

Salt balance (salt influx = salt efflux)

$1925 + 1.75\,LR_i = 7\,LR_i + 980 \rightarrow LR_i = 180$ mm
Total irrigation requirement $I = I_c + LR_i = 1\,100 + 180 = 1\,280$ mm
$LF = 180/1\,280 = 0.14$ (14%)

Had there been no rainfall, the leaching percentage would have to be much higher as in this case it would have to satisfy the following balance:

$I\;EC_i = LR_i\;EC_p \rightarrow (1\,655 + LR_i)1.75 = 7\,LR_i \rightarrow LR_i = 552$ mm
$I = I_c + LR_i = 1\,655 + 552 = 2\,207$ mm; $LF = 552/2\,207 = 0.25$ (25%)

When there is no contribution by rainfall, the leaching fraction could also have been calculated as $EC_i/EC_p = 1.75/7 = 0.25$ (25%)

15.5   Irrigation water quality

The salinisation/sodification hazards posed by irrigation water can be reasonably predicted on the basis of the content and types of the salts in the applied water. Irrigation development should not therefore be undertaken without prior analysis and appraisal of the irrigation water to be used. Three different hazards are distinguished, each corresponding with one of the three types of salinity problems identified in section 14.2:

* *salinity hazard*: danger of the applied irrigation causing soil salinisation and related osmotic problems

- *sodicity hazard*: the same with respect to soil sodification and dispersion problems
- *toxicity hazard*: the same with respect to toxicity problems.

A number of irrigation water quality appraisal systems have been developed. The system developed by the US Salinity Laboratory (1954) has been widely used but has now mostly been replaced by a new system developed by FAO (1985), which incorporates more recent research findings and experiences (for further updates see FAO 1997 and 2002a). The criteria adopted generally apply to crops grown under regular farming conditions. For most crops, considerably higher salinities can be permitted when special soil and water management practices are being followed. Some tolerant fodder crops have even been successfully grown with irrigation water salinity as high as EC = 10–20 dS/m (which compares to EC = 40–50 dS/m for sea water). Highly tolerant halophyte plants are abundant in nature but few of these are attractive as farm crops. Research to enhance salt tolerance through breeding is being conducted in a number of countries, which may lead to new salinity control approaches.

15.5.1   Salinity hazard

This hazard may be diagnosed on the basis of the EC-value of the irrigation water. The relevance of this parameter stems from the fact that there is a strong causal relationship between the EC_e of the soil and the EC_i of the applied irrigation water (Box 15.2). The shown relationships are based on field observations, lysimeter research and modelling. The EC_e values represent the weighted average EC_e over the main rootzone, taking into account the normal decrease in root density and water uptake with depth.

Few crops experience osmotic problems as long as $EC_e < 2$ dS/m. Based on this limit, the relationships suggest that for a conservatively low leaching percentage of 10%, the safe limit for the salinity hazard of the irrigation water is around $EC_i = 0.7–0.8$ dS/m. The FAO guidelines (Table 15.1) are based on an irrigation regime with a leaching percentage of 15% (so $EC_e \sim 2–2\frac{1}{2} EC_i$). This shows that the lower FAO limit of $EC_i = 0.7$ dS/m is of the same order as derived above.

Box 15.2   Salinity relationships for irrigated land in the (semi) arid zone

On the basis of the earlier established relationships $EC_{fc} – 2EC_e$ (section 15.4.1) and $LF = EC_i/EC_p = EC_i/EC_{fc}$ (Eq. 15.3), it may be shown that $EC_e/EC_i = 1/(2LF)$ which is called the *concentration factor* (concentration of the applied irrigation in the rootzone). For the normal range of $LF = 0.1–0.2$, the concentration factor would theoretically be expected to be in the range of 2.5–5. Observed concentration factors given in Figure 15.7 deviate from this range due to the interference by rainfall, leaching efficiency, non-equilibrium conditions and other factors.

Leaching fractions may readily be calculated by comparing ionic concentrations of the irrigation water and drainage water. This may best be done for the Cl- ion ($LF = Cl_i/Cl_d$) as this ion is less likely to be involved in precipitation/dissolution reactions. EC-values may be used when such reactions pose no significant interference ($LF = EC_i/EC_d$) during a rainless period.

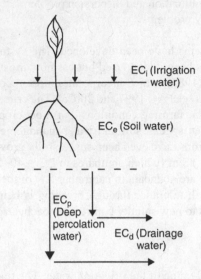

Approximate Relationships (FAO 1985)

Leaching fraction (LF)	$\frac{1}{2LF}$	EC_e/EC_i Arid zone	EC_e/EC_i Semi-arid zone
0.1 (10%)	5	3.0	2.5
0.2 (20%)	2.5	1.7	1.5
0.3 (30%)	1.67	1.3	1.2
0.4 (40%)	1.25	1.0	1.0

$EC_d \approx EC_p \approx 2 \times EC_e$

For LF = 0.1 (10%) $EC_d \approx$ 5–6 EC_i

LF = 0.2 (20%) $EC_d \approx$ 3–3½ EC_i

LF = 0.3 (30%) $EC_d \approx$ 2½ EC_i

Figure 15.7 Salinity relationships for irrigated land in the (semi) arid zone

15.5.2   Sodicity hazard

In the FAO guidelines, the sodicity hazard is appraised on the basis of the following two main diagnostic parameters (Table 15.1):

- *the EC_i value:* used to appraise whether the irrigation water is not too pure because this may enhance soil dispersion (section 14.2.3); such water may also over-leach the soils (too much leaching of semi-soluble Ca and Mg salts)
- *the SAR_i:* the relevance of this parameter stems from the close correlation between SAR_i → SAR_e → ESP of the soil. As shown, higher SAR_i values can be tolerated the higher the salt concentration (EC_i value) of the irrigation water. The relationship between the SAR_e and the corresponding ESP of the soil is shown in Figure 14.3. At a low leaching percentage of 10%, the saturation extract would be about 2.5–3.0 times as concentrated as the irrigation (Figure 15.7), so that $SAR_e = \sqrt{\text{(concentration factor)}}$ $SAR_i = \sqrt{(2.5 \text{ to } 3.0)}$ $SAR_i = 1.6$–1.7 SAR_i.

Residual Sodium Carbonate (RSC)

The relationship between SAR and ESP is complicated by the semi-soluble nature of the Ca and Mg salts in the soil. When the $(HCO_3 + CO_3)$ concentration of the irrigation water is high, Ca and Mg may precipitate as carbonates (e.g., in between irrigations when the soil dries out and the solubility limits of the Ca/Mg-carbonates are exceeded). Ca and Mg may also come into solution when the soil solution becomes diluted. This particular aspect of the sodicity hazard may be diagnosed by what is termed the *residual sodium carbonate* parameter (RSC) of the irrigation water (Eaton 1950, all concentrations in meq/l):

$$RSC = (HCO_3^- + CO_3^-) - (Ca^{++} + Mg^{++})$$ Eq. 15.7

Table 15.1 Guidelines for irrigation water quality appraisal (FAO 1985)

Potential Irrigation Problem	Degree of restriction on use		
	None	Slight to Moderate	Severe
Salinity *(affects crop water availability)*			
EC_i in dS/m	<0.7	0.7–3.0	>3.0
TDS in mg/l*	<450	450–2000	>2000
Infiltration *(affects infiltration rate of water into the soil, evaluated by using EC_i and SAR_i together)*			
$SAR_i = 0–3$ and EC_i = >0.7		0.7–0.2	<0.2
$= 3–6$ = >1.2		1.2–0.3	<0.3
$= 6–12$ = >1.9		1.9–0.5	<0.5
$= 12–20$ = >2.9		2.9–1.3	<1.3
$= 20–40$ = >5.0		5.0–2.9	<2.9
Specific Ion Toxicity *(affects sensitive crops);*			
Sodium			
surface irrigation (SAR_i)	<3	3–9	>9
sprinkler irrigation (Na in meq/l)	<3	>3	
Chloride (Cl in meq/l)			
surface irrigation	<4	4–10	>10
sprinkler irrigation	<3	>3	
Boron (B in mg/l)	<0.7	0.7–3.0	>3.0

* NB.TDS is widely used in the Colorado River Basin to indicate "salt" loads in the Colorado River (USBR 2017)

Example:

irrigation water with $HCO_3 = 5.2$, $CO_3 = 0.2$, Ca = 3.3 and Mg = 1.6 meq/l,
so RSC = (5.2 + 0.2) – (3.3 + 1.6) = 0.5 meq/l

Criteria:

RSC < 1.25 meq/l   : no problems to be expected
RSC = 1.25–2.50 meq/l : moderate problems to be expected
RSC > 2.50 meq/l   : severe problems to be expected.

The older version of the FAO guidelines (FAO 1976) included an assessment of the Ca precipitation hazard on the basis of an adjusted SAR parameter. When this parameter was found to over-estimate this hazard, it was replaced by a new adjusted SAR parameter. The use of this new parameter, however, is optional as its applicability is still under debate and its contribution to the improvement of the assessment seems quite marginal. The RSC parameter is still widely considered to be a reasonably solid parameter for assessing the Ca precipitation hazard (van Hoorn 2002).

15.5.3 Toxicity hazards

As described in section 14.2.2, the diagnosis of general toxicity caused by high salinity is covered indirectly by the EC_e parameter (and the corresponding irrigation water appraisal is thus covered by the EC_i parameter). Similarly, the SAR_i parameter covers the appraisal of Na-toxicity. In the FAO guidelines some toxicity is also explicitly appraised (Table 15.1):

- *Na-toxicity*: diagnosed by the SAR_i parameter and the Na concentration; given criteria apply to sensitive crops only (woody perennials)
- *Cl-toxicity*: diagnosed by the Cl concentration of the irrigation water; given criteria apply to sensitive crops only (woody perennials). Water which poses a Cl-toxicity hazard, normally also far exceeds the limits of the salinity hazard
- *B-toxicity*: diagnosed by the B-concentration of the irrigation water; given criteria apply mostly to sensitive crops only.

15.5.4   Examples of irrigation water quality appraisal

The application of the above methods and criteria is described in this section. The test results of the waters is shown in Table 15.2.

The four waters are all from NW India. Water 1 is canal water (abstracted from the Bakra reservoir system), waters 2 and 3 are both from tubewells (water 2 from Punjab, water 3 from Haryana) while water 4 is from a large drain in Haryana which was partly being reused.

Water 1 (canal water): this canal water is obviously of good quality for irrigation, posing no serious problems at all. With this water, the soil salinity levels, even at minimal leaching, will remain well below the tolerance levels of even rather sensitive crops while the drainage water may generally be readily reused.

Waters 2 and 3 (tubewell water): the somewhat higher salinity levels of both waters should only be of some concern under conditions were there is minimal leaching (under-irrigation, low rainfall, poor drainability), otherwise they pose no problems. The sodicity hazards are of more concern, especially in the long-term. The moderately high SAR values

Table 15.2 Analysis results of three irrigation waters (1–3) and a drainage reuse water (4)

Measurement	ion	water 1	water 2	water 3	water 4
EC_i (dS/m)		0.3	2,1	2.0	3.2
pH		7.8	7.6	8.9	9.3
Cations (meq/l)	Ca	1.5	2.3	2.2	2.6
	Mg	0.6	2.1	2.6	2.2
	Na	0.7	19.1	15.2	25.0
Anions (meq/l)	Cl	0.5	5.7	6.7	5.0
	SO_4	0.7	7.0	2.8	10.2
	CO_3	nil	nil	1.8	3.2
	HCO_3	1.7	6.6	8.2	8.8
SAR_i		0.7	12.9	9.8	16.1
RSC (meq/l)		−0.4	2.2	5.2	7.2

would seem to be quite acceptable in view of the prevailing EC values and the to be expected soil solute concentrations. The high RSC value of water 3, however, indicates that there is a considerable risk that part of the Ca cations will over time precipitate from the soil solution, that the SAR values of the soil solution and the related ESP values will increase and that the soil structure of the upper soil layer will deteriorate. The diagnosed sodicity hazard may be minimized by using this water only on the more calcareous soils and/or using it in conjunction with the canal water.

Both tubewell waters obviously also pose Na and Cl toxicity risks, especially the former. The water should not be used for sprinkler irrigation while these waters also should not be used in an undiluted form for surface irrigation of sensitive crops.

Water 4 (drainage water): all of the above restrictions on the irrigation use of the two tubewell waters, apply to an even greater extent to the reuse of this drainage water. The water may be used for tolerant crops and with sufficient over-irrigation/good leaching also for crops that are more sensitive. Limited use during the rainy season in higher rainfall areas should also be possible. Even then, adapted cropping and irrigation practices should be followed (e.g., short irrigation intervals to keep the soil solution well diluted). In all other cases, the water can only be used in conjunction, or be mixed, with better quality water.

Note

1 The older term alkalisation is also still commonly used to describe these same processes.

Chapter 16

Drainage of irrigated land

Although, by its nature, irrigated land will generally feature in a more arid climatic/physiographic setting than land used for rainfed agriculture, the occurrence of excess rainfall is still quite common. Rainfall in arid climates falls typically as intense storms, creating surface drainage requirements, which are easily 5 to 10 times greater than those due to the irrigation losses. For this reason, design discharges for surface drainage systems for irrigated land may, in most cases, be based solely on rainfall. This also applies to subsurface drainage when the drainage is exclusively for the control of excess rainfall (as may for example be the case with supplemental irrigation in semi-humid climates). Under the typical arid to semi-arid setting of most irrigated land where subsurface drainage is mainly needed for salinity control, different design criteria apply.

16.1 Waterlogging and salinity

Irrigation induced salinisation of irrigated land usually expresses itself in the form of the twin-problems of *waterlogging and salinity*. Waterlogging is to be expected in all cases where the subsurface drainage capacity of the irrigated land cannot cope with the irrigation and rainfall losses to the groundwater, and therefore, watertables rise into the upper soil layers (Figure 16.1). Waterlogging hinders leaching of the rootzone while it enhances capillary salinisation conditions and easily leads to salinisation of the soil as well. Changes in land use may also lead to rises in watertable and land salinisation (tree clearing for ranching has led to considerable dry land salinisation in some parts of Australia, see Box 14.1).

The generic occurrence of waterlogging and salinity is generally restricted to irrigation in the arid and semi-arid zones of the tropics with annual rainfall < 600–700 mm. Due to the low rainfall and the high evapotranspiration, the land and water resources in arid zones are often already naturally salinised. Observed watertable rises mostly range from 30–70 cm per year while the problems typically start to become critical after a time lapse of 10–30 years from the onset of the irrigation development. Estimates are that worldwide, the productivity of some 20–30 million ha of irrigated land in the (semi) arid zone are severely affected by irrigation induced soil salinity.

The problem of waterlogging and salinity of irrigated land should in principle always firstly be combated by reducing the causal high irrigation water losses. Effective measures are canal lining (see Giroud and Plusquellec 2020 for more information on actual water losses from lined canals) and improved field irrigation, as canal seepage and on-farm deep percolation losses are generally the two main sources of irrigation induced groundwater recharge. Improved drainage will however often also be necessary to ensure that watertables remain deep enough to establish a favourable vertical leaching in the upper soil layers and to minimize the capillary salinisation.

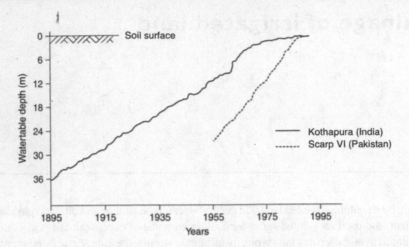

Figure 16.1 Irrigation induced rise of the watertable (van Achthoven et al. 2000, Vlotman et al. 1994)

Well drainage is usually the drainage technology of choice when the groundwater is relatively fresh and can readily be used for irrigation. Much of the formerly existing large-scale waterlogging and salinity in the semi-arid parts of South Asia (e.g., Pakistan) has been brought under control by tubewell development. When the groundwater is too saline for direct reuse, well drainage may still be the lowest cost technology but as pipe drainage usually generates a less saline effluent, the latter may be preferred when there are constraints on the disposal of saline drainage water.

Land that is waterlogged and salinised can generally not be reclaimed by improved surface drainage (such land needs subsurface drainage). Improved surface drainage, however, can be an effective preventive measure as it will collect and dispose of excess water which may otherwise contribute to the rise of the watertable. Effective surface drainage can also significantly reduce the subsurface drainage load and as such reduce the subsurface drainage costs or even obviate the need for subsurface drainage.

While in most countries the classic waterlogging and salinisation problems may have been more or less brought under control, because of the widely emerging freshwater scarcity and salt disposal constraints, the salinity problems of irrigated agriculture in the arid zone remain of considerable concern. Although in some cases the conventionally applied remedial measures might still suffice, generally it is to be expected that sustainability in the arid zone irrigation will increasingly depend on the sophisticated water and salinity management practices as applied in the San Joaquin Valley, California and elsewhere. Salinity control is heavily reliant on precision irrigation, reuse of drainage water and final disposal to wetlands, on-farm evaporation ponds and solar evaporators as well as enforcement of new environmental regulations (see FAO 2002 and section 16.5.2).

16.2   Surface drainage

Figure 16.2 illustrates a modern layout of irrigated land (surface irrigation; the layout of sprinkler or trickle irrigated land would generally be similar to that of rainfed land). The layout is designed to promote controlled overland flow of irrigation water. Excess irrigation water and rainfall is collected at the lower end of the field by a surface drain, typically consisting of a shallow V-shaped passable ditch.

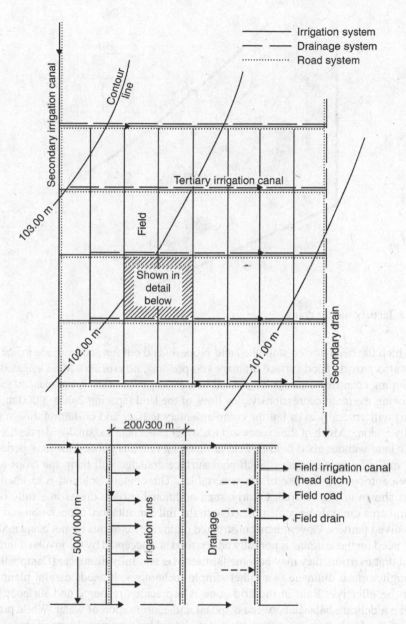

Figure 16.2 Modern surface irrigation layout

Length of irrigation runs are in the first place based on the needs of irrigation, and should also conform to the following requirements for surface drainage:

- minimum slope from 0.10–0.20% for clean furrows, to 0.20–0.25% for rough trashy furrows (see row drainage, section 9.2.2)
- drainage discharge in the furrow due to rainfall should not exceed its safe carrying capacity (which may lead to erosion and/or overtopping); this may pose the critical limit on furrow length.

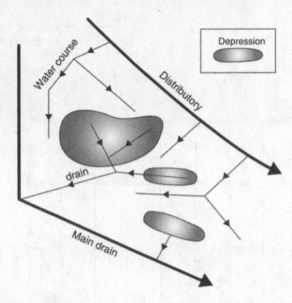

Figure 16.3 Tertiary surface drains

A field which has been graded/smoothed and properly laid out for good surface irrigation will normally also provide good surface drainage (no ponding, no erosion). Most irrigated land in the developing countries does not have the above drainage facilities. A low-density main system following the main depressions/valley lines of the land (spacing 2 000–5 000 m, density 2–5 m/ha) will generally exist but the complementary tertiary and on-farm drainage systems are usually lacking. Much of the excess surface water collects in the smaller depressions while in regular land without good field-to-field drainage, it remains ponded for long periods until it finally infiltrates or evaporates. Such poor surface drainage will harm the crops and may adversely contribute to the rise of the watertables. The obvious solution is to intensify the system as shown in Figure 16.3. Clearly, such additional surface drains in a fully occupied agricultural area can only be contemplated with the full consultation, cooperation and consent of the involved farmers. Government constructed drains are notorious for not being maintained when the need for these drains is not fully understood and accepted by the involved farmers (to the extent that overtime they may become farmed-over and fully disappear (Campbell 1994).

Although surface drainage is a rather simple technology, it needs careful planning and design to be effective. Rain in the arid zone is a precious resource and surface drainage must strike a delicate balance between disposal and conservation of water. While prolonged ponding of the land must be controlled, beneficial infiltration must be promoted. On-farm conservation of storm water may also help to reduce downstream flooding.

16.3   Pipe drainage systems

Subsurface drainage systems for salinity control in irrigated land should be capable of draining the recharge to the groundwater whilst maintaining a watertable regime that provides sufficient aeration and leaching and minimises capillary salinisation of the rootzone. Horizontal pipe drainage as described in Chapter 8 is widely used but vertical drainage using pumped wells also has its applications (see discussion in section 16.1). The design of a pipe drainage system, described in Chapter 11, requires the following decisions and information:

- type of system and layout of the field drains
- drain depth W below the soil surface (field drainage base)
- basic design criteria i.e., the design discharge q in combination with the watertable depth H to be maintained.

The above information, fed into an appropriate drain spacing formula, yields the required drain spacing L, (Figure 11.5). General reference is made to Chapters 8 and 11 since most of that discussion is also relevant to subsurface drainage for salinity control in irrigated land. Only the specific aspects have been included in this section.

16.3.1   Drain depth

Most factors influencing the drain depth W listed in section 11.6 also apply to pipe drainage for salinity control. In contrast to humid areas, the local/regional drainage base will seldom be a limitation. The drain depth may still be limited by soil conditions (unfavourable soil hydrological/mechanical conditions might occur at some depth), the available machinery (few machines go beyond 2.5 m) and cost effectiveness considerations.

To prevent capillary salinisation, the depth W should adhere to certain minimum depth criteria. The critical period for capillary salinisation occurs once the downward percolation of excess irrigation has stopped. This often coincides with the period after harvest time when irrigation is discontinued, or it could occur towards the end of a particularly long irrigation interval. The watertable will then have fallen back to near the drain depth W (slightly higher when head is needed to drain off seepage inflow into the area). These considerations lead to the minimum requirements:

- $W \geq D_c$ when there is no seepage
- $W \geq D_c +10–20$ cm when there is seepage inflow.

D_c-values for most soils are of the order of 100–150 cm rising to over 200 cm in very fine sandy or silty soil profiles (see section 15.2.1).

Box 16.1   Central Asian experience

On the basis of extensive research and experience the following guidelines for the determination of the drain depth have been formulated for the Central Asian region (Dukhovny and Umarov 2000):

Groundwater salinity (mg/l)	W/D_c
≥ 2 000–3 000	0.6
≥ 5 000–7 000	0.9
≥ 10 000–15 000	1.2

Where W is the drain depth and D_c the critical watertable depth (see section 15.2.1). These guidelines sensibly take into account the salinity of the groundwater.

Table 16.1 Least cost depth calculation, $ = USD, based on costs in Pakistan 1998

Description	Design 1	Design 2	Design 3	Design 4
Depth (m)	1.5	2.0	2.5	3.0
Spacing (m)	100	155	200	235
Pipe ($/m)	1.8	1.8	2.1*	2.1*
Gravel ($/m)	1.2	1.2	1.8*	1.8*
Installation ($/m)	0.4	0.7	1.2	2.5
Total $/m	3.4	3.7	5.1	6.4
Length (m/ha)	100	65	50	42
Cost ($/ha)	340	240	255	270

*These large spacings require a larger pipe diameter, wider trench and more gravel

Table 16.2 Commonly applied drain depths (see also Table 11.6)

Temperate Zone: mostly for aeration and workability control	
UK	0.80–1.00 m
Holland/North Europe	1.20 m
Canada	1.00 m
France	1.20–1.50 m

Intermediate Case	
Turkey (irrigation projects)	1.50 m

Arid/Semi-Arid Zone: mostly for salinity control but also some aeration control	
Egypt	1.50 m
India (Haryana)	1.50–2.00 m
USA (arid West)	2.00–2.50 m
Pakistan	2.00–2.50 m
Central Asia	2.50–3.00 m

The cost effectiveness considerations are usually taken into account in the *least cost depth* calculations as in Table 16.1.

Deeper installation allows for wider spacing (less drain length per ha) but as the costs per unit length increase with installation depth, the lowest costs are usually found at some intermediate depth. In the case of Table 16.1, costs decrease sharply between W = 1.5 m and 2.0 m, then reach a minimum before starting to increase significantly when W > 2.5 m. The least cost depth is this case would be around 2.0–2.2 m.

Table 16.2 gives an overview of the applied drainpipe installation depths in various countries. The depths are clearly much deeper for drainage for salinity control in the arid zones as compared to drainage for aeration control in the humid temperate zones.

Considerable variation however occurs. In Egypt with year-round cropping/irrigation (so mostly downward water movement in the soil profile), W = 1.5 m suffices. In Pakistan deep drains (W = 2.5 m) were found to be difficult to install and in some cases to cause over-drainage. The consensus in Pakistan and in many other countries with similar conditions is now moving to an optimal drain depth of about W = 1.5–2.0 m. Central Asia is the only exception to this view. The large depths used there are based on Russian theories on capillary salinisation (Dukhovny and Umarov 2000). The loess type soils of the region have strong capillary characteristics but still the applied depths are widely considered to be excessive. In many cases deep drain depths are unnecessary for salinity control. They generate more environmentally harmfully saline effluent and a consensus is emerging around an optimal 1.5 m drain depth (see Smedema 2007 and the effluent disposal description in section 16.5.2).

16.3.2   Design criteria

The q-H package needed for calculating drain spacing may be formulated in a steady state or a non-steady state form. The latter formulation is closer to reality although both represent rather fictitious situations, the significance of which is statistical rather than physical. The use of these criteria is validated only by their good correlation with such response indicators as crop yield or rootzone salinity level. Subsurface drainage design for salinity control in irrigated land can normally be carried out quite adequately using a steady state formulation which reflects the average discharge and watertable level during the critical part of the season. This formulation allows the use of the simple steady state drain spacing formulae.

Watertable depth and watertable head

Figure 16.4 illustrates the common case of a permitted general rise of the watertable during the season with, super-imposed on it, the short-term fluctuations due to the periodic irrigations. Capillary salinisation is of little concern since the predominant water movement in the soil will be downward due to the percolation of excess irrigation (and rain). The watertable depth may therefore be considerably less than D_c. The average watertable should, however, remain deep enough to ensure sufficient rootzone aeration. Recommended criteria are set out in Table 16.3.

Given the drain depth W, the design watertable head (h) to be used in the steady state drain spacing equation may be determined as h = W-H with H to be taken from Table 16.3.

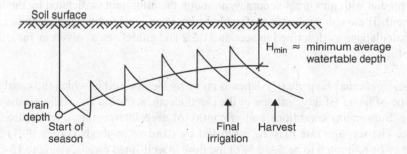

Figure 16.4 Watertable regime in irrigated land with subsurface drainage

Table 16.3 Suggested design watertable depths for irrigated land (after FAO 1980)

	Fine textured, moderately permeable soils	Light textured soils
Field crops	0.90–1.20 m	0.70–1.00 m
Vegetables	0.80–1.10 m	0.70–1.00 m
Tree crops	1.20–1.60 m	0.90–1.20 m

Note: the lower limits will meet the requirements of most crops; upper limits apply to deep rooting/high value crops

Design discharge

This rate may in principle be determined by estimating the drainage surplus in the groundwater balance:

$$q = (I + R + C + S - D)/T \qquad \text{Eq. 16.1}$$

where:

q = drainage discharge (mm/d)
I = deep percolation due to field irrigation (mm/period)
R = deep percolation due to rainfall (mm/period)
C = canal seepage (mm/period)
S = lateral/upward seepage (mm/period)
D = natural subsurface drainage (mm/period)
T = period in which the excess needs to be removed (days).

To arrive at matching q and H values, this balance should be applied over the same period to which H applies i.e., the period during which the watertable has risen to its maximum permitted level (generally the mid irrigation season). From then on all of the recharge to the groundwater has to be discharged without delay as no further rise of the watertable can be permitted. It is noted that the evapotranspiration (ET) is not an explicit component in the considered groundwater balance as it is assumed to be implicitly taken into account in the R and I estimate.

Deep percolation: the deep percolation due to rainfall (R) and field irrigation (I) is difficult to predict with any great accuracy, although the minimum is dictated by the leaching requirement. These values should preferably be based on local water balance studies. Excess rainfall calculations as described in section 11.5.2 and guidelines as given in Table 16.4 are also helpful.

Canal seepage: canal seepage (C) depends on many factors: soil in which the canal is excavated, type of lining (if any!), shape of the canal, depth, scouring and sedimentation conditions, etc. Substratum conditions and proximity of deep intercepting ditches also have an influence. The seepage rate may be estimated by standard methods (FAO, 1977) or more simply may be estimated to be 5–10 % of the flow in well-lined canals, rising to 15–30% for poorly lined and unlined canals.

Table 16.4 Guidelines for estimating deep percolation from field irrigation (FAO, 1980)

Irrigation system*	Conditions influencing percolation	Percolation as a percentage of the water delivered to the field	
		fine textured moderately permeable soils	light textured soils
Sprinkler	Daytime application, moderately strong wind	30	30
	Night-time application	25	25
Trickle		15	15
Basin	Poorly levelled and shaped	20	30
	Well levelled and shaped	30	40
Furrow	Poorly graded and sized	30	40
Border	Well graded and sized	25	35

* For irrigation efficiencies see Table 2.3

Seepage and natural drainage: the existence of the seepage inflow and natural drainage in an area may initially be assessed on the basis of the fall of the watertable during the dry season in combination with the salt concentration of the groundwater.

The rates of the seepage (S) into and the natural drainage (D) out of the project area may be approximated on the basis of the regional groundwater gradients and corresponding transmissivity values, using Darcy's Law. Best estimates are obtained by applying the groundwater modelling approach described further on.

Discharge period: this period may be taken equal to the one or two irrigation intervals over which the critical recharge is determined. The drain outflow will of course not be steady during an irrigation interval. It will rise sharply to a peak immediately after the irrigation and then recede gradually, in line with the fluctuation of the watertable. The recession flow will last longer for deep drains than for shallower drains. This may be taken into account by introducing the notion that the effective discharge period varies with the drain depth as indicated below (values applicable for values of H taken according to Table 16.3):

drain depth	W = 150 cm	effective period	50% of available period
	W = 200 cm		75%
	W = 250 cm		100%

The water balance and discharge period should also be considered over a longer time period (critical periods, seasons) as averages over these longer periods may well constitute more realistic design criteria.

Groundwater regime models

The conventional determination of the design discharge (q) as previously described is based on many assumptions and approximations, and reliable values can usually only be obtained when local drainage design experiences are available. Even then considerable doubt over the validity of the used criteria remains and cases abound where subsequent monitoring has revealed considerable over or under-design. The influences of lateral seepage and natural drainage are particularly difficult to assess.

The modelling approach as outlined by Boonstra et al. (1997) offers a highly recommendable alternative. It is based on the use of readily available groundwater regime models which predict watertable levels at nodal network points on the basis of inputted recharge (for details see section 21.3.3). These models can, however, also be used in the so-called inverse mode in which case the model estimates the net recharge at the nodal point on the basis of the observed watertable levels. Best estimates are obtained when a sufficiently long series of watertable records is available to run the model over a number of seasons/years (say 5–10 seasons/years). The estimated net recharges can then be used to establish drainable surpluses and design discharges. Application of the model in Pakistan allowed considerable adjustment to the existing design discharges (including a reduction of $q = 2.4$ mm/d to about 1.0 mm/d).

The model used in its normal mode, can also be used to compare the impacts of various watertable control alternatives (pipe vs. well drainage, improved surface drainage, canal lining, etc.) and as such help to establish optimal drainage designs and water management plans. For Pakistan, this showed that blanket drainage (uniform coverage of the entire area with drainage systems) may not always be necessary and may well be replaced by judiciously designed landscape drainage (a low cost/low density system of depression drains, reinforcing/assisting the natural drainage sinks of the landscape).

Salinity vs aeration control

While salinity control is almost always the main objective of drainage of irrigated land in the arid and semi-arid zone, there is usually also a need for some aeration control during prolonged periods of rain in the monsoon season. The latter applies especially to semi-arid zones with annual rainfall between 400–500 and 600–700 mm. For areas with rainfall > 600–700 mm, aeration control generally becomes the critical drainage objective since as mentioned earlier (see section 16.1) under such rainfall conditions, soil salinisation is generally not a generic problem (here the drainage requirements of the semi-arid zone merge with those of the semi-humid zone, see Chapter 19).

The drainage requirements for salinity control and for aeration control are in fact quite different. For salinity control it usually suffices to have on average a deep enough watertable for leaching and control of capillary salinisation during most of the season/year. Some intermediate salinisation is of little concern as long as the predominant water movement in the soil profile during the crop-season is downwards. This calls for a drainage system with a deep drainage base (W) while the drainage intensity (h/q) may be low (Table 16.5). Drainage systems for aeration control should have fairly high drainage intensities in order to rapidly lower the watertables after the cessation of rain and keep the waterlogging of the root zone to within the tolerated time limits. The drainage base depth may be rather shallow as long as it is compatible with the high h/q requirement.

The drainage requirements for aeration control in warm climates have been poorly researched. It is well-known that crops in warm climates are more sensitive to air deficiency than in temperate climates and that even a few days of waterlogging can do severe harm. The criteria given in Table 16.5 are indicative only. For drainage design in the semi-arid zones where salinity and aeration control appear to be required, analyses should be made for both objectives with the design to be based on meeting the most critical requirement. Agro-hydrological models described in section 21.4 can be of great help and value in these analyses.

Table 16.5 Indicative drainage design criteria for different climatic zones

	Aeration control		Salinity control
	Temperate zone	Semi-arid zone	Arid/semi-arid zone
q (mm/d)	5–10	5	1–3
H(m)	0.5	0.5	1.0
W(m)	1.0	1.5	2.0
h = W-H(m)	0.5	1.0	1.0
h/q (day)	50–100	200	>500

16.3.3 Layout patterns

The comparison between the use of pipes and ditches for subsurface drainage in section 8.2 also largely applies to their use in irrigated land. The dense system of irrigation canals, however, leads to a great many crossings of ditches that are avoided in an underground pipe system. Ditches within the field interfere severely with the field irrigation so that in general they should only be used as widely spaced boundary ditches for large fields (spacing of 200–300m). The maintenance of ditches should also be a major consideration as weed growth in the ditches is often rapid and abundant (especially various reed species). For these and other reasons (accessibility, public health) ditch drainage is seldom applied anymore. The present trend is clearly in favour of pipe drainage using layouts such as those detailed in Figure 16.5 (DRI staff 2001, EPADP/RWS 2000, Vlotman et al. 1994).

Both singular and composite pipe systems may be used, the choice depending on the weighting given to the factors listed in section 8.1. The major advantage of composite systems is that the number of open drains and crossings with irrigation canals are reduced. Areas of land up to 100 ha or more may be drained wholly under the ground, enabling the open drains to be spaced 1 000–2 000 m apart.

In Egypt a modified layout is used in areas with wet rice in the rotation (cotton-maize-rice rotation, Figure 16.5a, b). As in the standard layout rice farmers are likely to block the drains in order to keep their rice fields under water, upstream non-rice fields would not receive adequate drainage. The modified layout allows for a more independent water management at the block level (see also section 20.2.1).

Drainage laterals should preferably run in the same direction as the surface irrigation runs as irrigation water crossing a recently backfilled trench easily leads to piping. Solutions to piping were described in section 8.7.4 and essentially consist of preventing water from coming into direct contact with the trench backfill either by using bunds bordering the trench or by constructing a broad based ridge over the trench. These should be retained until the trench fill has stabilised and consolidated. In the layout where the laterals run along the contour intersecting the irrigation, the avoidance of direct contact between the trench fill and the irrigation water is not possible. In these circumstances the backfill should be thoroughly compacted.

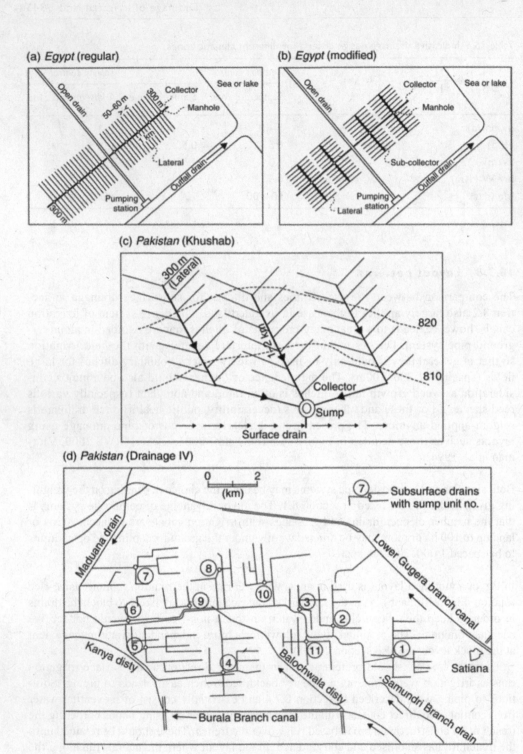

Figure 16.5 Examples of composite pipe system layouts

A layout pattern with extended laterals is an alternative to the composite system and is appropriate for both sloping and flat lands (Box 16.2).

Gravity outlets as are common for aeration drainage in the temperate zone can rarely be applied with the deep drainage systems used in the drainage for salinity control in the arid zone. It is, however, possible in Egypt as the drain depth is only 1.5 m and water levels in the receiving canals are kept low by pumping. In Pakistan, the composite pipe systems almost always discharge into a sump, from where the water is pumped into a shallow main drain. Deepening of these main drains to allow for gravity discharge would be costly, also considering the high maintenance costs (such deep drains would often cut into unstable fine sandy subsoil and be subject to much caving-in and sloughing of side slopes).

Box 16.2   Extended laterals

The extended lateral system is an interesting alternative layout (Boumans 1987). It is a singular system (laterals discharge directly into open drains) which, just like composite systems, can drain large areas entirely underground but is often less costly than composite drainage. The laterals may be up to 1 000 m long when provided with the regular maintenance/inspection facilities as deemed prudent. The system is best applied when the extended lateral drains can be aligned with the natural slope, but the system is still applicable in nearly flat land, using minimal drain slopes of 0.05%.

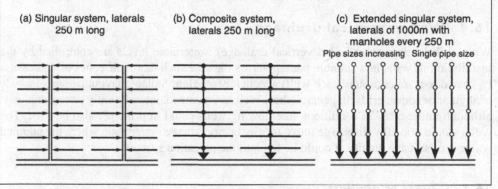

(a) Singular system, laterals 250 m long

(b) Composite system, laterals 250 m long

(c) Extended singular system, laterals of 1000m with manholes every 250 m

Pipe sizes increasing   Single pipe size

16.3.4   Pipe diameter

Pipe drains in irrigated land should normally be designed to carry the peak discharge which occurs for a short duration after each irrigation. This value may be determined as the discharge of the installed drainage system under its maximum watertable head (h_{max}). For the situation depicted in Figure 16.4, it may generally be assumed that $h_{max} = W - \frac{1}{2}H_{min}$ and the design discharge for pipe flow may then be estimated by solving the drain spacing formula inversely for q. For drain depths $W = 2.0$ m and $H_{min} = 1.0$ to 1.2 m, this roughly works out as q_{design} for pipe diameter $= 1.5 \times q_{design}$ for drain spacing.

Lateral design may further be based on the assumption that the entire area served by a field drain will be irrigated within one day (so that the peak inflow will occur nearly simultaneously along the entire length). However, this assumption would lead to over-design for collector drains. The irrigation supply is normally rotated among fields/farms and the larger

Table 16.6 Area reduction factors for collector drains in irrigated land (adapted from FAO, 1980)

% collector area irrigated in one day	q_{design}
50–100%	100% (= q_{design} for lateral drains)
30–50%	90%
20–30%	85%
10–20%	80%
< 10%	70%

the area considered, the less likely it is that the entire area will be irrigated on the same day. An area of 100 ha or more served by one collector drain may for example be irrigated in five days at 20 ha per day. So, the inflow into the collector will be at the peak rate from only 20% of the area while from the remaining areas it will be at a lower rate. Table 16.6 presents guidelines for estimating the design discharge rate for collector drains.

In semi-arid areas, rainfall discharge requirements may be more critical for the pipe capacity than those for excess irrigation water (see also the discussion on salinity and aeration control in section 16.3.2). In such cases, capacity may have to be based on rainfall intensities (following procedures similar to those described in section 20.2.1).

16.4 Well or vertical drainage

With well drainage (also called vertical drainage), watertable levels are controlled by the installation of wells in a suitable pattern across the area such that their cones of depression (= drawdown zones around each well) adequately overlap. Suitable layout patterns include triangular or square grids. In general, these wells are pumped, intermittently or continuously, although under artesian conditions free flowing wells could in principle also be used. The wells would normally discharge into a nearby open drainage system but, where the pumped water is of suitable quality, it could be fed into the irrigation system.

16.4.1 Types of aquifers

The applicability of well drainage depends especially on the geohydrological situation of the area. Two types of aquifer may be encountered.

Confined/Semi-confined aquifers

Where an area is underlain by artesian water as in Figure 16.6, well drainage may be used to take some pressure off this water and be used to reduce the upward seepage of this artesian water through the confining layer *(aquitard)* into the overlying soil layer *(leaky aquifer)*. Free flowing wells may be used where the artesian pressure reaches above the surface, although an effective solution to the seepage problem will generally require pumped wells. Through pumping the pressure in the aquifer may be reduced to the extent that sufficient downward flow through the aquitard occurs to control the phreatic watertable (thereby eliminating the need for watertable control by means of horizontal drainage).

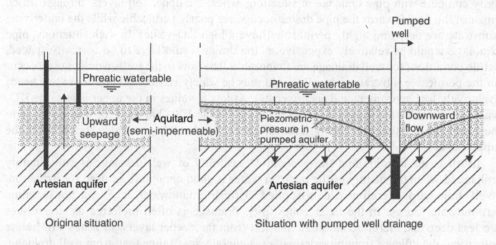

Figure 16.6 Relief of pressure on artesian water by means of well drainage

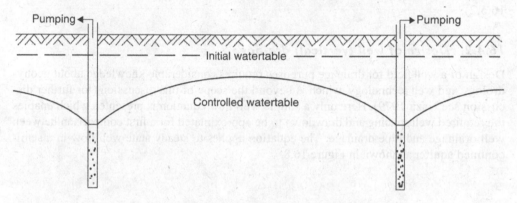

Figure 16.7 Watertable control by well drainage in an unconfined aquifer

Unconfined/Phreatic aquifers

The functioning of well drainage in this case (Figure 16.7) is similar to that of horizontal subsurface drainage and both systems could generally be alternatively applied. In humid climates with high drainage loads, operational costs of well drainage would generally be too high to be competitive with horizontal drainage. Well drainage, in order to allow wide spacings, establishes pronounced sinks around the wells which locally maintain watertables unnecessarily deep for aeration purposes and may lead to local over-drainage. In the humid zone, well drainage is only incidentally used e.g., for enclosed depressions which cannot be drained by horizontal systems.

Well drainage is most applicable for salinity control of irrigated land in the arid zone where drainage loads are rather low and deep watertables are desirable. For this type of drainage, capital costs of well drainage are usually less than those of pipe drainage. Fuel and other operational expenses can, however, be quite high. In general, well drainage can only compete with pipe drainage in situations where the upper soil layers, through which most of the flow towards the pipe drains occurs, are poorly permeable while the underlying substrata are deep and highly permeable (have a high KD-value). In such situations, pipe drainage would be relatively expensive as the drains would have to be narrowly spaced, while conditions for well drainage are favourable. The flow to the wells would mostly occur in the permeable substrata and wells could thus be widely spaced and operate at low heads (relatively low investment and energy costs). At low K values in the upper layers, the KD-value of the substrata should as a rule be at least some 100–200 m²/d for well drainage to be competitive, rising to > 500 m²/d for the case of moderately permeable conditions in the upper layers.

In all cases it holds that the economic feasibility of well drainage is considerably enhanced when the quality of the pumped water is good enough to be used for irrigation. Well drainage is therefore widely applied in fresh groundwater zones but not in saline groundwater zones. In the latter zones, pipe drainage is often preferred as streamlines go less deep and therefore mobilize less salt from the deeper layers. In zones with saline substrata, the effluent from pipe drainage is generally less saline than from well drainage, making it easier for the drainage water to be reused (e.g., when mixed with the irrigation water) or to dispose of it in an environmentally acceptable manner (see also section 16.5.2).

16.4.2   Design of well (vertical) drainage

Design of a well field for drainage purposes requires considerable knowledge about geohydrology and well technology, which is beyond the scope of this discussion (for further discussion see Nazir 1979). Here only a simple well flow equation is presented which enables the required well spacing and drawdown to be approximated for a first comparison between well drainage and pipe drainage. The equation applies to steady state well flow in a semi-confined aquifer as shown in Figure 16.8.

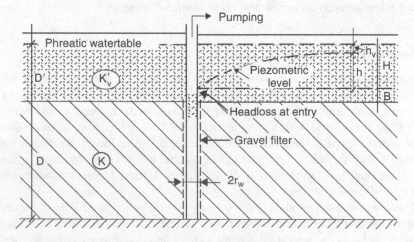

Figure 16.8 Notations used for the well spacing formula of Eq. 16.2

It is assumed that the deep percolation through the aquitard is uniform although in reality it decreases somewhat with increasing distance from the well:

$$H - cq = \frac{\pi r_e^2 q}{2\pi KD}\left[\ln \frac{r_e}{r_w} - 0.5\right]$$ Eq. 16.2

where:

H   = the head i.e., the difference in level between the watertable and the piezometric level adjacent to the well (m)

c   = the hydraulic resistance of the aquitard calculated as $c = D'/K'_v$ (in days) where D' is the thickness of the aquitard below the watertable (m) and K'_v = the vertical conductivity in the aquitard (m/d),

q   = the steady uniform rate of percolation of phreatic groundwater through the aquitard (m/d)

r_e = the radius of influence of the well (m), i.e., well spacing (m)

r_w = the radius of the well incl. the gravel filter (m)

KD = the transmissivity (see also K-D value) of the aquifer (m²/d)

This equation also roughly applies to situations with a unconfined/phreatic groundwater as depicted in Figure 16.7 when pumping is carried out in a highly permeable substratum (as would normally be the case for well drainage).

Example: annual drainable surplus = 320 mm and annual operating time excl. repairs and maintenance = 290 days, which gives q = 320/290 = 1.1 mm/d

Aquitard: D' = 13 m, K'_v = 0.033 m/d → c = 13/0.033 = 394 days. Aquifer: K = 8 m/d and D = 25 m → KD = 200 m²/d. Diameter of well, including filter: 2 r_w = 0.4 m. Head lost in vertical flow through aquitard: h_v = c.q = 0.43 m. Drawdown adjacent to well: h = D' – h_v – B = 13.0 – 0.43 – 3 = 9.57 m (B indicated in Figure 16.8).

From Eq. 16.2 the well spacing is calculated:

$$h = \frac{\pi r_e^2 q}{2\pi\pi D}\left[\ln \frac{r_e}{r_w} - 0.5\right] \rightarrow 9.57 = \frac{r_e^2 0.0011}{2(200)}\left[\ln \frac{r_e}{0.2} - 0.5\right] \rightarrow r_e = 670 \text{m}$$

Drained area: A = πr_e^2 = 141 ha; well discharge Q = Aq = 1551 m³/d = 18 l/sec.

16.5   Main drainage

Main drains in an irrigated area as a rule serve the dual function of collecting the subsurface drainage as well as surface drainage water. To fulfil the first function by gravity, the drains would have to be deep, often 2.5–3.5 m below the soil surface which is seldom feasible. As already explained in section 16.3.3 most subsurface drainage for salinity control in the arid zone therefore requires pumped discharge.

Due to the prevailing high storm intensities of the (semi) arid zone, the storm discharge requirements in irrigated areas generally far exceed the irrigation waste disposal and the subsurface discharge requirements. These latter relatively minor sources of excess water can normally be neglected, with the main drainage requirements being based solely on the storm water discharge.

16.5.1   Design discharge

Design discharges for main drains can generally be determined following the procedures and methods described in Chapter 12 with the method described in section 12.6 (area reduction formula) often being most relevant to irrigated land. The rainfall-discharge relationship to be used depends very much on the applied type of field irrigation and is for example quite different for furrow irrigation as compared to basin irrigation (basin irrigation can provide a considerable amount of temporary storage). Storage capacities in the soil will of course vary greatly, depending on when the field was last irrigated, but these field-to-field differences will average out when larger areas are considered. Box 16.3 provides some relevant values.

Box 16.3   Applied formulae and criteria

(based on $Q = q0\ A^{\alpha}$, Eq. 12.5, section 12.6)

Egypt: $q_0 = 0.8$ l/sec/ha; $\alpha = 1.0$

India $q_0 = 1$–3 cusec/sq mile = 0.1–0.3 l/s/ha; $\alpha = 2/3$–$3/4$

Pakistan, North: $q_0 = 2$–3 cusec/sq mile = 0.2–0.4 l/s/ha; $\alpha = 5/6$

               South: $q_0 = 1$–2 cusec/sq mile = 0.1–0.2 l/s/ha; $\alpha = 5/6$

Sudan: $q_0 = 3$ cusec/1000 feddan = 0.2 l/s/ha; $\alpha = 2/3$

USA: $q_0 = 10$–20 cfs/1000 acre = 0.75–1.5 l/s/ha; $\alpha = 4/5$

16.5.2   Disposal of saline drainage water

Drainage water from irrigated land in the arid zone is likely to have a fairly high salinity which may create problems for its environmentally acceptable disposal, especially when there is no ready access to natural or otherwise acceptable salt sinks (seas, salt lakes, playas, etc.). For such schemes, one of the following modes of disposal or disposal management options may be considered (see also Smedema and Shiati 2002 and USBR 2017).

Down the river: this is the natural mode of disposal, which was traditionally used in almost all basins until the limits of the downstream salinity were reached. In some cases, reaching these limits can be prevented by enhancing the river flow during critical low flow periods e.g., by changing the reservoir operation rules or by limiting the upstream water diversions, by reducing the saline water disposal during low flow periods or by otherwise adjusting the salinity disposal to the dilution capacity of the receiving river.

Evaporation ponds: evaporation ponds are widely used throughout the arid zones. Typically, these are natural depressions in the landscape towards which the drainage water can be easily directed. They are usually located in desert areas outside the irrigated perimeters, either at the edges or at the lower end of the irrigation systems. Small constructed evaporation ponds e.g., serving individual farms, may also be found within irrigated schemes.

Limiting the saline effluent generation: although salt balances need to be maintained, some options for limiting the disposal flow are generally available. *Reuse of drainage water* is such an option, although it is clearly not a long-term solution as salts are being stored somewhere in the basin and limits will eventually be reached. Reuse of drainage water is extensively applied in the Nile Delta, Egypt (Abdel-Shafi and Mansour 2013). The same applies to not meeting the leaching/drainage requirements. Improving the irrigation efficiencies helps by leaving more water in the river and also by reducing the drainage volumes but as the latter will have a higher salinity (see discussion in section 15.4.1 on drainage water salinity), this will generally be of only limited relief to the downstream salinity.

Limiting the salt mobilisation: this is a highly effective and desirable measure with no negative side effects. Ideally, the fossil and other resident salts stored in the basin should not be mobilised but as irrigation development almost inevitable leads to some rise of the watertable and consequently to the creation of new piezometric gradients and generation of new groundwater flows, the pick-up of some resident salts is often unavoidable. When mobilised, the salts may however be prevented from reaching the river by the installation of interception drains.

Land use planning: land use planning in the basin can help in various ways to limit the river water abstraction, the generation of saline drainage flows, the salt mobilization or otherwise help to control downstream river salinity. Less water demanding crops may be grown, irrigated land with uncontrollably high saline drainage rates may be retired or converted to rainfed cropping, land with a high salinity may be left un-reclaimed and unproductive depressions may be designated as salt sinks (*the dry drainage solution*).

Outfall drains: the construction of special drains for the collection and transport of saline effluent to a natural salt sink (usually the sea) may ultimately be required for basins which have no ready access to such sinks. Temporary solutions may be appropriate as intermediate steps but as indicated above, most of these solutions have a limited capacity and do not fully maintain or restore the salt balance and therefore do not assure the long-term sustainability of the irrigated agriculture in the basin.

Other options: disposal by means of bores to shallow saline aquifers is practised in the southern part of the Murray-Darling Basin, Australia. The salinity of the effluent is typically 1 000–2 000 mg/l and that of the receiving aquifer 25 000 mg/l and higher. Injection into safe deep aquifers and desalinisation are options under debate but are not applied in practice. The opportunities for deep injection seem to be rather limited while desalinisation will only become a serious option when the costs of disposal approach the costs of desalinisation (about US$ 0.5 per m^3 for saline drainage water in 2004, also considering the value of the produced freshwater). In most basins this point has not yet been reached.

Another option is disposal by means of *serial biological concentration of salts* using plants to concentrate saline drainage water. Drainage effluent from land respectively with (1) regular crops, (2) salt tolerant crops, (3) salt tolerant trees/bushes, and (4) halophytes is used sequentially and finally disposed to an evaporator (small/lined evaporation pond). Each next reuse area is only a fraction of the source area. In the process the drainage water is reduced in volume, but its salt concentration increases to the extent that all the salt will end up in solid form in the evaporator. This concept is seen as an alternative to regular evaporation ponds and was the subject of experiments in the San Joaquin valley in California (USBR 1999, FAO 2002).

These practices were only applied at a limited scale and it is unclear whether they will eventually provide answers to the identified challenges. Clear is, however, that not all arid zone countries facing salinity related sustainability problems have the economic means and the technical capacity to adopt the San Joaquin type water management practices (Box 16.4). More research and best-practices development work remains to be done to achieve wide applicability and acceptability.

Box 16.4   Innovative San Joaquin water management

The San Joaquin water management practices, incl. serial biological use, were developed in response to specific local emergencies, i.e., limitations imposed on the drainage disposal to the San Francisco Bay due to selenium contamination of the groundwater and closure of the Kesterson drainage outfall. Similar practices may ultimately be needed in more irrigated areas in the arid zone but in the interim, less demanding practices are used/experimented with in some parts of Australia/Murray-Darling (introduction of disposal permits), Egypt (drainage water reuse, directly and by blending), India/Haryana (shallow watertable management, salt disposal during seasons of high river discharge) and Pakistan (drainage water reuse for saline agriculture).

Part VI

Special topics

Part V

Chapter 17

Seepage and interception

Seepage refers to local waterlogging problems caused by groundwater inflow from a higher lying outside source. The causal groundwater flow is referred to as the *seepage flow* and the affected area as the *seep zone*. The source may be a hilly range but also a terrace at a higher elevation. The scale can be rather small and local (scattered seep spots of < 0.5–1.0 ha each at the foot of a slope) but also large and regional (seepage into a valley from a surrounding mountain range and from irrigated areas located on the higher plains adjacent to river valleys (MDBA 2015). Seepage is a natural phenomenon as in all the above examples but may also be manmade (seepage from an elevated canal or reservoir, seepage from an upslope irrigated area, seepage into a deep lying polder, etc.). The distance between the source and the seep area can be small (< 100 m to 1 km) but also quite large (> 10 km) while the seepage flow may travel through various layers (for regional seepage flows easily up to 50–100 m depth). Diagnosis of seepage problems therefore often requires considerable insight into the local and regional geohydrological conditions.

The required drainage measures may involve the collection and disposal of the excess water in the seep areas but also the interception of the causal seepage flow at some convenient point upstream of the affected area. Some seepage problems may also be solved by source control (lining of leaking canals, reducing deep percolation in upslope irrigated areas, etc.). Solutions to seepage problems are usually quite specific and need to be tailor made. A number of common seepage problems with relevance to land drainage are described in this Chapter.

17.1 Drainage systems for sloping land

The hydraulic gradients that are associated with natural groundwater flow are generally small (< 1%). Those associated with artificial drainage are often somewhat higher although they seldom exceed 2–5% (this would apply to a parallel subsurface drainage system with 20–50 m spacing and a mid-spacing watertable head of 0.5 m). The down slope directed gravitational gradients generated by the elevation differences in sloping land are often also of this order and therefore may be expected to have a significant influence on the groundwater flow pattern within such land.

Field drains on sloping land may be aligned down the slopes (*longitudinal drainage*) or across the slope (*transverse drainage*). In fact, these are the extremes and in between there are endless varieties of oblique alignments of the drains relative to the contours. The systems may be of the singular or of the composite type, usually of the latter (Figure 17.1).

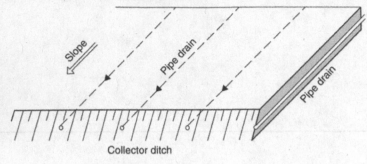

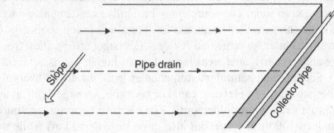

Figure 17.1 Subsurface drainage systems for sloping lands

17.1.1   Longitudinal drainage

In this case, the gradient generated by the pipe drains acts in a direction parallel to the contours while the gravitational gradient generated by the slope of the land acts in a direction perpendicular to the contours. The water will thus flow at an angle towards the drains, the degree of which depends on the relative strength of these two gradients. On mild slopes, the performance of the system (measured in terms of drain discharge and mid-spacing watertable control) is, however, virtually the same as that of a similar system in a horizontal field. Formulae described in Chapter 11 may be used to calculate the required drain spacing. When the gradient generated by the slope is of the same order or exceeds those generated by the drainage system, the flow direction becomes strongly oriented downslope. Consequently, much of the flow in the lower part of the field will either enter directly into the collector (Figure 17.1a) or when there is no cross-slope interceptor at the lower field boundary, pass on into a lower field. This makes longitudinal drainage less effective on steeper slopes ($\sim > 2$–5%).

17.1.2   Transverse drainage

In the case of transverse drainage, gradients generated by the slope and by the drainage system reinforce each other in the flow towards the downslope drain, but oppose each other in the flow towards the upslope drain. This results in a flow pattern and watertable shape between the drains as depicted in Figure 17.2.

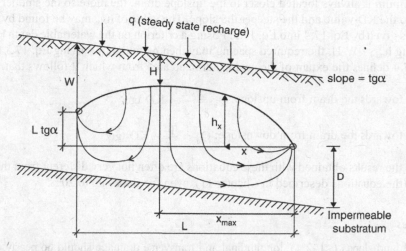

Figure 17.2 Flow pattern and watertable shape between two drains in a transverse drainage system

Under the simplifying Dupuit-Forchheimer assumption of horizontal flow through a layer with uniform KD[1], it holds that:

$$Q_x = -KD\frac{dh_x}{dx}$$

$$\frac{dQ_x}{dx} = q$$

$$\frac{d^2h_x}{dx^2} = \frac{-q}{KD}$$

Eq. 17.1

Integration of Eq 17.1 gives

$$h_x = \frac{-q}{2KD}x^2 + Ax + B$$

Eq. 17.2

Since at x = 0, h_x = 0 and at x = L, h_x = L tgα, it follows that:

$$B = 0 \text{ and } A = \frac{qL}{2KD} + tg\alpha$$

$$h_x = \frac{q}{2KD}(L - x)x + tg\alpha\, x$$

Eq. 17.3

For x_{max}, where h_x attains its maximum value (h_{max}), it holds that $dh_x/dx = 0$.

$$\frac{dh_x}{dx} = \frac{-q}{KD}x_{max} + \frac{q}{2KD}L + tg\alpha = 0$$

$$x_{max} = \frac{L}{2} + \frac{KD}{q}tg\alpha$$

Eq. 17.4

The maximum is always located closer to the upslope drain, the more so the smaller q is in relation to the KD-value and the steeper the slope. The value of h_{max} may be found by inserting x_{max} as given by Eq. 17.4 into Eq. 17.3. Posing a criterion on the watertable depth (H) and calculating $h_{max} \doteq$ W-H, the required spacing may then be found by solving Eq. 17.3 for L.[2]

Eq. 17.4 defines the extent of the upslope drained area from which it follows that:

flow towards the drain from upslope: $Q_u = \dfrac{qL}{2} + KD\ tg\alpha$

flow towards the drain from downslope: $Q_d = \dfrac{qL}{2} - KD\ tg\alpha$

Actually, the results obtained with these equations are often not very different from those for flat land (the equations described in Chapter 11), even for slopes up to 20%.

Guidelines

For small land slopes (<1–2%), longitudinal and transverse drainage should be nearly equally efficient in controlling the groundwater. On steeper slopes there will always be a significant downslope flow component and here transverse drainage should be used as cross-slope drains are most efficient at intercepting the downslope component. Field drains in a transverse system may be aligned at a slight angle to the contours (~ 10°) to provide a gradient for flow in the drain lines. The collector, usually a pipe as in Figure 17.1, runs down the slope at a maximum gradient.

17.2   Interception

Interception refers to the capturing of a drainage flow on a slope at some point along its flow path. Two common cases of subsurface interception are described in this section.

17.2.1   Interception of seepage down the slope

This case is depicted in Figure 17.3. A seepage flow (Q in m³/d per m slope width), for example generated by excess water on the upper part of the slope, moves downslope through a permeable upper soil layer overlying an impermeable substratum. To protect downslope land from becoming waterlogged, an interceptor drain may be installed across the slope. This may either be a ditch or a pipe drain.

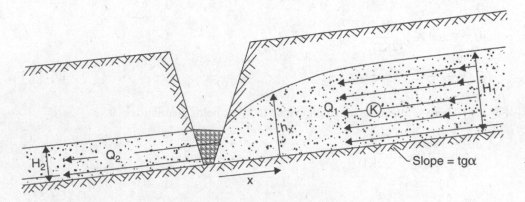

Figure 17.3 Interception of seepage flow down the slope

Under the assumption of parallel flow (streamlines parallel with the impermeable substratum), the downslope watertable establishes itself at the water level in the interceptor. It then holds that:

Upslope of interceptor: $Q_1 = KH_1 tg\alpha$ Eq. 17.5

Downslope of interceptor: $Q_2 = K H_2 tg\alpha$ Eq. 17.6

Intercepted flow: $Q_1 - Q_2 = K (H_1 - H_2) tg\alpha = \dfrac{H_1 - H_2}{H_1} Q_1$ Eq. 17.7

The difference in level between the upstream flow depth (H_1) and the water level in the interceptor (H_2) thus determines how much of the seepage will be intercepted. When the water level in the interceptor is maintained at $0.75H_1$, 25% of the flow will be intercepted; when at $0.5H_1$, 50% will be intercepted and so on. For full interception, the drainage base should be at or in the impermeable substratum, meaning in practise excavating the drain well into this substratum. When pipe drains are installed in a trench excavated into the impermeable substratum and then backfilled, there is a danger that considerable flow may pass over the drain without being intercepted. This so-called *bridging* should be minimised either by backfilling the trench with gravel or by installing the pipes just on top of the impermeable base rather than into it. In general, bridging is liable to become a problem on slopes steeper than 3–4%.

The flow in the drawdown zone immediately upslope of the interceptor, may be described as (Dupuit-Forchheimer assumptions):

$$Q_1 = K h_x \left(tg\alpha + \frac{dh_x}{dx} \right)$$ Eq. 17.8

which, when combined with Eq. 17.5, yields:

$$KH_1 tg\alpha = K h_x \left(tg\alpha + \frac{dh_x}{dx} \right)$$

$$dx = \frac{h_x}{(H_1 - h)tg\alpha} dh_x$$ Eq. 17.9

Integration of Eq. 17.9 gives:

$$x = -\frac{H_1}{tg\alpha} \ln(H_1 - h_x) + \frac{H_1 - h_x}{tg\alpha} + C$$

For $x = 0, h_x = H_2$ so $C = \dfrac{H_1}{tg\alpha} \ln(H_1 - H_2) - \dfrac{H_1 - H_2}{tg\alpha}$

$$x = \frac{1}{tg\alpha} \left[H_1 \ln \frac{H_1 - H_2}{H_1 - h_x} + (H_2 - h_x) \right]$$ Eq. 17.10

This shows that the watertable upslope of the interceptor approaches the undisturbed watertable as an asymptote. Defining the effective drawdown zone as the zone in which the drawdown is at least 10%, the upslope extent of this zone (x_{eff}) may be found by solving Eq. 17.10 for $h_x = 0.9 H_1$:

$$x_{eff} = \frac{1}{tg\alpha}\left[H_1 \ln\frac{H_1 - H_2}{0.1H_1} + (H_2 - 0.9H_1)\right]$$

For $H_2 = 0$ (100% interception): $x_{eff} = \dfrac{1.4H_1}{tg\alpha}$

$H_2 = 0.5 H_1$ (50% interception): $x_{eff} = \dfrac{1.2H_1}{tg\alpha}$

The parallel flow assumption underlying all these relationships is most realistic for an interceptor ditch cutting deep into the flow zone (with the ditch bed at the impermeable substratum as in Figure 17.3). In situations where a pipe interceptor is installed well above the impermeable substratum as in Figure 17.4, additional head will be lost due to radial flow towards the pipe, rendering these equations less accurate. Under these circumstances some water will also flow towards the pipe from the downslope side (to provide head for this, the watertable just downslope of the interceptor must be somewhat above the water level in the interceptor).

17.2.2   Interception of canal seepage

Waterlogging due to canal seepage as shown in Figure 17.5 may be controlled by the installation of interceptor drains in the toe of the embankment. Such drains are most effective

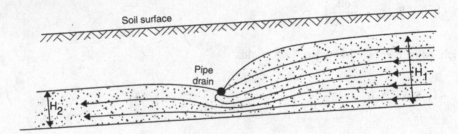

Figure 17.4 Pipe interceptor with pronounced radial flow

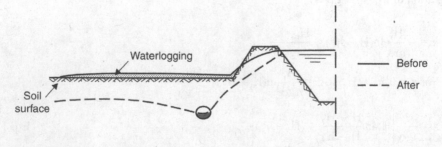

Figure 17.5 Interception of canal seepage

when installed on the valley/downslope side of the canal. The intercepted water may either be discharged by gravity into a nearby deep open drain or be discharged via a pumped sump into a nearby irrigation canal. Since the water is usually of good quality, it should be reused wherever possible.

The depth and location of the interceptor drain relative to the canal can best be determined by trial and error or by using a suitable numerical groundwater flow model (see section 21.3). Deeply installed drains are most effective in controlling the waterlogging but also induce more seepage. Research work on the performance of interceptor drains in Pakistan showed that the installation of interceptor drains may induce so much extra seepage as to render the net interception and the overall impact to be too small to be cost effective (Bhutta and Wolters 2000).

17.3   Natural drainage of river valleys

Figure 17.6 depicts a situation where seepage flow (Q in m³/d per m width) from the hills bordering a valley, enters directly into the water-bearing strata underlying the valley floor (the latter sloping towards the river). In the valley the groundwater is recharged by irrigation water losses and/or by rainfall (recharge q in m/d). The groundwater in the valley drains naturally towards the incised river.

Assuming horizontal flow (all streamlines parallel to the top of the impermeable substratum), it holds that:

$$(L-x)q + Q = Kh_x \frac{dh_x}{dx}$$ Eq. 17.11

Integration over the limits $x = 0$, $h_x = D$ and $x = x$, $h_x = h_x$ yields:

$$h_x^2 = \frac{q}{K}(2L-x)x + \frac{2Q}{K}x + D^2$$ Eq. 17.12

Eq. 17.12 indicates that two flows occur which are super-imposed: a uniform seepage flow from the hills (represented by the linear term) and a non-uniform flow generated by the

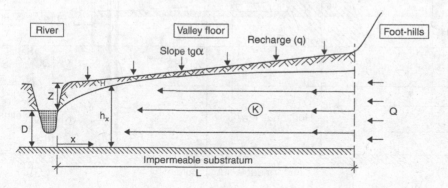

Figure 17.6 Natural groundwater flow in a river valley

recharge in the valley (quadratic term). The watertable has a curvature that becomes increasingly pronounced towards the river (note that Eq. 17.12 does not account for the radial resistance near the river which would have made the curvature near the river even more pronounced).

At minimum watertable depth location (where the watertable is closest to the valley floor surface), it holds that $dh_x/dx = tg\ \alpha$ which, applied to Eq. 17.12, enables its location to be determined. The critical high watertable zones occur typically in the middle to lower parts of the slope (Figure 17.6), their exact location depending on the relative values of D, Z, tg α, K, q and Q. Subsurface drainage may be required in this zone (longitudinal or transverse, see section 6.1) while the areas above and below would normally be adequately drained naturally.

17.4 Seepage into a polder

By their very nature, involving a difference between the inner and outer drainage base (Figure 10.7), polders are prone to seepage inflow. On the basis of the nature of the inflow, two cases may be distinguished.

17.4.1 Semi-confined flow

Figure 17.7 illustrates a fairly common situation where a polder is underlain by a semi-confined aquifer which is recharged from outside. Inside the polder the water in the aquifer is under over-pressure (as indicated by the piezometer readings), creating a head difference for upward flow through the overlying semi-permeable layer to the phreatic groundwater.

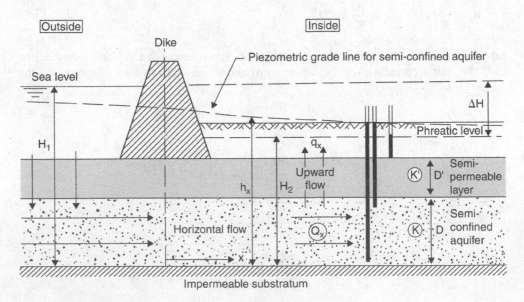

Figure 17.7 Seepage into a polder from a semi-confined aquifer

Under steady state conditions, the head of the semi-confined water (h_x) decreases in the flow direction in a manner as depicted by the piezometric grade line. The upward flow from the semi-confined aquifer to the phreatic groundwater is described by:

$$q_x = K' \frac{h_x - H_2}{D'} = \frac{h_x - H_2}{c}$$

Eq. 17.13

where:
q_x = upward seepage flow (m/d)
h_x = head (m)
H_2 = head (m)
K' = vertical hydraulic conductivity of the semi-permeable layer (m/d)
D' = thickness (m) of the semi-permeable layer (often combined into c = D'/K', *the hydraulic resistance* of the semi-permeable layer, in days).

The horizontal seepage flow through the aquifer (Q_x in m³/d per m width) is given by;

$$Q_x = -KD \frac{dh_x}{dx}$$

Eq. 17.14

This flow decreases in the flow direction, the rate of decrease being equal to the rate of upward seepage:

$$\frac{dQ_x}{dx} = -q_x$$

Eq. 17.15

Combining Eqs. 17.13, 17.14 and 17.15 results in:

$$Q_x = cKD \frac{d^2Q_x}{dx^2}$$

Eq. 17.16

Where the semi-confined aquifer extends infinitely both outside and inside the polder, Eq. 17.16 may be solved to show that (ILRI 1972):

$$Q_x = Q_0 e^{-x/\sqrt{cKD}} \quad \text{where; } Q_0 = \frac{\Delta H}{2}\sqrt{\frac{KD}{c}}$$

Eq. 17.17

$$q_x = q_0 e^{-x/\sqrt{cKD}} \quad \text{where; } q_0 = \frac{\Delta H}{2c}$$

Eq. 17.18

$$h_x = H_2 + \frac{\Delta H}{2c} e^{-x/\sqrt{cKD}}$$

Eq. 17.19

Defining the effective extent (x_{eff}) of the seepage zone inside the polder as the zone in which $q_x > 0.1q_0$, it follows from Eq. 17.18 that:

$$x_{eff} = -\ln(0.1)\sqrt{cKD} = 2.3\sqrt{cKD}$$

Eq. 17.20

The piezometric grade line forms an asymptote to both the levels H_1 (outside the polder) and H_2 (inside). Half of the driving head ΔH is used outside the polder, half inside. Where the semi-permeable layer does not extend outside the polder, the full head is available for the seepage flow into the polder (so the head in Eq. 17.17, 17.18 and 17.19 becomes ΔH instead of ½ΔH).

17.4.2   Phreatic flow

In this case, the seepage takes the form of phreatic groundwater flow as shown in Figure 17.8. The inflow into the polder may be estimated by the *flow-net method* or by modelling (see section 21.3.3).

An analytical solution has also been derived by conformal mapping in which the B/D-ratio is used to characterise the geometrical situation. The inflow is given by:

$$Q_0 = K\Delta H\beta \qquad\qquad\qquad\qquad \text{Eq. 17.21}$$

where:
Q_0 = seepage flow passing beneath the dike into the polder (m³/d per m length of dike)
K = hydraulic conductivity of the flow zone (m/d)
ΔH = head (m)
β = geometry factor, function of B/D (see Figure 17.8 and Figure 17.9)

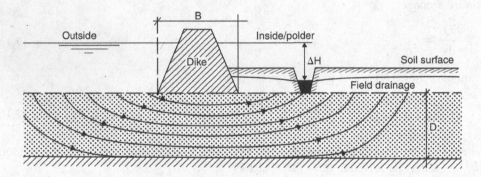

Figure 17.8 Seepage into a polder by phreatic flow

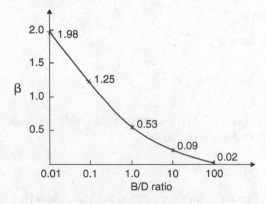

Figure 17.9 The geometry factor β as a function of B/D (ILRI, 1972)

Examples

Semi-confined flow (with semi-permeable layer not extending outside the polder): $D' = 5$ m, $K' = 0.01$ m/d, $c = 500$ days; $D = 200$ m, $K = 25$ m/d, $KD = 5000$ m²/d, $\Delta H = 2$ m. Using Eq. 17.17 and Eq. 17.20:

$$Q_0 = 2\sqrt{\frac{5000}{500}} = 6.3 \text{ m}^3/\text{d per meter length of dike}$$

$$x_{eff} = 2.3\sqrt{500 \times 5000} = 3636 \text{ m}$$

Phreatic flow: $B = 50$ m, $D = 200$ m, $B/D = 0.25$ which gives $\beta = 0.87$ (Figure 17.9). $K = 10$ m/d, $H = 2$ m. Applying Eq. 17.21: $Q_0 = 10 \times 2 \times 0.87 = 17.4$ m³/d per m length of dike.

Polder: square polder of 10×10 km, area 10 000 ha, dike length 40 km. Seepage load for the semi-confined case: $40\ 000 \times 6.3 = 252\ 000$ m³/d $= 2.5$ mm/d and for the phreatic case: $40\ 000 \times 17.4 = 696\ 000$ m³/d $- 7.0$ mm/d. The seep area in the semi-confined case would essentially be restricted to a 3 600 m wide strip inside and adjacent to the dike. In the phreatic flow case, the seep area may be determined by the *flow-net method* or by modelling (usually restricted to a narrower strip, 2 to 3 times D or in this case 400–600 m). There will obviously be overlap and inter-action of seepage flow in the corners where the seepage flow will be from different directions which are neglected in these simple calculations.

17.5   Seep zones and springs

Seep zones typically occur in situations such as the one depicted in Figure 17.10 where at the foot of a hill slope the hydraulic grade line of the phreatic seepage flow emerges onto the surface (*day-lighting*). This is likely to occur where slopes change abruptly from steep to flat, although much depends upon the soil, geological conditions and stratification.

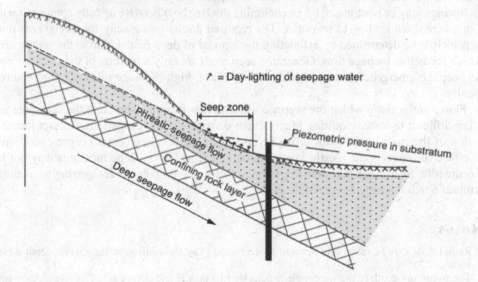

Figure 17.10 Seep zones and springs at the foot of hill slopes

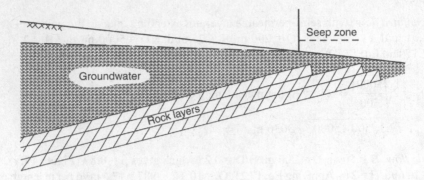

Figure 17.11 Seep zone due to daylighting of rock layers

In addition, there may be a seepage flow through deeper strata from the hills emerging in the valley. Springs arise when this seepage creates (semi) confined groundwater conditions under the valley from which water spills to the valley floor through fissures, faults or other leaks in the overlying layers. Typically, a fault line occurs along the foot of hills, giving rise to a spring line. Seep zones are also commonly caused by upward sloping and/ or day-lighting rock layers which force moving groundwater to the surface (Figure 17.11). Naturally, seep areas often abound with reeds, rushes and similar wetland indicator plants.

Seepage zones as in Figure 17.10 and Figure 17.11 may in principle be drained by intercepting the causal seepage flow above the wet spot, just about where the watertable emerges onto the surface of the land. The interceptor drains (ditch or pipe drain) should be deep enough to cut well into the layer transporting water. As described in section 17.2.1 the deeper the drain is installed in this layer, the more water it will intercept and for full interception the drain should reach to the base of the water transporting layer.

Springs may be neutralised by an encircling ditch or be otherwise directly connected with a drainage ditch leading to an outlet. The required discharge capacity for a seep area may in principle be determined by estimating the amount of deep percolation in the source area which feeds this seepage flow. Generally, seep areas are only a fraction of the source areas. As seepage emergence is often highly non-uniform, high drainage rates may be required locally.

Flows paths along which the seepage water moves from the source to the seep area are often difficult to trace, requiring in any event considerable investigations. Correct identification of the main water transporting layers (often just a vein or crack) is pre-conditional to effective interception. Costly investigation and remedial drainage measures may not be worthwhile. Seep areas may sometimes be used for rice growing or for grazing/haymaking without much drainage being required.

Notes

1 Radial flow may be taken into account by replacing D by the equivalent Hooghoudt depth d (see section 7.3).
2 The minimum depth to the watertable is actually less than H and occurs at ½ L where dh/dx = tgα; for gentle slopes the difference is negligible.

Chapter 18

Reclamation and drainage of unripened soils

Unripened soils are encountered in the reclamation of sea, lakebeds, marshes and swamps, floodplains, coastal forelands, etc. The overall appearance of these soils after emergence from under water is that of soft mud. Following drainage certain changes take place in the mud, which together are referred to as *ripening* of the soil. Ripening precedes the common soil formation processes leading to full development of the soil profiles.

Much of the knowledge about the reclamation of unripened soils has been gained during the impoldering of four polders in the former Zuiderzee in the Netherlands. This work started in 1927 with the construction of the closure dam which over time converted the former sea-arm into a freshwater lake (IJsselmeer) due to several rivers flowing into it. While the first polders were reclaimed under saline water conditions the latter three were reclaimed under brackish to freshwater conditions. Various reclamation experiences from elsewhere around the globe have also been included in the discussions.

18.1 The soil ripening processes

Soil ripening of fresh sub-aqueous sediments starts with the loosely packed, over-saturated mud losing (permanently) part of its excess water, initially mostly by evaporative drying, later also by drainage. The watertable falls and the soil above the watertable becomes exposed to capillary forces that pull the soil particles into a closer packing which is enhanced by the fact that they no longer float, supported by the buoyant forces, but become subjected to gravity forces and particle pressures. In this process the soil particles are also spatially rearranged, all in all resulting in a permanent reduction of the soil volume, being partly shrinkage (caused by the capillary forces) and partly settlement (caused by the increase in particle pressures). This induces subsidence, cracking and in the end, structural development of the soil, together comprising the *physical ripening* of the soil. The higher the colloid content of the soil the more intensive the ripening and in fact sandy sediments, lacking colloids, hardly ripen at all.

18.1.1 Physical ripening

A prominent feature of the physical ripening process is the change in phase composition (Figure 18.1). The pore volume (V_{pores}) reduces markedly and consequently the bulk density (BD) of the soil increases as the soil ripens. As long as the soil remains saturated, this

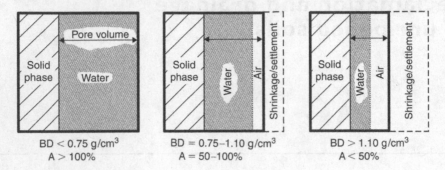

$$BD < 0.75 \text{ g/cm}^3 \qquad BD = 0.75-1.10 \text{ g/cm}^3 \qquad BD > 1.10 \text{ g/cm}^3$$
$$A > 100\% \qquad\qquad A = 50-100\% \qquad\qquad A < 50\%$$

Figure 18.1 Changes in the phase composition of the soil during the ripening of the soil (data apply to a clay soil)

decrease in pore space shows as a decrease in the water content of the soil as measured by the *A-value* (= water content of the saturated un-ripened soil in weight % (%w) i.e., g water per 100 g dry soil). Under these conditions, it holds that:

$$BD = \frac{100}{100/\gamma_s + A}$$ Eq. 18.1

$$\% \text{ pores} = BD \times A$$ Eq. 18.2

where γ_s = specific density of the solid phase and BD are both in g/cm³. Taking for the average specific density of the soil mineral and of the organic material (OM) respectively 2.65 and 1.45 g/cm³, it may be shown that γ_s = 2.65 – (%OM/100) (2.65–1.45).

The consistency of the soil changes as the soil ripens, from nearly liquid to firm/hard. For field assessment, consistency grade scales have been developed enabling the experienced soil surveyor to arrive at fair estimates of the current stage of ripening of the soil on the basis of a simple hand kneading of the soil (Pons and Zonneveld 1965).

Progress of the ripening with depth and time

The ripening of a mud soil normally starts at the surface, initiated by evaporative drying, from where it slowly extends to the deeper soil layers. The rate of ripening at all depths decreases as the ripening of the soil progresses while the rate of extension into the soil also decreases as the ripening reaches deeper in the soil. Both trends are clearly shown in Figure 18.2. The first trend is an inherent characteristic of the ripening process while the second trend is mostly due to the slower rate of drying deeper in the soil. Full ripening of a soil profile to 1.5–2.0 m depth may take centuries and typically, subsoils remain unripened long after reclamation. Ripening generally does not extend below the regional drainage base or below the deepest fall of the watertable as e.g., reached at the end of a dry period.

Since much of the water-loss during the initial ripening period is due to evaporative drying, it generally holds that the ripening progresses more during dry years than during wet years. Good drainage, however, is most essential for rapid and deep ripening.

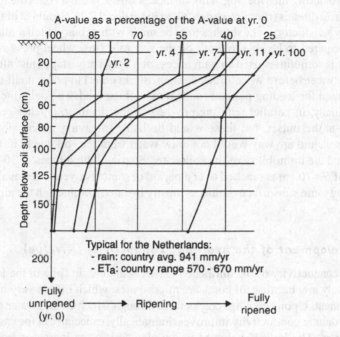

Figure 18.2 The progress of the ripening in the soil profile with time and with depth (light clay soil in the Zuiderzee polder, the Netherlands)

18.1.2   Other ripening processes

In parallel with the physical ripening, various chemical and biological changes occur in the soil (respectively referred to as the chemical and biological ripening of the soil).

The *chemical ripening* comprises the oxidation processes and the adjustments in the cation composition of the adsorption complex of the soil. The oxidation of the sulphides found in marine sediments, which may lead to the formation of acid-sulphate soils, is of particular concern (section 18.3). Many marine soils will also be salt affected and will have to be leached before they can be used to grow crops.

The *biological ripening* of the soil involves the development of aerobic microbiological life within the soil (Figure 18.2). This may take time, especially the development of an adequate nitrification capacity in the soil. Inoculation of the soil may be required to establish an effective nodule type of bacteria population for the fixation of atmospheric nitrogen.

18.2   Reclamation and drainage methods

Reclamation and drainage of fresh sub-aqueous sediments are closely related. During the initial ripening period they are in fact largely synonymous. During later stages, drainage develops its own particular characteristics.

18.2.1   Early reclamation stage

During this period, the hydraulic conductivity of the still unripened soil is usually very low and drainage only entails maintaining the surface clear of ponded water. As such, however, it performs a very useful function in that it maximises the evaporation of water from within the soil, thus promoting the ripening. This surface drainage is most conveniently undertaken using parallel field ditches (the system described in section 9.2.1). Spacings of the order of 10 m are generally satisfactory. The ditches may be made with a rotary ditcher although special equipment is required to be able to work on the soft soils (low wheel pressures). Amphibious equipment is sometimes used to gain access at a very early stage (just after emergence from under the water before wheel/track equipment can enter) to press small drainage pathways into the mud for leading ponded (rain) water off the land. The field ditches should be deepened gradually, in parallel with the progress of the ripening. Ditches could of course be made deep at the outset, but these would be liable to caving in and sloughing in the unripened subsoil, and anyway would not draw water from the soil due to its low hydraulic conductivity and the immobility of the soil water. Starting with ditches of 30–40 cm depth, a ditch depth of 60–70 cm is reached over a period of some five years. Gradually the ditches start performing some subsurface drainage, mostly by interflow through the already partially ripened top layer.

18.2.2   Development of the hydraulic conductivity (K-value)

The hydraulic conductivity of the unripened clayey sediment, in spite of the large pore volume, is extremely low because all pores are micro-pores, which offer a very high resistance to water movement. Upon ripening, cracks form through which excess water can move rapidly and the hydraulic conductivity improves dramatically, especially in the case of clay soils which crack most. The bulk K-value of heavy clay soils often increases from only a few mm/d in the unripened sediment to several m/d in the cracked, ripening soil (the hydraulic conductivity of the soil mass between the cracks remains low). When the crack volume represents irreversible shrinkage, the cracks remain intact when the soil is re-wetted. This is common with the ripening of fresh marine sediment in the Netherlands. Ripening cracks in some soils, however, are known to close again by swelling in the rainy season, probably due to differences in the clay mineralogy.

In tropical coastal swamps, roots of the cleared mangrove vegetation may leave large well-preserved channels, which confer on the soil a good hydraulic conductivity. The ripening history also has an influence. Naturally pre-ripened sediments generally develop less cracking upon reclamation, resulting in lower hydraulic conductivity values than fully unripened sediments. Future K-values, therefore, depend on a number of factors, the influence of which are not well understood, making predictions difficult. Much depends also on the soil structural development taking place during the later stages of the physical ripening, parallel and integrated with the chemical and biological ripening of the soil. In general, very little improvement of the hydraulic conductivity is to be expected due to ripening cracking when the existing bulk density (BD) values of the soil have attained values of 1.5-1.6 g/cm^3. More ripening cracking is generally to be expected in humid climates than in semi-humid/semi-arid climates since in the latter climates some ripening may already have occurred prior to reclamation during the prolonged dry periods.

18.2.3   Advanced reclamation stage

By the time the ripening front has extended well into the subsoil (say to some 60–70 cm depth) pipe drains may be installed if desired. In the Netherlands this was done normally after some five years of ripening, using drain depths of 1.0–1.2 m. These pipe drains function initially very much as the drains described in section 9.4 (Figure 9.14). The excess rainwater sinks in vertically through the cracks, then flows laterally over the impermeable unripened subsoil to the trenches and finally vertically down through the backfill to the pipes. The trenches should be backfilled with well-ripened soil and when this is done carefully and under dry conditions, experience indicates that fine textured soils generally do not need a pipe envelope. Earlier installation of pipes is generally not advisable as firstly their effectiveness is restricted by the limited ripened depth, secondly the pipe lines may become disturbed and mis-aligned by the considerable subsidence still to occur, while thirdly, due to this subsidence, the drain depth would also in time become too shallow. Even after five years the pipes should be laid at some 10–20 cm over-depth to allow for subsequent subsidence of the overlying soil.

The spacing to be used depends very much upon how the K-value develops upon ripening. When a good number of stable ripening cracks become established, spacing up to 30–50 m may well be used for clay soils. Where considerable further cracking is expected, the pipe drains may be extra widely spaced, maintaining for the time being some of the field ditches. In the spacing calculations, the deeper unripened subsoil should in general be treated as an impermeable base for the subsurface flow. Pipe drainage design for date palms in the Karun delta in southern Iran takes into account the fact that the soil below 1.5 m, the regional drainage base, will remain unripened and virtually impermeable.

18.2.4 Drainage, evaporation and ripening

Good drainage enhances the ripening considerably. Drainage can, however, only remove the soil water above field capacity (FC). In the case of unripened soils most of the pore water (the pores being almost all micro-pores) is held below field capacity, and so drainage involves mostly the removal of water from the surface and the cracks. The removal of pore water from unripened soil depends mostly on evaporative drying, which of course is most effective when all the non-pore water has already been removed by drainage. Evaporation losses from bare soil diminish in time due to the formation of a protecting mulch layer (section 15.2.2), although in this case the reduction in the rate of loss is partly offset by the formed cracks. Through the cracks the subsoil is able to continue to lose water by evaporation. This occurs also, and to a much larger extent, when there is vegetation on the land.

Vigorous vegetation with a deep and extensive root system which is able to thoroughly deplete the rootzone, helps to promote and to extend the ripening to deeper layers. Mainly for this reason, the land to be reclaimed in the Netherlands is sown with reed (broadcast by aircraft). This has the additional advantage that the land will not be infested by all kinds of weeds. The first crops follow this reed, usually colza (rape seed), followed by wheat and barley. Alfalfa is also useful during this early cropping period as it is deep rooting and adds nitrogen to the soil.

The ripening is considerably retarded, often even stagnating at a shallow depth, when there is seepage inflow into the area. Intercepting the seepage inflow can accelerate the ripening in such areas. Tropical lowland soils used for wet rice cultivation will generally remain unripened except for a shallow top layer, which periodically dries out during the off-season. The application of irrigation will in general of course also retard the rate of soil ripening.

18.3   Acid sulphate soils

Seawater is normally rich in sulphates, as are marine and brackish water sediments. In the presence of fresh organic matter, the sulphates in the unripened/non-aerated sediments change into sulphides by microbiological reduction. These then combine with the Fe ions to FeS (ferrous sulphide), giving these sediments their typically dark blue appearance. In time, the FeS will turn into pyrite (FeS_2) with soil colours becoming somewhat greyer. The pyrite may accumulate, seabed soils usually having a higher concentration than tidal lands. Although river water is usually low in sulphur, low-lying basin sediments may become enriched by accumulation. For all sediments, it holds that they contain more sulphur the higher their clay content.

18.3.1   Acidification processes

Upon reclamation, drainage and ripening of these sediments, air will enter the soil and the sulphides will become oxidized through one of the following reactions:

$$4\,FeS + 10H_2O + 9O_2 \rightarrow 4Fe(OH)_3 + 4H_2SO_4$$

$$4\,FeS_2 + 14H_2O + 15O_2 \rightarrow 4Fe(OH)_3 + 8H_2SO_4$$

$$12\,FeS_2 + 30H_2O + 45O_2 + 4K^+ \rightarrow 4KFe_3(OH)_6(SO_4)_2 + 16H_2SO_4 + 4H^+$$

(jarosite)

The oxidation of the FeS is a rapid process but some forms of aged FeS_2 are rather inert to oxidation, especially when the soil reaction is alkaline. The oxidation processes are promoted by the sulphur-oxidising bacteria in the soil while the role of the Fe ions can be taken over by Al ions, in the latter case leading to the formation of $Al(OH)_3$. The sulphuric acid (H_2SO_4) may react with the $Fe(OH)_3$ or the $Al(OH)_3$ to form salts which, because they are composed of weak bases and a strong acid, will readily hydrolyse to form some Fe/Al hydroxide-sulphate compounds and again sulphuric acid. The soil reaction may become strongly acidic with the pH values falling as low as 2–3, although pH = 4 is more common as some of the hydrogen ions formed will normally be leached from the soil. Under such conditions, crops suffer from a lack of phosphorus (becomes unavailable at low pH) and from toxicity (Al toxicity for dry land crops; Fe toxicity for wetland rice). The generally poor physical conditions of the soils also limit crop growth. Where such soils develop, the land is often abandoned.

18.3.2   Neutralisation and reclamation

In calcareous soils the sulphuric acid is neutralised and the pH values remain neutral when the above described sulphide oxidation processes take place in the ripening soil:

$$H_2SO_4 + CaCO_3 \rightarrow CaSO_4 + H_2O + CO_2$$

This neutralisation may also be achieved by adding lime to the soil. The lime requirements are high although they may sometimes be somewhat reduced by applying some leaching before liming of the soil (e.g., waiting for one to two rainy seasons to pass or by leaching the soil by irrigation). For clay soils, up to 50 tonnes $CaCO_3$ per ha may be required although lighter textured soils can be neutralised using less. Deep ploughing may sometimes be used to bring up calcareous subsoil to mix with the acidic top layer. Flooding with base rich water (i.e., water high in Ca and Mg ions) is also useful as these cations can replace H ions on the complex and subsequently leach these replaced H ions from the rootzone. Sea water may be used for this purpose, provided that the resulting salinisation/sodification of the soil can be readily corrected afterwards.

The acidification can be stopped/prevented by maintaining high watertables (keeping the soil under water, out of contact with air). In theory watertables may be lowered gradually at such a rate that leaching, alone or in combination with liming, maintains the balance between H production and removal/neutralisation, but few examples of successful application of this approach exist. For crops like grass (temperate climates) and rice (tropical climates) shallow rootzones suffice and the subsoils may be left un-reclaimed or under water. Of course, there will be little leaching under such high watertable conditions and the alternative (low water-tables maintained by deep drains) might be preferable when neutralisation can be achieved at low costs. Pipe drains should generally not be installed until the neutralisation has reached an advanced state as the pipes are liable to become clogged by iron ochre when installed too soon.

18.3.3 Diagnosis

Comparing the lime and the pyrite content of the sediment can be used to roughly assess the chances of acid sulphate soil formation during the ripening/drainage of muddy sediments. There is a clear hazard when the lime concentration of the sediment is equal to or less than the pyrite concentration (both expressed in meq/100 g dry soil with the pyrite converted into SO_4). The unripened sediment may also be left exposed to aeration and drying, and the pH development be investigated over a period of one to two months. To sustain the acidification processes, the sample should be kept moist and the temperature should not be too low. The changes in colour of the sample are often already indicative (yellow jarosite mottling).[1] A number of rapid evaluation tests have also been developed (ILRI, 1973). Pyrite accumulation is especially to be expected in tropical mangrove swamps due to the abundance of organic matter. Reclamation in such areas almost universally faces the problem of acid sulphate soil formation.

18.4 Subsidence prediction

The shrinkage and settlement of unripened soil upon drying (and the underlying compression/re-arrangement of the soil skeleton) is essentially irreversible. Some of the original volume is usually regained by swelling once the soils are rewetted. The irreversible part of this will be seen as a long-term gradual decrease of the soil volume, superimposed on which shorter cycled reversible changes in soil volume may occur (Figure 18.3).

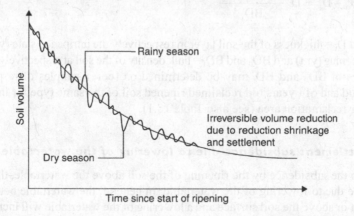

Figure 18.3 Reversible and irreversible soil volume reduction during the ripening of clay soils

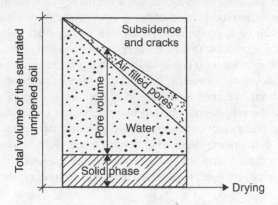

Figure 18.4 Changes in phase composition of the unripened soil upon drying

The soil volume reduction reveals itself in the field by subsidence of the soil surface *(vertical contraction* of the soil skeleton) and by the formation of cracks *(horizontal contraction)*. In unripened soils, subsidence and crack formation will occupy most of the soil volume vacated by the water when the soil ripens, the air-filled pore space often increasing very little, see Figure 18.4 (this contrasts with non-shrinking ripened soils which have rigid skeletons and water lost from the pores will be wholly replaced by air).

18.4.1 Ripening subsidence

Cracks are filled up from the flanks and/or from the surface and in the long run all ripening shrinkage and settlement translates itself into a lowering of the soil surface (subsidence). The total subsidence due to ripening may thus be estimated as:

$$D_t = \frac{BD_0}{BD_t} D_0$$

Eq. 18.3

$$\Delta D = D_0 - D_t = D_0 \frac{BD_t - BD_0}{BD_t}$$

Eq. 18.4

where: D_0 and D_t = thickness of the soil layer in respectively the unripened state (yr 0) and after 't' years of ripening (yr t) and BD_0 and BD_t = bulk density of the soil at respectively yr o and yr t.

The values of BD_0 and BD_t may be determined on (core) samples of respectively the unripened mud and of t years old reclaimed/ripened soil of the same type for instance taken from an older reclamation area (see also Table 18.1).

18.4.2 Settlement subsidence due to lowering of the watertable

In addition to the subsidence by the ripening of the soil above the watertable, there will also be subsidence due to lowering of the watertable. In general, the watertable before reclamation will be at or above the soil surface and a lowering of the watertable will increase particle pressures throughout the whole of the soil (Figure 18.5). This settlement subsidence in the soil may be estimated according to Terzaghi (ILRI 1974) as:

$$\Delta z = \frac{D}{c} \ln \frac{\sigma_p + \Delta\sigma_p}{\sigma_p}$$
Eq. 18.5

where:
Δz = the settlement subsidence (cm)
σ_p = the average particle pressure in the soil layer considered (g/cm²)
$\Delta\sigma_p$ = increase in particle pressure due to the lowering of the watertable (g/cm²)
D = thickness of the soil layer considered (cm)
c = consolidation coefficient (to be determined by means of a laboratory test).

Under hydrostatic conditions, σ_w is positive below and negative above the watertable, its value corresponding to the heights respectively below and above the watertable. The value of σ_s in a fully saturated soil at depth H below the soil surface is:

$$\sigma_s = H\left(\frac{\% \text{pores}}{100} + BD\right)$$
Eq. 18.6

It may generally be assumed for an unripened soil that as the watertable is lowered, the soil above the watertable remains (nearly) saturated. The value of the particle pressure σ_p follows from $\sigma_p = \sigma_s - \sigma_w$ which gives, again for a fully saturated soil:

watertable at soil surface (Figure 18.5a):

$$\sigma_p = H\left(\frac{\% \text{pores}}{100} + BD - 1\right)$$

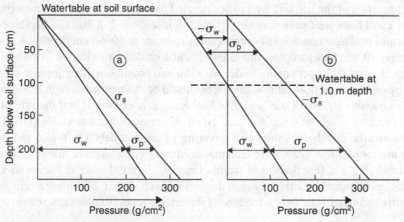

σ_w = Pressure in the water
σ_p = Pressure on the soil particles
$\sigma_s = \sigma_w + \sigma_p$ = Total pressure exerted by the soil and water

Figure 18.5 Changes in the pressures in the soil due to a lowering of the watertable (numerical values based on unripened clay soil)

Table 18.1 Average values of the physical soil constants for settlement subsidence prediction

	Specific density solid phase (g/cm³)	% pores	Bulk Density BD (g/cm³)	c Consolidation coefficient
Sand: loose packing	2.65	45	1.45	100
dense packing	2.65	40	1.60	150
Loam, ripened	2.65	50	1.30	25
Clay: ripened	2.65	60	1.00	20
unripened	2.65	80	0.50	8–10
Peaty clay, half-ripened (50% OM)	2.00	60	0.80	6–8
Peat, drained	1.40–1.60	80–90	0.15–0.30	4–5

watertable at depth W below the soil surface (Figure 18.5b):

$$\sigma_p = H \left(\frac{\% \, pores}{100} + BD + \frac{W-H}{H} \right)$$

The value of σ_p in Eq. 18.5 may thus be readily determined given the values of the soil constants. Ideally, these should be determined for each case but for rough estimates, the values given in Table 18.1 may be used. The value of $\Delta\sigma_p$ (in g/cm²) under these saturated conditions is numerically equal to the fall of the watertable in cm. For details on the applied calculation methods see ILRI (1973).

18.4.3   Oxidation of peat soils

Additional subsidence is to be expected when peat/peaty soils are reclaimed/drained, as the improved aeration of the soil will lead to accelerated oxidation of the organic material. Under temperate conditions peat soils may be expected to lose some 1–2 mm soil depth per annum under normal drainage conditions (watertable in summer at 50–60 cm depth). At the higher temperatures of the (sub) tropics, the rates of soil loss due to oxidation are always much higher, up to several cm per annum under well-drained conditions. To minimise subsidence/ soil loss by oxidation of peat soils, watertables should be maintained as high as is compatible with the (agricultural) use of the land. For Malaysia, it is estimated that the rate of subsidence will increase by about 4 mm for each 10 cm lowering of the watertable (DID/LAWOO 1996). On swamp/marsh soils the usual covering of organic debris will also mostly disappear in a matter of a few years after reclamation/drainage, the organic matter content of the topsoil stabilising at a few percent in warm climates, slightly higher in temperate climates. The oxidation of organic matter deeper in the soil is much slower, becoming negligible in the permanently saturated zone below the lowest watertable level (summer/dry season level).

18.4.4   Experiences in the Netherlands

No great accuracy can be expected from the predictions of subsidence based on Eq. 18.4 and Eq. 18.5, the accuracy in fact depending almost entirely upon the accuracy with which the final bulk density value of the ripened/half-ripened soil can be estimated. The Terzaghi formula (Eq. 18.5) was developed for instantaneous loading while in land reclamation the loading due to lowering of the watertable is rather gradual. Best results are obtained when the coefficient c is determined in the laboratory under loading conditions rather similar to those existing in the field. The Terzaghi formula also supposes the increase in load to be fairly uniform across the area considered and the pressures in the groundwater at all times to correspond to the prevailing hydrostatic pressures. This last condition will prevail in freely draining soils though in clay soils of low hydraulic conductivity, excess soil water pressures may continue to exist for a considerable time after the lowering of the watertable.

Rates of subsidence

The ripening subsidence may be calculated with Eq. 18.4 at any time after the start of the reclamation. In contrast, Eq. 18.5 only enables the final settlement subsidence to be predicted. In peat soils, some 90% of the final subsidence is often reached within two years of the initial reclamation. In total, peat layers may reduce to some 60–70% of their original thickness before reclamation/drainage (ignoring the loss due to oxidation which continues). Conditions, however, vary considerably depending upon the prevailing conditions during the formation of the peat. Peat may, for example, have an extremely low bulk density (and consequently subside considerably upon reclamation) or it may be rather firm (higher bulk density) either due to the nature of the peat itself or to preceding ripening and settlement during historically low rainfall/better drainage periods.

The settlement in clay soils is much slower than in peat soils. Much of the subsidence in fresh sub-aqueous mud is due to ripening, the rate of which depends to a great extent upon the prevailing drainage and weather conditions. Typically, ten years after reclamation, some 50% of the final subsidence will have occurred. The ripening subsidence is usually restricted to a depth of about 100–150 cm (= average lowland watertable depth). The settlement subsidence affects a much deeper layer. In the Netherlands it is estimated that during the first 100 years after reclamation, the total subsidence of the subsoil below 150 cm (post-subsidence), amounted for peat, clay, sandy clay and sand respectively to some 10, 7½, 5 and 2½ cm per m soil layer. The subsidence continues beyond this period but at such a slow rate that it only needs to be taken into account in very long-term planning (periods of 500 to 1 000 years).

Effects

The subsidence affects the planning of the drainage system for the reclaimed area. Starting from the present levels (pre-subsidence), the soil map and the subsidence calculations enable future levels and lie of the land to be predicted, based on which canal alignments and locations of outlets may be planned. Medium term (10 to 20 years) and long-term (~ 100 years) subsidence predictions should be made and be related to those engineering works with a life (technical and/or economic) of the same order of magnitude. Separate calculations based on detailed site investigations should be made for all-important structures.

The subsidence of peat soils is roughly proportional to its original thickness and so the post-reclamation topography of peat land often closely reflects the relief of the underlying substratum. Much depends also on the drainage conditions. In the Netherlands, for example, high watertable levels are often maintained in peat land during the summer to stop subsidence and oxidation (see also the discussion on controlled drainage in Chapter 20).

Note

1 The popular name *cat clays* for acid sulphate soils stems from this mottling which give the soils an appearance rather similar to cat-droppings.

Chapter 19

Drainage of rice lands

Rice is one of the principal food crops of the semi-tropical/tropical zone. It is distinct from almost all other important crops in that it has a well-developed internal conduit system for aerating the roots from the parts growing above the ground, which facility allows rice to grow well under anaerobic waterlogged conditions. Rice is therefore usually the crop of choice for (semi) humid tropical lowland plains where waterlogging conditions (high watertables and/or surface ponding) prevail naturally during much of the wet season or may readily be created by the retention of rain within bunded fields and/or by supplying some additional irrigation water. In the hydro-topographically higher or lower parts of the plains that are less suitable for wet rice culture, other cultures (upland rice and deep water/floating rice, see Figure 19.1) may be grown.

Wet rice may be established by transplanting from a nursery or by direct seeding. The transplanting is usually done in a puddled saturated soil. Puddling establishes a soft soil for easy transplanting and creates a poorly permeable bed in the fields. Under direct seeding, the seed is broadcast either in a dry seedbed (*dry seeding*) or on a wet soil (*wet seeding*). The water layer is established after the seeding. *Upland rice* is grown very much like normal dry foot crops. *Deep water rice* varieties have a stem growth ability which can keep up with a flood water rise of several cm's per day which allows it to survive in several meters of water. The yields of the latter cultures are however as a rule much lower than those of wet rice. Various intermediate type rice cultures also exist, e.g., rice planted in receding flood water and maturing on the residual soil moisture.

Wet rice requires sufficient surface drainage to keep the water depth on the fields under control. Prolonged periods of excess depth as may occur under uncontrolled conditions during the rainy season can greatly reduce yields. Subsurface drainage is applied in some countries but the need for this type of drainage is not yet generally accepted.

19.1 Surface drainage

Almost all wet rice in the developing countries is grown in areas with historically developed layouts with small (0.10–0.25 ha) and irregularly shaped fields as shown in Figure 19.2. The fields are formed or subdivided into small basins by the construction of small bunds. When irrigated, the water is supplied from a higher lying secondary/tertiary canal while the drainage is towards a lower lying drain.

Internal drainage systems within a block of fields are almost always lacking and the collection of the drainage water within the block depends mostly on field-to-field movement of the water (with the water either overflowing the bunds or passing through dedicated openings). The density of the existing main drainage infrastructure is usually not more than 5–10 m per ha which may be conceived of as an equivalent spacing of 1 000–2 000 m.

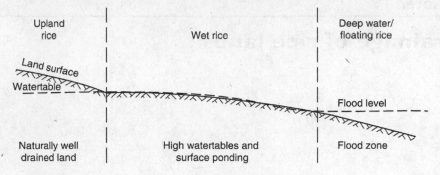

Figure 19.1 Rice cultures for different hydro-topographic zones

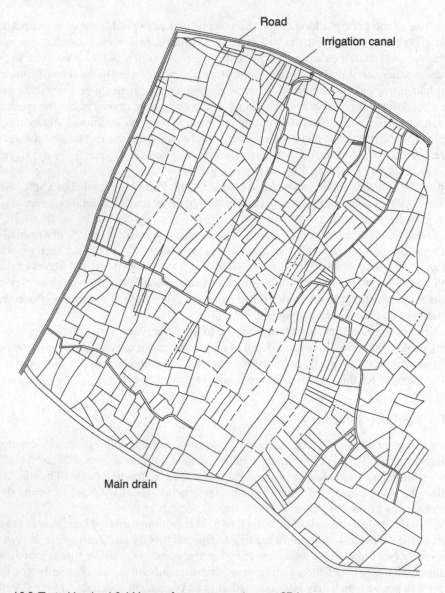

Figure 19.2 Typical lowland field layout for wet rice culture in SE Asia

Fields located at mid-spacing therefore are some 500–1 000 m from a drain and the field-to-field drainage has to travel this distance to reach it.

Modern layouts with good sized rationally dimensioned fields and fully developed internal irrigation, drainage and road systems can be found in most of the rice growing areas in the developed countries (Australia, Spain, USA, etc.). Japan has implemented large scale land consolidation and modernization programs, aimed at arriving at standard 3 000 m² fields (30 × 100 m) with irrigation, drainage and road facilities running along the short ends. Modern layouts may also be found in some recently developed areas in the developing countries (often polder type reclamation of marshes, coastal foreland, etc.). Some developing countries have also made a modest start with land consolidation and the stepwise modernisation of the traditional layouts (Taiwan, Thailand).

Drainage needs

Although over-irrigation may contribute, rainfall is almost always the most critical source of excess water on the rice fields. Therefore, the drainage requirements are usually derived from the rainfall falling directly on the rice fields. Runoff from surrounding uplands should be dealt with separately (intercepted and by-passed around or through the rice area without entering the fields, see earlier discussion in section 10.2.3).

The design rainfall should in principle be based on an analysis of the expected drainage process and of the costs and benefits, although in practice it is usually based on local experience and rules of thumb. For most situations, the critical rainfall duration is taken at 2–3 days and the frequency at 5 years with an exceptional 10 years frequency for expected high damage cases.

Water depths

The water layer on the wet rice fields establishes the desired waterlogged growth conditions although it also helps in the weed control. The modern short stem high yielding varieties (HYV's) generally demand stricter water depth control requirements than the traditional varieties. Optimal water depths for modern rice farming are generally assumed to be of the order of 5–10 cm.

As shown in Figure 19.3 and in Table 19.1, the sensitivity to excess depth varies per crop stage with the crop being most sensitive during its panicle-formation stage.

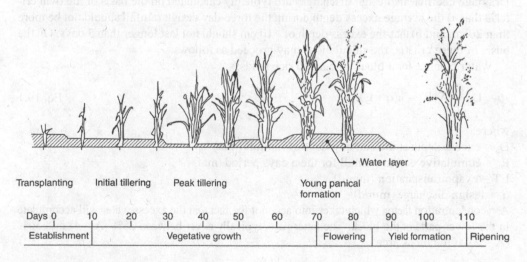

Figure 19.3 Stages of wet rice development

Table 19.1 Yield reduction % due to submergence (IR 30 variety Philippines, Bhuiyan and Undan in ILRI 1987)

Growth Stage	Crop height (cm)	Period of submergence in days		
		1	*3*	*5*
2 weeks after transplanting (early tilling)	30	25	61	84
		7	4	12
4 weeks after transplanting (mid/peak tilling)	48	25	38	95
		0	8	5
6 weeks after transplanting (panicle initiation	68	74	94	100
		10	8	11

Note: first line applying to full submergence and second line to half height submergence

Yield reductions due to excess depth generally remain quite limited provided the rice plants are not fully submerged for more than one day (warm period) to two days (moderately warm period). Submergence by muddy water is usually more harmful than by clear water as the mud may (partially) block the stomata and impose constraints on respiration and photosynthesis. Strong winds during full or partly submergence may also cause extra damage, especially when this occurs during the seedling stage. Wet seeding requires careful land preparation since even short periods of submergence of the young small seedlings can do considerable harm.

Good terminal (end of season) drainage enhances the uniformity of the ripening and creates improved harvesting conditions (especially for mechanical harvesting). The uniformity and thoroughness of the water removal in this case is more important than the discharge capacity. This also applies to any desired intermediate draining of the water layer e.g., in the mid-season drainage as is practised in Japan.

Drainage coefficient

Drainage coefficients/design discharges are typically calculated on the basis of the twin criteria that a) the average excess depth during the three-day design rainfall should not be more than 20 cm and b) that the excess depth of >10 cm should not last longer than 3 days. On the basis of these criteria, the calculations may proceed as follows:

Water balance for a block of bunded rice fields:

$$D_n = A\{R_n - n(q + ET)\}$$

Eq. 19.1

where:
D_n = excess depth at the end of n days of rainfall (mm)
R_n = cumulative design rainfall for the n days period (mm)
ET = evapotranspiration (mm/d)
q = design discharge (mm/d)
A = concentration factor which takes into account the fact that the excess water will accumulate in the lower parts of the block. The latter parts typically cover between 25% (A = 4) and 50% (A = 2) of the block area

Applying Eq. 19.1 for n = 1: $D_1 = A\{R_1 - (q + ET)\}$
$\qquad\qquad\qquad$ n = 2: $D_2 = A\{R_2 - 2(q + ET)\}$
$\qquad\qquad\qquad$ n = 3: $D_3 = A\{R_3 - 3(q + ET)\}$

First criterion: $(D_1 + D_2)/2 = \frac{1}{2}A(R_1 + R_2 - 3q - 3ET) < 20$ cm (200mm)

$$q = 1/3R_1 + 1/3R_2 - 400/(3A) - ET \qquad\qquad\qquad \text{Eq. 19.2}$$

Second criterion: $D_3 = A\{R_3 - 3(q + ET)\} = 10$ cm (100 mm)

$$q = 1/3R_3 - ET - 100/(3A) \qquad\qquad\qquad \text{Eq. 19.3}$$

Example: $R_1 = 110$ mm, $R_2 = 155$ mm, $R_3 = 180$ mm, ET = 5 mm, A = 3, applying Eq. 19.2 gives q = 40 mm/ d = 4.6 l/s/ha while applying Eq. 19.3 gives q = 44 mm/d = 5.1 l/s/ha of which the latter value is most critical.

The drainage coefficient may also be estimated by the graphical procedure shown in Figure 19.4. The design rainfall is represented by the rainfall duration curve for the chosen frequency. The water depth on the rice field at the beginning of the rainfall is assumed to be at the desired level (5–10 cm). As the discharge initially is less than the rainfall, an excess depth starts to build up in the fields, which may not exceed its prescribed maximum (100 mm in this example). Graphically this condition means that the discharge curve, starting at 100 mm

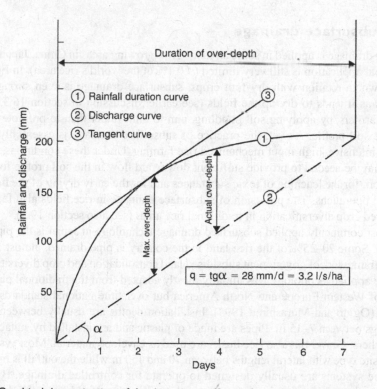

Figure 19.4 Graphical determination of the drainage coefficient for wet rice fields

permitted storage value on the vertical axis, should be tangential to the rainfall line. The build-up and depletion of excess depth can also be read from the graph and if found not to meet the requirements, a steeper discharge curve (greater q-value) should be chosen. It should be noted that in the case presented in Figure 19.4, it is assumed that the excess water will be uniformly distributed over the considered area ($A = 1$).

Applied values

Surface drainage systems for rice are designed for a design discharge range from as low as 1.0 l/s/ha (India /Pakistan) to as high as 10–15 l/s/ha (Japan). Calculated drainage coefficients for most significant rice countries in SE Asia are between these extremes:

- Malaysia: design rainfall 3 days, 1×5 years, q = 6.5 l/s/ha
- Philippines: design rainfall 2 days, 1×5 years, q = 6.7 l/s/ha
- Thailand: design rainfall 3 days, 1×10 years, q = 5.3 l/s/ha

In most of these countries, the applied design discharges are somewhat less than the calculated requirements while the actual discharge capacities, considering the generally poor state of maintenance of the drainage systems, are again less than the design values. It should also be emphasised that the drainage coefficients described apply to the design of the tertiary systems. For the design of the main systems covering larger areas (sub-basin and basin), area reduction factors as described in section 12.6 may be applied.

19.2 Subsurface drainage

Subsurface drainage is applied in some advanced rice growing areas in China, Japan and Korea but its global application is still very limited (< 0.1% of the world's rice area). In Egypt where rice is grown in rotation with dry-foot crops, subsurface drainage is even strongly resisted by farmers as it tends to dry up the fields (see earlier discussion in section 16.3.3). Indeed, most rice farmers, by applying soil puddling, aim to limit rather than to promote downward flow in the soil profile. As yet, the practice of subsurface drainage is essentially limited to areas with intensive, high input mechanised rice farming. Under these conditions, subsurface drainage may be needed to provide sufficient downward flow in the soil profile for enhanced root aeration, for the leaching of toxic substances and for the early drying of the fields for the mechanical operations. The provision of subsurface drainage in rice fields also facilitates the often-desired crop diversification in traditional rice areas (see also section 19.3).

The most commonly applied subsurface drainage technology in Japan is the pipe drainage technology. Some 20–25% of the rice land in the country is pipe drained, almost all installed within the framework of government subsidised land consolidation and crop diversification programs. The applied technology was initially mostly adapted from the traditional pipe drainage countries of Western Europe and North America but over time national standards have been developed (Ogino and Murashima 1993). Installation depths are usually between 50–70 cm and spacings between 7–15 m. Pipes are made of plastic and are installed by suitably adapted, small trenchers. Gravel but also rice husks are used as envelope material. Most systems are of the composite type with lateral lengths between 70 and 150 m while the outfall is by gravity.

The pipe systems are usually designed to operate for controlled drainage. By means of valves in the lateral junction boxes and/or in the collector outfall boxes, the drainage flow can be turned on and off (with the surface water entering the system through the outfall boxes). The design discharges are usually of the order of 20–30 mm/d. The systems are also designed to dry out the fields after the removal of the water layer for the mid-season and terminal drying of the fields. The systems also provide regular watertable control for the upland crops (beans, wheat) cultivated in rotation with the rice crop.

Mole drainage of rice fields is practised on a limited scale in some advanced rice cultivation areas in South China (depth 0.4–0.5 m, spacing 3–5 m, bamboo end-pipes). The outflow can roughly be controlled by raising/lowering the water levels in the outfall ditches. The installed mole drainage has allowed the introduction of rice/wheat rotations and an increase in the cropping intensity under continuous rice cropping. Mole drains, usually in combination with pipe drains (see Figure 9.13), are also used in Japan for the drainage of heavy clay soils.

19.3   Crop diversification

Crop diversification programs are in progress in many traditional rice-cultivating countries in the humid tropics, especially in SE Asia. The general aim of these programs is to promote the cultivation of more dry foot/upland crops and to replace the rice-rice rotation by a rice-upland crop rotation. A rice-upland crop rotation with the latter grown during the dry season generally requires no additional drainage measures. Growing upland crops during the wet season will, however, almost always require improved drainage.

The current cultivation of upland crops during the wet season is mostly restricted to the higher lying, better drained parts of the lowlands such as the riverine levee lands and bordering terrace lands. The wet season cultivation of upland crops in lowlands is exceptional and is mostly restricted to high value crops grown on raised beds in the home yards. Large scale cultivation of upland crops during the rainy season in rice fields requires extensive drainage measures. Under the prevailing high rainfall conditions, subsurface drainage systems can generally not provide adequate aeration control of the rootzone and the installation of such systems in the humid tropical lowlands is widely considered to be technically/economically not feasible (ILRI 1987). The Sorjan cropping system (Figure 19.5), developed by Indonesian farmers, i.e., a system that constructs alternate deep sinks and raised beds and is viable in both flood-prone and drought-prone area will be more appropriate. The construction of these systems is, however, highly labour intensive while also the cropping of such systems is only suited to small-scale labour-intensive farming. Cambered bed systems as described in section 9.1 would also be suitable for upland crops (e.g., applied for sugarcane) but not for wet rice.

Figure 19.5 Sorjan bedding system (left) and traditional rice paddy fields with nursery beds in the centre on the right

19.4   Flood control

In most humid tropical lowlands, land use is well adapted to the prevailing characteristics of the natural hydrological regime, specifically to the expected annual flooding of the land. The flooding is generally restricted to rather well-defined periods and develops slowly over a span of time, which patterns are taken into account in the cropping calendars. Full flood control is often technically/economically not feasible and often even undesirable as it has a number of adverse effects (blockage of the natural drainage by embankments, inadequate recharge of aquifers due to the containment of flood water, loss of fish spawning grounds). Significant progress in agricultural development is possible with only partial flood control.

Partial flood control can be practised at the village level by construction of small polders at low cost and with the full participation of the beneficiaries. These polders can be developed into multi-purpose water management units by providing them with inlet facilities for irrigation water and outlet facilities for excess rainwater (NEDECO 1991). Together with the progress made in the breeding of short duration rice varieties, partial flood control enables farmers to switch from one to two rice crops per year. In the course of time, with increasing development of the region/country, these small polders can be consolidated into larger polders with increasingly higher and stronger surrounding embankments, which provides better assurance of early flood protection and induces farmers to raise the level of their input spending. Ultimately, at some future date, these developments may lead to full protection of most of the flood prone area.

The above approach to flood control in tropical lowlands is aptly illustrated by the developments in the water management for rice growing in the Mekong Delta. As shown in Figure 19.6, the construction of low embankments in combination with low lift pumping,

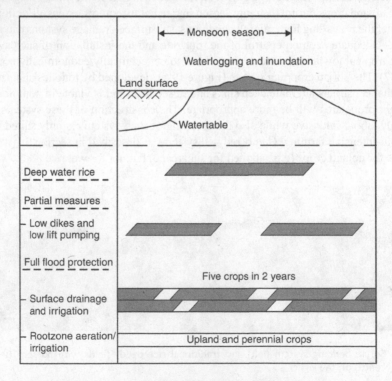

Figure 19.6 Water management for rice in the Mekong Delta (Vietnam)

allowed farmers to shift from the traditional one deep water rice crop to two improved local wet rice crops with a 100–110 days growing period, per year. The low embankments protect the land just long enough for the first crop to be harvested before the high floods arrive and the embankments are overtopped. The second crop is planted when the floodwater starts to recede. When planted early, the crop can mature before the onset of the dry season (with the pumping helping to achieve early evacuation of the flood water and also providing supplemental irrigation water in case the residual water does last till the end of the season). Under full flood control (higher land or in future), year-round rice cropping can be practised when irrigation and surface drainage is provided while even upland/perennial crops can be grown when sufficient root zone aeration during the wet season can be provided.

Chapter 20

Controlled drainage

Traditionally drainage systems, whether surface or subsurface, were designed to evacuate excess water within a specified period (2–7 days depending on the climate), with the aim of preventing waterlogging and/or flooding. With the emergence of water shortages and water quality and environmental concerns, cases have emerged where drainage needs to be managed, only discharging when necessary. Such drainage is broadly referred to as *controlled drainage*. By maintaining high watertables, water savings of up to 40% during the rice-growing season were achieved in Egypt. Controlled drainage prevents unnecessary mobilisation of nitrogen and phosphorous (Figure 1.10), pesticide residues and other undesirable elements (i.e., selenium) and helps to maintain downstream water quality.

Controlled drainage may also be applied at a basin scale and this application may well have greater potential than it does at a field scale (Evans et al. 1996). It may involve improved reuse of upstream drainage water in the downstream areas. In areas where deep outlets have been constructed to provide drainage for the lowest land in the basin, over-drainage of the higher land is typically the result. Over-drainage frequently occurs in soils with higher permeabilities and low water holding capacities. The drought problems experienced in some higher lying forest and nature reserve areas in the Netherlands is a typical example.

From the above it should be clear that controlled drainage is an important instrument for *Integrated Water Resources Management* (IWRM). Ample evidence from around the world shows that controlling water quantity and quality can result in higher yields and less negative environmental impacts. There are also some disadvantages to controlled drainage (biological clogging of pipe drains due to prolonged submergence, reduced hydraulic conductivity around drains in swelling clay soils and sloughing of ditch banks when water levels in drains are lowered), but these can either be mitigated or are outweighed by the advantages.

Controlled drainage requires more control structures and a higher density of sub-collector drains, resulting in higher capital investment than with conventional drainage systems. It also requires more management, preferably done by the farmer, and O&M costs are also expected to be slightly higher. However, as was shown in Egypt (DRP/DRI 2001), these costs were recovered by the farmer in one to two growing seasons with rice cultivation because of savings in pumping and water costs.

20.1   Issues and developments

Netherlands/Europe: in the late 1990s the Dutch, renowned for their traditional fight against water, changed their paradigm from fighting against water to accommodating water and nature. A shift in thinking from disposing of water as quickly as possible, to one of retaining water as long as possible, providing water room for expansion during the periods of flooding and the re-creation of natural wetlands and habitats along the rivers, streams and drains, accepting some waterlogging for short periods. This compensates for the encroachments (e.g., for industrial and urban development) in the traditional flood plains along the rivers and for the higher and more rapid runoff from upstream areas. The natural meandering of some small streams which in the past were canalised are now being restored, both for water management and environmental reasons. Conflict resolution, compromises, the polluter pays, and acceptable economic damage have become key phrases of the IWRM in the Netherlands. They are also the guiding policy elements for the European Union Water Framework Directive (Lallana et al. 2002). Considerable attention is being paid to advanced monitoring systems (water quality and quantity) and the use of GIS coupled with simulation models, to arrive at satisfactorily sustainable water management solutions.

Australia: irrigators across Australia have used subsurface drainage to manage waterlogging and salinity since the 1930s. Some 35 000 hectares of horizontal (pipe) drainage and about 64 000 ha of vertical drainage (tubewells) have been installed. It is believed that another 65 000 ha still need drainage in one form or another (Christen and Ayars 2001). The main problem with existing drainage systems is that they often drain more water than is required (over-drainage), reducing irrigation efficiency and removing too much salty water. This then causes disposal difficulties, especially increasing the salinity of the receiving rivers and streams. It is estimated that up to 40 times more salt was removed in some sub-surface drainage schemes than was brought in by the irrigation water. The imbalance occurred because rising watertables mobilised and then in effect mined the native salts stored deep in the soil. Increased recognition of these problems has persuaded irrigators to manage drainage systems more effectively. The ideal subsurface drainage system should allow irrigators to grow the best crop possible while minimising irrigation water loss and drainage salt loads.

North America: controlled drainage and sub-irrigation have been recommended as a sustainable agricultural management practice in the USA and in Canada. Controlled drainage regulates pipe discharge to provide storage of rainfall and minimize drain discharge, losses of nitrogen and other agricultural chemicals. The crop can use the stored water with its nitrate during dry periods in the growing season, which would otherwise leach from the crop root zone. Controlled drainage combined with sub-irrigation has been shown to improve corn and soybean yields (Skaggs and van Schilfgaarde 1999).

20.2   Design considerations

In view of the foregoing, the following objectives of controlled drainage have been formulated:

- Maintain a watertable regime that will provide adequate rootzone aeration but will not remove more water than is necessary. The aim being to improve water availability by retaining water in the soil profile for plants to use
- Mobilise as little as possible of the salts below the root zone, preferably by removing salt only from within the root zone, maintaining salt levels low enough for acceptable crop production. This implies maintaining the smallest possible net downward water movement through the root zone on a seasonal basis
- Mobilise as few solutes as possible, other than the salts mentioned above, to reduce downstream impacts.

Different types of drainage system lead to varying responses in terms of watertable position, soil salinity level and drainage volume. Depending on the nature of the drainage process, the layout of the system and the soil conditions in the area, the systems should have a control option at the tertiary level (farm level) and/or at a secondary or primary drainage system level. Site-specific conditions should be used to develop designs rather than accepting traditional regional rules of thumb, which can lead to over-designed systems. Computer models such as DRAINMOD and SWAP have controlled drainage options which can be used to simulate different system designs and operational strategies (Chapter 21).

Drainage design should also include acceptable means for disposing of drainage water, such as licensed disposal to rivers, reuse (preferably at local level), serial biological use (also referred to as IFDM system; Integrated Farm Drainage Management, FAO 2002), storage in evaporation ponds, or disposal to sea via dedicated outfall drains (i.e., Pakistan, see section 16.5.2). Farmers should be able to check water quality in their disposal drains by simple means, using for instance, multi-purpose handheld water quality meters (e.g., oxygen, salinity, temperature and other indicators combined in one instrument), and have easy access to dedicated (government or private) laboratories.

When designing a drainage system, recycling or reuse of water should be considered in the design right from the beginning and be discussed with the farmers. Although it may take some doing to re-design some of the existing drainage systems, benefits in water savings, water availability, and reduction of negative downstream impacts should outweigh the extra costs.

To meet the objective of minimising downstream water quality impacts, limits on the emission of certain substances need to be set at control points (see Table 1.5). These limits need to be based on maximum protection for all living creatures, achievability (technique, time, money) and acceptability. They should be set *As Low As Reasonably Achievable* (known as the ALARA principle), which implies that the applied technologies are essentially the best available and are deemed economically feasible/viable. Furthermore, in setting standards according to the Water Framework Directive of the European Union, the precautionary principle is used, which for water management means:

- *Pollution reduction*: pollution should be minimised irrespective of the type of substances concerned
- *No-deterioration principle*: harmful substances should not increase in the environment.

A major requirement for successful application of controlled drainage is that soil conditions are such that local head differences and different watertable regimes between adjacent fields can be maintained. This is possible in the Nile Delta in Egypt, but not in the various parts of the Indus plains where the transmissivity of the underlying aquifer is so high that even drawdown of tubewells is measured in centimetres rather than decimetres. Controlled drainage in the Indus plain requires an approach on a much larger scale than in the Nile Delta.

20.2.1 Layout and technical provisions

In irrigated land, the drainage system needs to match up with the irrigation system that is already in place. More generally, controlled drainage needs to match up with the existing natural drainage and consider multi-functional objectives on an equal footing during the initial planning stages. Figure 20.1 shows (a) a common drainage design, (b) an ideal drainage design layout without regard for reuse by farmers and (c) an optimised design from the reuse and the final disposal point of view.

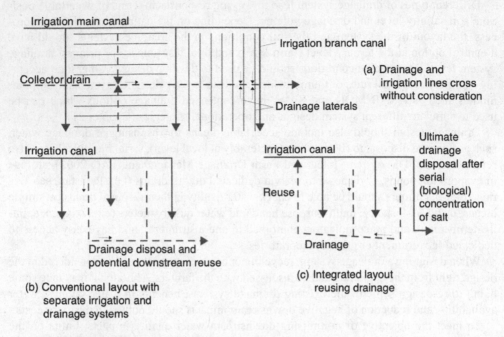

Figure 20.1 Traditional drainage and controlled drainage layout

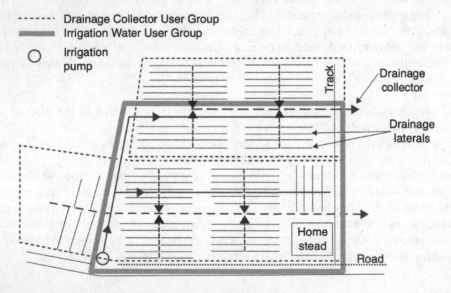

Figure 20.2 Mismatch between hydrological systems and water user groups

Figure 20.2 illustrates the problem of matching existing drainage and irrigation systems into a single unit for the formation of water user groups.

Figure 20.3 shows how in the existing system of Figure 20.2, it is possible to close part of the drainage system because of the existence of closing devices in manholes at the collector junctions (see Figure 8.11b). Gates are installed in the manholes to close-off certain

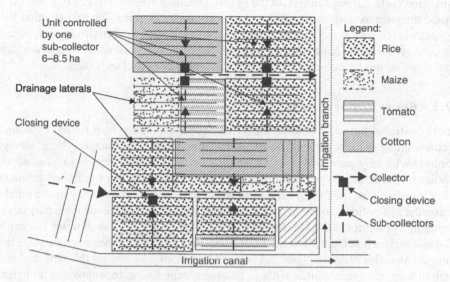

Figure 20.3 Controlled drainage in Egypt

sections of the subsurface drainage system. This would not have been possible without the willingness of farmers to consolidate cropping patterns, such that rice was grown together in a single drainage sub-unit. In traditionally designed systems, only a very small proportion of the total drained area allows such type of operation. Even if farmers in a larger area would agree to crop consolidation, the traditional drainage design does not allow partial closure without affecting upstream and adjacent areas. It should be kept in mind that the partial closure of sub-sections of the drainage system is only effective in medium to heavy textured soils where even over short distances an appreciable difference in watertable head is possible. Such a system would not work when drains are installed in a highly permeable layer with hydraulic conductivities of 5–50 m/d.

Most existing drainage designs are based on the design engineer's perceptions, while controlled drainage designs must rely on farmer management and provide much greater flexibility in closing off certain sections. Re-cycling the drainage water at specific locations should be considered and facilities to accommodate this should be built into the drainage and irrigation system. In Egypt, it is estimated that this will make a traditional subsurface drainage system 15–20% more expensive due to extra sub-collectors and manholes. However, when considering the farm budget, it has been shown that farmers can recover these extra costs in 3–5 years, depending on the type of crops grown and the local and world market prices for the locally grown agricultural produce.

20.2.2   Drain depth

Since drains continue to abstract water from the soil/land as long as the water level in the drains is below the watertable, drain depth is of critical significance to controlled drainage. Deep drains continue to remove water much longer than shallower drains and deep drains have in some instances led to over-drainage. Drains in the Mardan project (Pakistan) installed at 2.5 m depth have been provided with control devices in response to farmer's complaints of over-drainage and irrigation water shortage. Based on an economic analysis of construction costs, amongst others, at the Fourth Drainage Project (Pakistan), it was decided that pipe drainage as deep as possible and as wide as practical was the best solution for the design. However, with water shortages, scarcity, social (O&M) and environmental concerns, other cost-benefit items have been included in the economic evaluation. Shallower drains, spaced more narrowly, are now deemed more acceptable, as well as feasible.

20.2.3   Reuse arrangements

Farmers control of drainage water at the tertiary system level can be achieved by designing and constructing options for supplemental irrigation from the drainage system: designing a manhole with easy access for the farmer (Figure 20.4). Allow farmers to have access to manholes of the drainage system for pumping water, rather than the traditional approach of making it difficult for the farmer (and vandals) to have access. If farmers can use manholes beneficially, self-policing to maintain a good operational status of the manhole may develop.

Sump units commonly used in drainage design in Pakistan, dispose of water 1–2 m above the natural surface level, only to have it fall back into a stilling well box over the same height (Figure 20.5b). This elevation gain can be used beneficially by storing this potential energy in a small local reservoir, with or without geo-membrane lining, for supplemental irrigation or for the reclamation of (nearby) highly saline fields. Not all reuse locations are necessarily

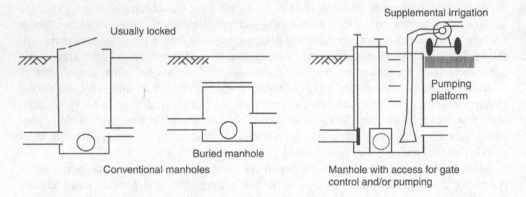

Figure 20.4 Farmer friendly manhole

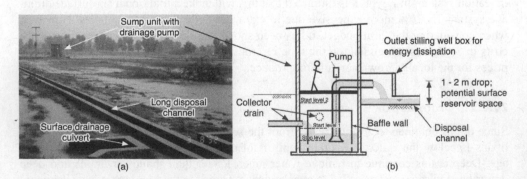

Figure 20.5 Sump unit; controlled drainage[1]

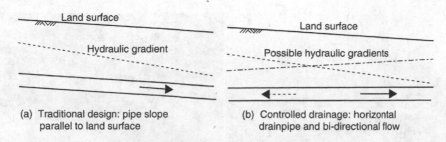

(a) Traditional design: pipe slope   (b) Controlled drainage: horizontal
    parallel to land surface   drainpipe and bi-directional flow

Figure 20.6 Horizontal drains to accommodate optimal location for reuse pumping

downstream from the drainage pump. If downstream reuse is not possible a canal/ditch to route the water back to an upstream point in the irrigation canal system is essential. Design engineers should consider locating the sump unit closer to the irrigation canal entrance to the area. Hence, rather than designing a long disposal channel as shown in Figure 20.5a, a small reservoir could be built on the abandoned area.

In a system with reuse options in mind (Figure 20.1c) it is possible to either reuse the water, or, when the water quality is too poor for reuse, the water is transported to the end of the irrigation system without further reuse, except perhaps in the case of serial biological use. Most existing irrigation systems have a disposal option at the end of the tertiary system and this should be used for drainage purposes only. The irrigation canal has a dual function in such a case: first for irrigation, and second as a surface drain. Poor quality water may be disposed of after serial biological use in evaporators. Evaporators are ponds that are designed in such a way that on a seasonal basis they evaporate all the remaining water. Evaporators impact less on the environment than evaporation ponds.

A drainage pump does not have to be located at the lowest point in the area. Subsurface pipe drains could be constructed horizontally (Figure 20.6b). The driving force for flow is the hydraulic head difference, not the slope of the pipe! Hence pumping at one end of the drain will cause this necessary head difference. Therefore, subsurface drainage pumping units could be located at more practical locations to stimulate reuse of water.

Irrigation designers should consider bi-directional flow in open ditches, and design structures in such a way that they allow for drainage water reuse through the tertiary irrigation canal system. Local drainage water supply[2] may be routed through the irrigation system during closure periods (cyclic irrigation of good and poor-quality water), or during peak water demand periods (i.e., blending of irrigation and drainage water) when many canals operate under a rotation scheme. To accommodate both irrigation and drainage water at the same time (reuse during peak crop water demand periods) larger capacity tertiary canals may be needed. Two-directional flow in irrigation canals best applies with land slopes of 0.001 or less. Spoil from canal maintenance may be used to raise banks along the gravity irrigation canals to increase storage capacity of the canal and to allow reverse flow from pumps or small reservoirs downstream.

20.2.4 Discharge control and watertable management

A key element required of controlled drainage is discharge control in time and quantity. Time of discharge of the drainage system can be controlled by simple open and shut structures or by more sophisticated adjustable weirs (Figure 20.7).

(a) Adjustable automatic weir with flow measurement, viewed from upstream (photograph W.F. Vlotman)
(b) Adjustable automatic weir, viewed from downstream (photograph W.F. Vlotman)
(c) Adjustable weir/gate in Dutch polder (photograph W.F. Vlotman)
(d) V-shape weir in an urban setting (photograph W.F. Vlotman)

Figure 20.7 Controlled drainage and flow measurement in surface drains in the Netherlands

Structures to control the rate and total quantity of discharge, as well as to measure flow, are more common with irrigation systems (Clemmens et al. 2002), but are especially designed for use with (surface/shallow) drainage systems. An example some of these is the inlet box to control surface runoff, thereby reducing peak flows in open drains (Figure 20.8). Some of these inlet boxes have not been successful because their operation was not explained properly to farmers, who were generally more concerned with the rapid evacuation of water from their fields, rather than allowing a more controlled discharge approach to take place; and, as a consequence, farmers enlarged the openings increasing peak flows in the open drains.

Structures for watertable control

In order to properly manage the watertable, as well as to control the mobilisation of solutes, it may be necessary to measure flows in drainage systems. The same structures used in the irrigation system can be used in surface drains but since drains tend to carry more floating material than irrigation canals, structures that easily pass debris need to be chosen. Flow measurement in subsurface drainage systems requires special structures in the manholes or at the outlets of the subsurface drains into the open collector drains (Figure 20.9).

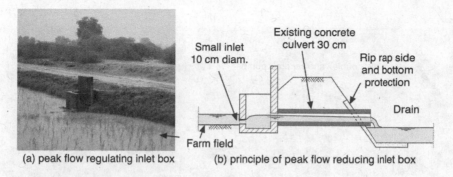

Small inlet
10 cm diam.

Existing concrete
culvert 30 cm

Rip rap side
and bottom
protection

Drain

Farm field

(a) peak flow regulating inlet box

(b) principle of peak flow reducing inlet box

Figure 20.8 Peak flow control structure

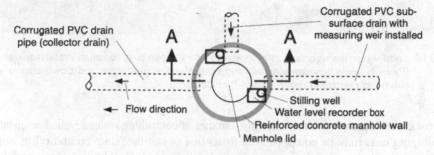

Corrugated PVC drain
pipe (collector drain)

Corrugated PVC sub-
surface drain with
measuring weir installed

A   A

Flow direction

Stilling well
Water level recorder box
Reinforced concrete manhole wall
Manhole lid

Top-view manhole

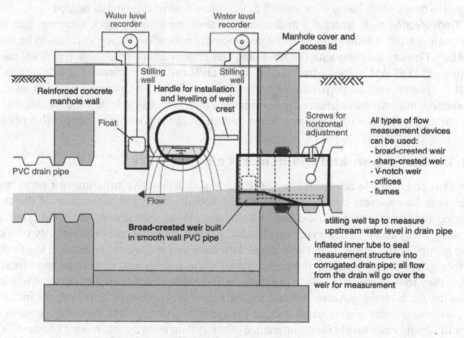

Water level
recorder

Water level
recorder

Manhole cover and
access lid

Reinforced concrete
manhole wall

Stilling
well

Stilling
well

Handle for installation
and levelling of weir
crest

Float

Screws for
horizontal
adjustment

PVC drain pipe

Flow

Broad-crested weir built
in smooth wall PVC pipe

All types of flow
measuement devices
can be used:
- broad-crested weir
- sharp-crested weir
- V-notch weir
- orifices
- flumes

stilling well tap to measure
upstream water level in drain pipe

Inflated inner tube to seal
measurement structure into
corrugated drain pipe; all flow
from the drain will go over the
weir for measurement

Cross-section A - A

Figure 20.9 Flow measurement for subsurface drains using a weir that can be inserted into the drainpipe

Figure 20.10 Local water management in polder; windmill for pumping (Bosman Watermanagement (Piershil) www.bosman-water.nl) and control gates (upstream and downstream) with culvert to control drainage (photo P. Hildering)

In most cases discharge control is simply a matter of controlling water levels. For instance, sump pumping units may be equipped with floats that switch the pump on and off at certain predetermined levels. Small scale automated (windmill) pumps for local sub-drainage and water level control are often used in the Netherlands, including double vane windmills (Figure 20.10). One vane keeps the windmill facing into the direction of the wind, while the second vane is connected to floats which cause the windmill to turn into or out of the wind as needed.

Traditionally, most structures in drainage systems are made of steel, concrete, cast iron or stainless steel. Advances in material development now allow these structures to be made of High-Density Polyethylene (HDPE), and design can easily be adjusted to local needs through CAD/CAM systems that control the HDPE cutting machines. The advantages of HDPE structures are no corrosion, high chemical resistance, no coatings needed to increase protection, minimal maintenance, lightweight and therefore easy to install and handle, and no expensive mould (as with concrete), while custom design is possible at competitive prices.

20.3   Operation and maintenance by farmers

Whereas considerable advances have been made with innovative structures for water management, their success depends largely on the social setting and acceptance of the engineering innovations by the users and stakeholders. Water user groups are often promoted as being a key factor in achieving this but the success, viability and sustainability of water user groups at the tertiary level will depend on clear and immediate benefits to the farmer. Before governments promote the formation of water user groups to save on maintenance costs, they should consider how to promote the formation of water user groups as a win-win situation for both the government and the farmer. If this cannot be achieved, for instance, because the benefit is downstream from the farmers property/field, then the government should remain responsible for maintenance of the system components having (downstream) environmental impacts.

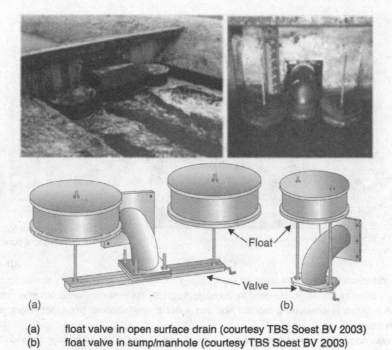

(a)   float valve in open surface drain (courtesy TBS Soest BV 2003)
(b)   float valve in sump/manhole (courtesy TBS Soest BV 2003)

Figure 20.11 Local sub-drainage control structures; float valves

The positive impacts on O&M of farmer-operated irrigation schemes are well documented, but farmer operated drainage systems are less common. Only if drainage design is fully integrated with the irrigation system through clever reuse options, will farmers begin to feel responsible for maintaining and operating the drainage system. Creating a sense of ownership is important. Lack of ownership seems to play a role at secondary water management level too, as has been observed in Pakistan (Vlotman et al. 1994): drainage tubewells that discharged into irrigation canals were well maintained, but those discharging into open surface drains were not. This situation is probably a case of the pump owners seeing no direct benefit from pumping. There seemed to them to be no economic incentive to keep the drainage pump going.

For salinity control it is not necessary to drain year-round provided a net downward water movement is maintained through the root zone over the growing season. The period over which this water movement is necessary will depend on climate, land use, watertable depth and water quality. Temporary high watertables can provide supplemental irrigation during the peak water demand period. The farmer should have the option to monitor and manage watertable depths and obtain instant feedback about the necessity to drain or not to drain. The watertable monitor (Figure 20.12a) pioneered in Australia, provides an excellent example of a solution (Vlotman et al. 2003). This 'traffic light' in the field comprises an observation well with a matching float and indicator rod. Colours in this case indicate the level of action needed by the farmer. The length of each coloured section depends primarily on crop and soil type and needs to be adjusted based on local experience.

A similar approach, but more elaborately executed, is used in the Netherlands (Figure 20.12b–d) to inform the public about the state of drought in forest and other nature

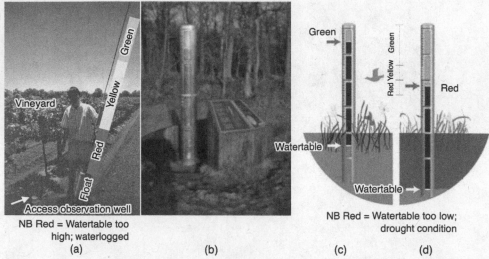

NB Red = Watertable too   NB Red = Watertable too low;
   high; waterlogged   drought condition
         (a)   (b)   (c)   (d)

Green: Watertable is at the right level: no waterlogging and/or danger of secondary salinity and no drought
         condition for grassland. No need for pumping. In gravity systems manhole gates may remain closed.
Yellow: A problem is developing, monitor soil and water quality, observe crop/plant response. Drain as
         needed.
Red: Waterlogging and/or salinity danger! Drain. Pump. Reuse. Store in surface reservoir. Order less
         irrigation water and allow the plants to lower the watertable through transpiration.

Figure 20.12 Watertable monitors

reserves. Because the purpose is different the colour red and green are reversed on the marker column when compared to the agricultural version: a high watertable is desirable in this case.

Pumping systems should be no larger than can be maintained by the local farmers and/or user groups. For instance, if changing parts of the pump, or the pump itself, means that a large crane needs to be hired, it cannot be called 'manageable' by the farmer (e.g., Figure 20.5). Moreover, spare parts should be locally available. Hence the smaller the (drainage) area that is controlled by a single farmer, or by a water user group, the easier it is for the O&M of the system to be under local/private control. Ideally, the farmer should be able to use his transportable irrigation pump for drainage (Figure 20.4) or use a multi-purpose transportable submersible pump (Figure 8.12).

A higher degree of sustainability will be achieved if the water user group can implement maintenance of a drainage pump without having to contract the job out. Yet, paradoxically, economics of scale dictates generally that water user groups should be large enough to be legally and financially in a position to levy fees and contract (maintenance) work out as and when required. For instance, in the Netherlands, Water Boards reduced in number from more than 3 500 in 1850 to less than 50 in 2004, a development in parallel with socio-economic developments and also stimulated by a series of national flood disasters. Economic and social development of a country will dictate which size of water user group is optimal; either a small or a large water user groups may be appropriate.

20.4   Controlled drainage at the regional level

A number of cases of controlled drainage at a larger scale than the field level have already been described, although the applied interventions were not always classified as such (compartmentalisation of polders in section 10.2.1 and of sloping basins in section 10.2.3). Controlled drainage applies in particular to basins in which part of the basin (usually the more elevated part) may be subject to drought because of the influence of the natural drainage exercised on it by the other, lower lying part. In the past drainage designs, often unwittingly, enhanced the natural drainage of the elevated parts. Controlled drainage aims to avoid such over-drainage and to conserve water in the drought prone areas.

Controlled drainage at the regional level has long been an accepted practice in the Dutch polders (the above mentioned compartmentalisation, Figure 20.10) and also in almost all of the sloping drainage basins in the eastern and southern part of the Netherlands. These latter basins typically vary between 30 000 and 50 000 ha with elevation differences between the upper and lower parts typically ranging from 30 to 50 m. The upper areas are mostly occupied by light textured drought prone soils with the land being used for forestry and mixed cropping. Water conservation is practised by aligning drainage canals as far as possible

(a)   Solar powered adjustable weir, Betuwe Polder District, photo W. F. Vlotman
Photos Bergschehoek-KWTwaterbeheersing, the Netherlands:
(b)   Solar powered tilting weir; solar panel on green control box
(c)   floating-crest weir, Berkenwoude
(d)   Thomson-V weir under construction

Figure 20.13 Adjustable, automatic weirs, solar powered and SCADA[3] controlled

along the contours and by installing weirs at various control points along the (natural or artificial) main drains. Major weirs (e.g., Figure 20.7b) are usually encountered at the breaks in
slope of all the major longitudinal streams/canals while these are complemented by smaller
weirs in many of the secondary/tertiary elements of the drainage canal system (e.g., Figure 20.7). Depending on their location or function these may have fixed or moveable crest
levels and function as drop structures too. All these weirs are operated by the responsible
public drainage authorities. Modern weirs may operate using solar energy and can be controlled remotely (Figure 20.13). In addition, there are many farmer-operated small stop-log
weirs in the on-farm collector and field ditches. The weir density, considering all different
types, may vary from one weir per 300 to 1 000 ha (which compares to a density of some
5 000 ha in a compartmentalised polder area).

The above controlled drainage design was originally adopted more for the purposes of
protecting the downstream/lower lands from the frequent winter floods rather than for their
water conservation benefits for the higher land and the 'win-win' feature of this design concept has only recently been fully recognised. The original design also included canalisation
of the natural streams and longitudinal alignments of the drainage canals. As mentioned
earlier (see section 20.1) some of these design features are now often being re-engineered to
their original state. It is also noted that drainage control in the surface system does not necessarily reduce the natural groundwater seepage from the higher to the lower lands. Recent
research indicates that the improvements to water conservation (especially in terms of maintaining sufficiently high inland watertables in the drought prone areas) in some basins may
actually be quite small (van Bakel 2003).

The changes in focus of drainage objectives stems in part from the European Water Framework Directive (EU WFD, or just WFD) which has the aim of establishing a framework for the
protection of inland surface waters, transitional waters, coastal waters and groundwater, which
prevents further deterioration and protects and enhances the status of aquatic ecosystems.
Transitional waters are bodies of surface water in the vicinity of river mouths which are partly
saline in character as a result of their proximity to coastal waters, but which are substantially
influenced by freshwater flows. Drainage is not specifically mentioned in the WFD, but since
most drains discharge their water into river and coastal systems and also interact with groundwater, the implications of the WFD on drainage design, operation and management may be
obvious. The WFD was established in 2000, is expected to be accepted in national legislature
by 2004 and the first milestone as far as implementation is concerned is in 2015.

Notes

1 In 2017, at the 13th International Drainage Workshop, it was observed that this drainage system (i.e.,
   this particular sump unit of the Fourth Drainage Project, completed in 1994) appears to be not functioning as intended; salinity patches observed in 1994 are similar in 2004 and 2018 (Google Earth);
   a shortage of water for leaching is the most obvious cause (Vlotman 2017).
2 To control salt accumulation the drainage system should be designed for minimally 1 mm/d. However, drainage water 'supply' may be much higher: maximum drainage discharges during the season
   may vary from 2.5 mm/d in Pakistan (Vlotman et al. 1994) to more than 8 mm/d in Egypt (DRI
   1998).
3 Supervisory Control and Data Acquisition (SCADA) is a control system architecture that uses computers, networked data communications and graphical user interfaces for high-level process supervisory management and may include other peripheral devices such as programmable logic controller
   (PLC) and discrete proportional-integral-derivative (PID) controllers to interface with the control
   centre computers and operators.

Chapter 21

Computer applications

Computer applications are widely used throughout the drainage sector. Most of them are successors of previously used conventional methods (described in Chapters 11–13) and some have introduced new methodologies and concepts. Many of the conventional methods are now implemented by computer programs, but description of these remain of value since they provide specific understanding which the computer programs do not provide. The computer programs on the other hand provide insights which conventional methods cannot give. Drainage engineers should therefore be familiar which both methodologies and use them inter-actively.

The availability of Digital Elevation Models and/or Maps (DEM) and LiDAR[1] imagery has increased significantly in both private and public domains. There are programs that derive contour lines from gridded elevation data and will produce natural drainage pathways. Google Earth may, for instance, be used during initial planning stages of drainage system as it can provide elevation profiles along imported lines, lines which may represent rivers, canals and drains. The Australian Water Observations from Space (WOfS; see www.nationalmap.gov.au a service provided by Geoscience Australia; look for "Landcover 25") web application provides historic data on where and how often water in the landscape has been observed, which may help in assessing drainage needs.

In this chapter, the computer applications have been grouped according to their field of application while for each group also the generic characteristics and a few well-known specific applications are described. In all cases, the applications are presented as examples and do not imply any endorsement. No software has been provided but ample reference is made to public sources where the programs and other software can be procured. For software, which can be downloaded from the Web, the pertinent sites have been indicated. All weblinks given were accessible September 2018. Although programs are grouped according to their (major) areas of application it is safe to observe that most currently available programs include most areas that were subjected to rainfall-runoff, channel and flood routing, flood mapping and flood recession analysis and therefore include some form of "natural" drainage. Most programs also make extensive use of GIS applications and advanced historic and real data management including calibration processes. When considering planning and design of drainage systems a web search is highly recommended checking for the latest versions of existing programs as well as for new programs.

21.1 Drainage design applications

This group covers the applications which perform specific design calculations (to solve the Hooghoudt formula, Manning/Strickler formula, etc.) but also packages of applications which can handle a number of related design calculations and other design tasks (design drawings, quantity tabulations, etc.). These latter applications are generally referred to as CAD packages (*computer aided design*). Most of the design applications were developed for internal use by engineering companies and are not publicly available. In many cases, however, comparable applications have also been developed in the public domain.

21.1.1   Field systems

Single purpose and limited package design programs for field drainage can be found in FAO (2007). DRAINMOD 6 provides a more all-encompassing drainage design solution including several water quality aspects. The FAO publication also covers the computer solution of the earlier mentioned Toksöz-Kirkham drain spacing equation (see sections 11.3.2 and 11.3.3) which is analytically the most complete description of the steady state flow to parallel subsurface drains for complicated situations with multi-layered anisotropic soil conditions. Computer programs for solving this equation are available with some engineering companies but hitherto not within the public domain.

21.1.2   Canal systems

The relationships between conventional methods of hydraulic design of drainage canal systems and structures (described in Chapter 13) and various pertinent features of these programs are described in this section.

Computers are well suited to perform many repetitive/interactive types of calculations. In conventional design, the burden of these calculations was typically reduced by the use of tables and graphical solutions (multi-factor graphs, nomographs, etc.). Computer programs are now readily available to help with the design of canals, culverts, bridges, weirs, etc., replacing e.g., the nomographs shown in Figure 13.3. Computers are ideally suited to perform the interactive type of calculations involved in most tidal outlet designs (replacing the tabular calculations shown in Table 13.4). Available programs can usually cope with a variety of outlet structures (pipe and box type culverts, horizontally or vertically hung doors and gates, etc.). They can also readily keep track of variously shaped and distributed inland storages.

Steady flow: these programs are based on the Manning/Strickler formula (see Eq. 13.3/13.4) and are used to evaluate how a designed or an existing system performs under higher and/or lower discharges than the assumed design discharge. The evaluations will show where the water levels or the flow velocities reach unacceptable values. These programs can also be used to evaluate the impact of poor or improved canal maintenance (re-calculations with lower and/or higher k_m values). The available programs can deal with tree type and/or with network type of canal patterns. An example of the publicly available programs is *HEC-RAS* which was developed by the US Army Corps of Engineers and is widely used in the USA to re-calculate/check water levels and grade lines for different design discharges under steady and un-steady state flow conditions. Much of this software is publicly available and can be readily procured through Water Resources Publications (WRP) and the US Army Corps of Engineers website (USACE) while enhanced commercial versions are available from Haestad Methods Inc. (See Box 21.1).

**Box 21.1   Procurement of hydrological software
(last accessed September 2018)**

- www.wrpllc.com. Water Resources Publications (WRP). Most of the hydrological software developed by US government Organisations and other public sector Organisations, as well as all publicly financed software developed by private Organisations is in principle publicly available at marginal costs (usually only covering the material and shipment costs, not the development)
- www.ess.co.at/GAIA/gaia_intro.html: Austria. GAIA: A Multi-Media Tool for Natural Resources Management and Environmental Education. Its primary objective is to build multi-media tools for environmental education and management, in a collaboration with 10 countries from Europe, Africa, Asia, and Latin America
- www.bentley.com/en/products/brands/haestad. Haestad computer programs are now available via Bentley.com and include: PondPack to assist in detention pond design; CulvertMaster assists in culvert design, while FlowMaster is a program for the design and analysis of pipes, ditches, open channels, weirs, orifices and inlets
- www.nrcs.usda.gov/wps/portal/nrcs/main/national/technical/tools/. Website of the USDA, Natural Resources Conservation Service (formerly Soil Conservation Service), includes TR-20 versions for various operating platforms (WinTR-20)
- www.bossintl.com/products/download/item/HEC-RAS.html. Boss International. TR-20 is part of a set of comprehensive hydrologic models offered commercially which includes also HEC-HMS, TR55 (WinTR-55), Rational Method (SCS 2003)
- www.stowa.nl: STOWA. A central register of hydrological software (Stichting Toegepast Onderzoek Waterbeheer, Utrecht, the Netherlands). Publications/manuals of SOBEK, and other computer programs. For more information contact Deltares at sobek.support@deltares.nl. (last accessed 2019)
- www.ars.usda.gov/pacific-west-area/riverside-ca/us-salinity-laboratory/docs/models/. Website with models of the United States Salinity Laboratory, Riverside, California, USA. Provides WATSUIT
- www.usgs.gov/products/software/overview. Library of the United States Geological Survey. A full range of software on climate change, environment, energy, mapping, Remote Sensing and water: www.usgs.gov/products/software/water.
- www.scisoftware.com: Site of the Scientific Software Group, with most of its software in the field of groundwater, surface water, soil and water pollution modelling
- http://fluidearth.net/default.aspx. Site with software of HR Wallingford Ltd, Oxfordshire, U.K (www.hrwallingford.com)
- www.hec.usace.army.mil/software/ Site with software of the Hydrologic Engineering Center (HEC), US Army Corps of Engineers (USACE)

Non-steady flow: various computer programs are now available to analyse a range of non-steady flow events as may occur in a system designed for steady flow. This refers to short duration high local inflows or abstractions, surge flows near weirs and pumping stations, sudden flow changes resulting from operational interferences (changes in gate/weir settings, etc.) and accidental canal blockages, wind set-up, etc. The programs are used to calculate level and velocity changes and backwater and drawdown curves. Most programs can also be used for flood routing calculations. The National Resources Conservation Service (NRCS) has adopted the Hydrologic Engineering Center's River Analysis System program, HEC-RAS, for simulation of both steady and non-steady state flow conditions to route hydrographs through existing river systems. Changes in tidal gate settings can be included such that non-steady, non-uniform flow, both rapidly varied and gradually varied conditions can be simulated. An important feature of HEC-RAS is the ability to import HEC-2 data. This feature makes it easy for a user to import existing HEC-2 data sets and start using HEC-RAS immediately. HEC-2 is a steady flow hydraulics model that was the predecessor to HEC-RAS. Thousands of hydraulic studies were done using HEC-2 in the 70's, 80's, and 90's.

Another example of the publicly available program is SOBEK (2019), a modelling suite for flood forecasting, optimisation of drainage systems, control of irrigation systems, sewer overflow design, river morphology, salt intrusion and surface water quality. The modules within the SOBEK modelling suite simulate the complex flows and the water related processes in almost any system. The modules represent phenomena and physical processes in an accurate way in one-dimensional (1D) network systems and two-dimensional (2D) horizontal grids.

21.1.3   Preparation of drawings and documents

Standard software can be used to facilitate various technical and administrative tasks associated with design.

Drawings: most design drawings can be readily digitised, and computer printed. This applies to the various types of situational drawings (topographic maps, layout maps, site maps, etc.), longitudinal and cross-sectional profiles of canals (showing existing and planned levels, sections and other relevant features) and various types of structural drawings. These drawings may be GIS based and can usually be easily scaled up and down to suit convenient use.

Tender documents, budgets, implementation plans: the computer has also amply proved its convenience and value in some of the design associated project activities. Tender specifications can now be prepared effortlessly by selecting relevant sections from available master text files. Various administrative and management programs can be used to prepare budgets and work programs.

21.2   Rainfall discharge models

A large number of computer models, which simulate the discharge generation and transport processes in drainage basins, are available. Some of these models not only simulate the discharge processes but also the underlying soil hydrological regimes. With the latter models, rather long periods (a series of years/seasons) can be simulated and drainage conditions/requirements of an area assessed (for details see section 21.4). Here, however, the description is limited to the deterministic modelling of single rainfall – discharge events for the purposes of estimating design discharges of small to medium sized agricultural drainage basins. Two models will be described briefly: the TR-20 model and the HEC-HMS model. These models have been widely tested, are well documented and readily available to the public.

21.2.1 TR-20 model

Technical Release No. 20 (TR-20) is a physically based watershed scale runoff event model. It computes direct runoff and develops hydrographs resulting from any synthetic or natural rainstorm. Developed hydrographs are routed through stream and valley reaches as well as through reservoirs. Hydrographs are combined from tributaries with those on the mainstream stem. Branching flow (diversions), and baseflow can also be accommodated. The model was developed by the US Soil Conservation Service (SCS) in 1964 in Fortran and a PC version became available in 1985. Since 1992 a Windows based (free) version is available while commercial software vendors have also developed user friendly and enhanced versions.

This model is essentially a computerised version of the SCS methods and procedures for estimating rainfall generated (peak) discharges from small sloping agricultural drainage basins (see section 12.5.2). Further details on the structure and application of the TR-20 models can be found in the National Engineering Handbook, Section 4 (Hydrology), SCS 1972, while for the procurement of the software reference is made to the USACE website.

21.2.2 HEC-HMS

This model was developed in the eighties by the US Army Corps of Engineers Hydrologic Engineering Centre, Davis, California. The HEC Hydrologic Modelling System (HEC-HMS) is designed to simulate the precipitation-runoff processes of dendritic watershed systems. It supersedes HEC-1 and provides a similar variety of options but represents a significant advancement in terms of both computer science and hydrologic engineering.

The program includes a variety of mathematical models for simulating precipitation, evapotranspiration, infiltration, excess precipitation transformation, baseflow, and open channel routing (for details on flood and open channel routing see Chow 1983 and other handbooks). A versatile parameter estimation option is also included to assist in calibrating the various models. In addition, outputs and mapping of the systems make extensive use of readily available Google Earth and other Remote Sensing based imagery.

For the rainfall-runoff conversion, a choice of routines is available, including the SCS CN-method (see also section 12.5.2) which is classified as a runoff-volume model. Seven runoff-volume models are included as well as seven direct-runoff models that include simulation of overland flow and interflow. Three baseflow models and eight routing models are included; for details see the technical reference manual that is available on-line. The routing modules can cope with a wide range of canal characteristics and on-stream and off-stream storages.

21.3 Ground water flow models

These models can be used to analyse complicated groundwater flow situations and patterns for which there are no simple analytical solutions available. Some of these cases were approximated in the past by means of semi-analytical solutions (often developed on the basis of conformal mapping techniques) using 'tele-deltos' paper, electrical analogue methods or by numerical and graphical solutions of the Laplace equation (respectively also known as the *relaxation method* and the *flow net method/squares method*). These conventional methods have now been largely superseded by computer models. These models are, in essence, all finite difference/element solutions of the Laplace equation. While the past solutions were essentially restricted to schematised cases of steady state two-dimensional flow, the computer models can readily handle complicated anisotropic multi-layer three-dimensional non-steady flow configurations.

21.3.1   Spreadsheet models

These models, which are a spreadsheet version of the earlier mentioned relaxation methods for solving the Laplace equation, have now largely been replaced by the model MODFLOW (see below). Spreadsheet models which do not need special software (standard spreadsheet programs suffice) still have their value in such steady state two-dimensional flow situations as the design of interceptor drains (described in section 17.2) and also for the analysis of seepage cases as described in section 17.4. The spreadsheet models can be used to analyse complicated steady state subsurface drainage flow to parallel drains as described in section 11.3, replacing e.g., the Toksöz-Kirkham drain spacing equation; for most simpler cases the Hooghoudt solution is quite adequate (For details see Olsthoorn 1985, and 1998).

21.3.2   MODFLOW and integrated programs

This is a finite-difference three-dimensional groundwater flow model originally developed by the U.S. Geological Survey (USGS). The flow zone is covered by a grid of varying shape and density, which can deal with a wide variety of up to three-dimensional flow configurations. Since its initial development, additional packages have been added which enhance its capabilities. Individual model layers can be confined, unconfined or a combination of the two. Separate packages allow simulation of abstractions or infiltrations, area recharge, evapotranspiration, drains and streams. In addition, MODFLOW has been integrated with models to simulate solute transport, particle tracking (flow path), as well as inverse modelling routines, automatic calibration programmes and GIS. Extensive documentation of MODFLOW is available on the website and includes descriptions of a variety of other computer programs linked with MODFLOW; see-https://water.usgs.gov/ogw/modflow/.

21.3.3   SGMP and SOURCE

The Standard Groundwater Model Program (SGMP) is a regional model which can be used to predict the effects of man's interference with existing groundwater systems (irrigation development, drainage development, groundwater abstractions, artificial recharge, etc.). The modelled area is divided into discrete polygons as shown in Figure 21.1 whereby each of these polygons represents an internal node (typically an area represented by one observation well) or an external node (posing a boundary condition). Conditions within the polygons are assumed uniform and shapes and sizes are chosen accordingly.

The Standard Groundwater Model Program (SGMP) model has been applied very successfully for the design of pipe and tubewell drainage systems for salinity control in the Indus Plain, Pakistan (see also the inverse mode application for estimating natural drainage described in section 16.3.2). For a detailed description see Boonstra and de Ridder 1990.

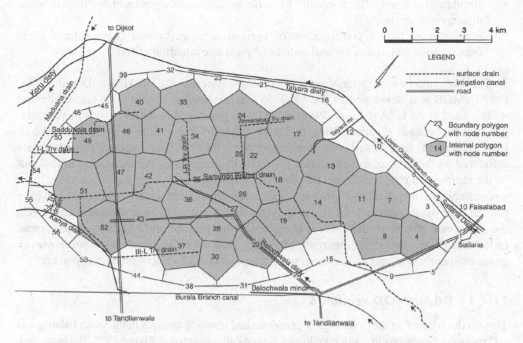

Figure 21.1 Polygon network of SGMP for the Fourth Drainage Project area Pakistan

Alternatives to SGMP are the groundwater interaction modules, protocols and guidelines that simulate the exchange of groundwater and salt between rivers and the underlying groundwater systems developed by eWater Ltd. This model (SOURCE) was formerly known as the Groundwater – Surface Water Interactions Tool (GSWIT) and is now part of the eWater SOURCE software (Source model 2018). eWater was a Cooperative Research Centre (CRC) until 30 June 2012 and continues as eWater Limited (Ltd), a public incorporated limited guarantee, not-for-profit Australian company.

21.4 Agrohydrological models

Models are now available which can reliably simulate the soil moisture, salinity and watertable regimes expected to prevail under given climatic, land use and water management conditions. Some of these models have crop growth modules by which crop yields can be predicted. Historical records are used for calibration and validation and the calibrated models can then be used for a wide range of prediction, evaluation and scenario assessment purposes. As such, they can greatly help with the diagnosis of drainage problems, with the establishment of the drainage needs/requirements and with the comparison of alternative drainage designs. Some typical drainage applications are:

- soil moisture/waterlogging simulations as described in section 5.2, but much more refined taking into account capillary rise and deep percolation
- simulation of achieved watertable control and/or crop yields for various drainage designs (different drain spacing, different drain depths, etc.)

- simulation of drain effluent quality for different drainage designs and or different water management regimes
- establishment and/or confirmation of optimal drainage criteria (e.g., optimal drain depth), using watertable control indices or yield as evaluation criteria.

The two models most widely used for drainage purpose are DRAINMOD and SWAP. DRAINMOD was developed in the 1960–70's at the North Carolina State University and widely used in the USA (Skaggs 1987). SWAP was developed in the 1970–80s in Wageningen, the Netherlands (Feddes 1988). In SWAP, the soil-water regime is simulated based on two fundamental laws of soil water movement while the crop responses and crop yields are estimated on the basis of the calculated ET rates. In comparison, the simulation of the soil-water regime and crop yields in DRAINMOD relies more on water balance concepts and well-established empirical relationships. Both models have been widely field-tested and have proven their worth. Although the use of these models is still very much restricted to research, the above examples show that there is also ample scope for their application in project preparation and project design (comparison of design alternatives; the models cannot, however, generate specific designs) as well as in water management (development of best practices).

21.4.1 DRAINMOD version 6

This model is based on the hydrological processes and resulting changes in the water balance and soil moisture conditions in a soil profile as schematically depicted in Figure 21.2. The water balance is predicted on an hour-by-hour basis. Complex transport equations are avoided by assuming a 'drained to equilibrium' state for the soil water distribution above the watertable. The model is, therefore, specifically applicable to soil profiles with shallow watertables as this assumption is less valid in the case of deep groundwatertables. Version 6.0 also includes routines for soil temperature modelling and considers freezing and thawing effects on drainage processes.

Inputs to the model include weather data, soil properties, crop variables and site parameters. Model results (infiltration, soil moisture conditions, evapotranspiration watertable regime, drainage rates, etc.) are available on a daily, monthly or annual basis, thus allowing for the evaluation of the effects of the year-by-year and seasonal variability of climatic data.

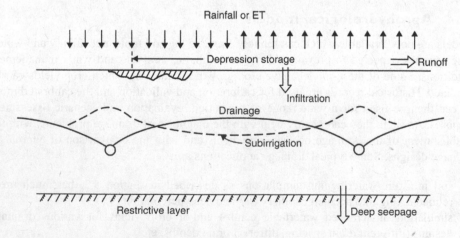

Figure 21.2 Schematic of DRAINMOD simulation

Optionally, the effects of water management system design on yields can be evaluated. Trafficability and planting dates are predicted while stress index methods are used to calculate yield response to excessive and deficient soil water conditions.

DRAINMOD is principally a one-dimensional model using the Kirkham equation to calculate drainage rate. Lateral and downslope seepage can be taken into account. The latest version incorporates the simulation of the movement of salts and nitrogen in shallow watertable soils. Soil salinity distribution, salt concentrations of drainage water and the effects of salinity on crop yield can be predicted. This also applies to the nitrogen concentrations in the soil profile and in the surface and subsurface drainage water. The solute transport module is based on the average daily soil water fluxes and water contents and considers the various dispersive, advective and chemical reactions of chemicals (nutrients and salts) while moving through soil pore space filled with water. For nitrogen, functional relationships are used to quantify rainfall deposition, fertiliser dissolution, net mineralisation, denitrification, plant uptake, runoff and drainage losses. For details see Workman et al. 1990. Detailed information can be obtained from the DRAINMOD home page at North Carolina State University:

www.bae.ncsu.edu/agricultural-water-management/drainmod

Last accessed Sep 2019; includes links to download the program.

21.4.2   SWAP

The Soil Water Atmosphere Plant (SWAP) model has evolved from SWATRE(R), SWACROP and other earlier versions. The considered hydrological processes have been schematically indicated in Figure 21.3. Vertical water movement in the unsaturated zone is simulated

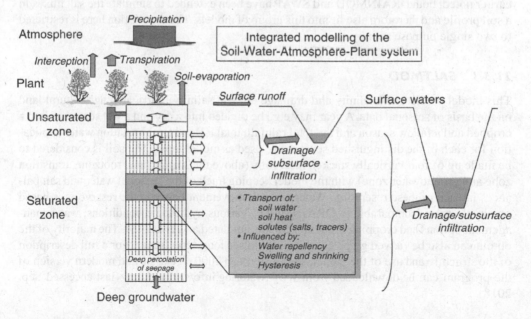

Figure 21.3 Schematisation of the soil water movement and other hydrological processes simulated by the SWAP model

based on the Richard's equation (see section 6.4). Root water extraction at various depths is calculated from potential transpiration, root density and possible reductions due to wet, dry or saline conditions. There may be ponded water and watertables may be shallow as well as deep. Conditions at the base of the considered soil profile may be defined by a water flux, pressure head, flux as a function of groundwater level or free drainage. Watertable control may be based on the Hooghoudt or Ernst drainage equations. The soil may be layered while spatial soil variability may be dealt with by using scaling factors.

Daily actual ET values are estimated on the basis of the Penman-Monteith ET values and prevailing water and/or salinity stresses, but other reference ET values and appropriate crop factors may also be used. Partitioning of the ET into transpiration rate and soil evaporation rate is based on either the leaf area index or the soil cover fraction. A choice can be made from three crop growth routines.

Simulation of salinity regimes was added in the 1980–90s, while the latest versions can also be used to simulate the movement of some nutrients (e.g., nitrogen) in the soil. Irrigation may be prescribed at fixed times or be scheduled according to a number of criteria (allowable stress, allowable depletion of total or readily available water, critical pressure head or water content at a certain depth). For the design of surface drainage and for the management of the local and regional drainage bases (open water levels) also a number of design and management options are available, including controlled drainage. For details see van Dam et al. 1997, Kroes and van Dam 2003, and visit www.swap.alterra.nl (last accessed Sep. 2018).

21.5 Salinity prediction models

This group includes the models that simulate soil salinity regimes under different water management conditions. They may be used to evaluate scenarios of water management. As earlier noted, both DRAINMOD and SWAP have been extended to simulate the salt fluxes in a soil profile and therefore also fit into this group of models. The description here is restricted to two single purpose salinity models.

21.5.1 SALTMOD

This model predicts soil salinity and drainage effluent salinity in irrigated agricultural land on the basis of seasonal data. A year may e.g., be divided into a dry and a wet season or into a cropped and a fallow season and the total rainfall, total ET and total irrigation water application for each of the distinguished seasons are used as model input. The soil is considered to be made up of four vertically stacked reservoirs (above the soil surface, rootzone, transition zone and groundwater zone) with the model keeping track of the seasonal water and salt balances in each of these reservoirs. Water and salt movement between the reservoirs is based on simple rules (comparable to DRAINMOD). Various hydrologic conditions, water management options and cropping schedules can be simulated and evaluated. The majority of the output can also be viewed graphically in order to see long-term trends. For a full description of the structure and use of the model, see Oosterbaan 2000. A classic and modern version of the program can be downloaded from www.waterlog.info/saltmod.htm (last accessed Sep. 2018).

Box 21.2 Tools and models available from the NRCS

- USDA/NRCS Hydraulic & Hydrology models (drainage):
- TR-19 (reservoir storage requirements analysis, 1967)
- WinTR-20, Version 3.20, (for Windows7 and earlier operating systems, 2017)
- WinTR-55, Version 1.00.10 (18.5 MB – last updated 2/7/2013) Windows 7 Version for 32- and 64-bit PCs (simplified flood peak and hydrograph development for small watersheds, SCS 2003, www.nrcs.usda.gov)
- SPAW Hydrology and Water Budgeting including the Soil Water Characteristics (SPAW is a water budgeting tool for farm fields, ponds and inundated wetlands).

21.5.2 WATSUIT

All the models described above only predict the salt concentration (EC-values) but not the composition of the solute. Sodicity hazards under these conditions may be evaluated based on the composition of the irrigation water (see section 15.5). This evaluation, however, does not fully take into account all composition changes and evaluation on the basis of a simulated composition of the soil solution would be preferable. The WATSUIT model developed by the US Soil Salinity Laboratory, Riverside California predicts the salinity, sodicity, and toxic-solute concentration. The model was developed in 1991 and updated in 2001 and was available Sep 2018.

Note

1 LIDAR, which stands for Light Detection and Ranging, is a remote sensing method that uses light in the form of a pulsed laser to measure ranges (variable distances) to the Earth. These light pulses, combined with other data recorded by the airborne system, generate precise, three-dimensional information about the shape of the Earth and its surface characteristics. Two types of LIDAR are topographic and bathymetric. Topographic LIDAR typically uses a near-infrared laser to map the land to make digital elevation models for use in geographic information systems, while bathymetric lidar uses water-penetrating green light to also measure seafloor and riverbed elevations.

Part VII

Management and New Developments

Chapter 22

Research and innovation

Research, innovation and recent developments in the field of drainage includes the use of new pipe/drain envelope materials, new tools in the field of precision agriculture that allow significant water quality control, and experiments with multi-level drainage systems.

Modern Land Drainage (MLD) is an extended approach to the traditional drainage design methods for rainfed agriculture in the humid temperate zone. It includes an extensive consideration of salinity control of irrigated land in (semi-) arid zones, drainage of rice land in the humid tropics, and advocates controlled drainage in the framework of integrated water resources management (IWRM). Institutional, management and maintenance aspects are included as well as the mitigation of adverse impacts of drainage interventions on the environment. Beyond Modern Land Drainage embraces the Triple Bottom Line (TBL), the triangle that considers interactions between social, environmental (or ecological) and financial aspects and extends it to consideration of drainage within the Water-Energy-Food Nexus (Vlotman 2014). These concepts were presented at the 13th International Drainage Workshop of the Working Group on Sustainable Drainage (WG-SDG) of the International Commission on Irrigation and Drainage (ICID), held in Ahwaz, Iran, March 2017.

Over the years it has become clear from worldwide experiences that economics and technical expertise are not the only key drivers of drainage development and that care for the natural physical and social/cultural environment will enhance the likelihood of sustainable water management and sustainable drainage systems (Darzi-Naftchali et al. 2017).

The drivers of sustainable environments are, amongst others, the Key Performance Indicators (KPIs) of Triple Bottom Line (TBL) frameworks that inform us how well we are doing. These KPIs are either oriented towards internal business performance or towards external impacts of water management organisations, including business by government departments. It is important to keep the internal and external KPIs separate such that mission, strategies and operational objectives of the organisation that is responsible for the drainage system are clear in the mind of all stakeholders. Drainage environmental KPIs are related to salinity, waterlogging and water quality while many others relate to the broader objective of Integrated Water Resources Management (IWRM).

22.1 Hydroluis pipe-envelope drainage

Sedimentation of pipe drains can be a serious problem during and after installation and work continues to come up with prevention of sedimentation as shown by Bahçeci et al. 2018. They developed the Hydroluis pipe-envelope combination drainage concept to prevent clogging, sedimentation and root invasion. A corrugated drainpipe with only three rows of perforation on the top of the drainpipe is covered for 2/3 by a non-perforated pipe with 8 mm of space between the two, forcing the water to enter between the inner and outer pipe two-thirds down; the drain water is flow upwards to the perforations on top of the inner pipe without carrying sediment. This system also showed no root penetration as the space between the two pipes is either fully filled with water or fully dry, neither condition being conducive to root penetration. Special arrangements are made in the trencher box to ensure that the inner and outer pipe are placed in the right orientation over the full length of the drain. The innovative pipe – envelope concept was tested on a 50 ha pilot area in Harran, Turkey, in 2015 and 2016.

22.2   Capiphon drain

Another example of a new development is the Capiphon drain (Figure 22.1b) which uses the capillary action of the Capiphon concept to both drain and supply water to the root zone; a new form of controlled drainage and irrigation (www.capiphon.com.au). This type of drainage is also known as wick drainage although this is different in its applications and configurations (Koerner 1994, Vlotman et al. 2000). It has been used mostly in urban settings and on small scale agricultural settings and has great potential for wider application in agricultural settings.

22.3   Precision agriculture for water quality control

Reconnaissance during the operation, management and maintenance (OMM) stages of the life cycle of a drainage system could be with the use of "swarmfarm" robots for precision application and control of drainage water quality (Figure 22.2). The idea is that farmers instead of large tractors and sprayers use a swarm of autonomous, collision-avoiding robots that can spray with accuracy and in the right quantity when using GPS and satellite linkage other farm inputs, such as soil type, moisture content, etc. Software is used to control the swarmfarm robot and adjust the intensity and concentration of the spraying. Clearly a variety of sensors can be added or included in the swarmfarm robots with salinity measurement

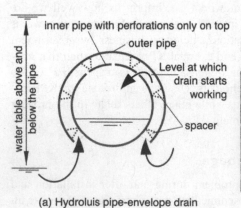

(a) Hydroluis pipe-envelope drain

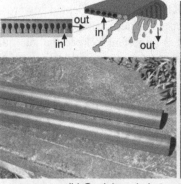

(b) Capiphon drain

Figure 22.1 Hydroluis and capiphon drains

Figure 22.2 Swarmfarm robots in action (www.swarmfarm.com)

(EM38, see section 14.5.1), soil moisture, temperature of the soil and a variety of chemical assessments with probes drawn through the top layers of the root zone (i.e., TDR probes, see section 14.5.2) can be performed.

22.4   Bi-level subsurface drainage

Installation of subsurface drains at two different levels (shallow and deep) with different spacings and varying number of shallow drains in between the deep drains, albeit at a different level, was investigated to manage 1 – watertable regime, 2 – drainage water quality, and 3 – soil salinity. A field scale land reclamation experiment was implemented in the Murrumbidgee Irrigation Area of New South Wales, Australia (Hornbuckle et al. 2007). A traditional single level drainage system and a bi-level drainage system was compared in the experiment in an irrigated field setting. The single level drainage system consisted of 1.8 m deep drains at 20 m spacing. This configuration is typical of subsurface drainage system design used in the area. The bi-level drainage system consisted of shallow closely spaced drains (3.3 m spacing at 0.75 m depth) underlain by deeper widely spaced drains (20 m spacing at 1.8 m depth). Data on drainage flows and salinity, watertable regime and soil salinity were collected over a 2-year period (2000–2002). The site had salinity levels well above the recommended level of 1.5 dS/m (EC_e) for wine grapes which were to be grown without yield loss (NB average soil salinities varied from EC_e = 10 dS/m to 20 + dS/m). For crop-tolerance levels see Table 14.1 and Table 15.1.

Results showed large reductions in soil salinity in both the bi-level and single level systems over the two-year period. Shallow soil layers (< 0.5m) were more effectively drained in the bi-level system. Between 0.5–1.0 m the two treatments performed similarly but below 1.0 m the single level system removed more salts (Figure 22.3). The shallow system was also more effective in controlling waterlogging; there was an 84% reduction in time the watertable was above 0.5 m, 67% reduction in time above 0.75 m and 41% reduction in time above 1m when compared with the single level deep drains. The drainage water from the shallow drains in the bi-level system was at a significant lower salinity than the deep drains and

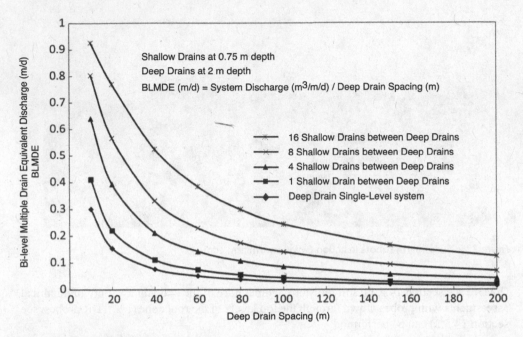

Figure 22.3 Equivalent system discharge for combinations of shallow and deep drains (after Hornbuckle et al. 2012).

because both systems had their own collector drains there is the potential to separate these flows and manage them differently; e.g., reuse.

Hornbuckle et al. 2012 subsequently developed a theoretical model to complement the field experiment but no further field observations are reported. The theory of bi-level design was developed by Kirkham 1949, 1957 and Kirkham et al. 1997 for a ponded water condition. Drainage behaviour is assumed to be ideal and the soil is considered to be homogenous; thus allowing application of Darcy's Law and the Laplace Equation.

Using the analytical solution for bi-level drainage situations with single and multiple shallow drains between deeper drains showed that for equivalent rates of total drainage, spacing between deep drains could be significantly increased by the use of shallow drains (Figure 22.3). It also demonstrated that flow paths and drainage rates from shallow and deep drains and the total system drainage could be significantly modified by changing the number of shallow drains. This information should be useful when considering various drainage configurations to meet the objectives of root zone salinity and waterlogging control and minimisation of drainage salt loads. Thus, the combination of shallow and deep drains offers a subsurface drainage alternative which can satisfy the agronomic, economic and environmental constraints of modern land drainage. No cost evaluation for construction of the different combinations was performed but is essential to assess the viability of bi-level systems.

Bi-level drainage was also performed in Pakistan using the principle of skimming wells (Nazir 1979). The Indus Plain has a deep unconfined aquifer with highly saline groundwater overlain by a freshwater layer. Initially farmers merely used tubewells and bores to pump water from the fresh shallow watertable, but when these started pumping saline water another solution was needed. When the pumps operate streamlines extend well below the level of the filter of the tubewell. In order to prevent water from deeper levels flowing towards the tubewell, two tubewells were installed. One tapped into the freshwater layer and one into the deeper more saline layers. When both pumps operated, water from the deeper well drew water from just below the shallow tubewell, allowing this to extract fresher water. In doing so fresh and saline water were separated and prevented from mixing.

Chapter 23

Institutional, organisational and financial arrangements

Drainage development and management usually requires cooperation between several parties. Most countries have adopted arrangements, which define the institutional relationships between the involved parties and provide the legal and administrative basis for system management. These arrangements are described in this Chapter.

23.1 Drainage laws

All drainage water will eventually end up in the streams and water bodies which comprise the natural and/or man-made disposal drainage system of the area. Most countries have laws stating that the communal streams are in principle under state ownership and control. Additional man-made drainage canals may have been installed and be owned by the local government or a public drainage organisation. The latter organisations may also manage the state-owned streams/water bodies in their service area.

In most countries, landowners have certain rights and/or are subjected to certain restrictions in the disposal of the excess water from their land to the communal systems. Depending on the prevailing law system and jurisprudence, more or less weight is given to the following basic rules:

- the upslope landowner is entitled to the natural discharge of excess water from his land inherent to its elevated position (Roman Law)
- excess water is considered a common enemy and both upslope and downslope landowners may take reasonable remedial measures (English Law)

Landowners who alter the natural drainage situation can generally be held responsible for the resulting damage. This encourages landowners to consult and co-operate with neighbouring owners and to make prior agreements.

The above rules generally deal adequately with the classic hydraulic/hydrological quantitative aspects of drainage. In most countries, however, drainage also has to meet various environmental standards dealing with conservation of biological and ecological values and the protection of the water quality. An ever-growing body of laws, regulation and jurisprudence deals with disputes arising from conflicts between the classical drainage laws and the more recently adopted environmental regulations.

23.2   Development and management models

Drainage development and management is generally distinctly different for the developed and the developing countries. These differences and related other features are described in this section.

23.2.1   Public/private good model

As main drainage systems generally have a wider public benefit than on-farm systems, the drainage development/management model, which is generally considered to be most appropriate, treats main drainage systems largely as a public good and the on-farm systems as a private good. Under this model, therefore, the responsibility for the development and management of the main systems rests primarily with the government or other public body (e.g., a public drainage organisation, see section 23.3) and that for the tertiary/on-farm systems with the farmers. Almost all existing drainage infrastructure in the developed countries has been established and is managed under this model. It should, however, be stressed that within the context of this model, governments have always played a highly proactive promotional role in the development of tertiary/on-farm drainage by providing for research, extension and technical assistance. Governments have also provided various types of direct and indirect subsidies (almost all on-farm drainage development in Europe and North America was until the 1960/70-ties subsidised at rates up to 50–70 %).

A straightforward application of the above described public/private good model in the developing countries will seldom be possible due to local weaknesses in management and financial capacities. In these countries, drainage development (including much of the tertiary/on-farm systems) is generally initiated, implemented and financed by the central government (Ministry of Agriculture/Irrigation) or by so-called authorities, bodies with statutory powers created by the (central) government. These organisations normally also operate and maintain the systems after their completion. The local/farmer participation in the various decision-making processes is usually very limited. As long as the public budgets are adequate, the above approach actually works quite well. The weaknesses of the approach appear when government commitments are not met, budgets fall short, systems are not adequately maintained and start to under-perform. This unfortunately, has been the fate of many drainage projects in the developing countries.

23.2.2   Participatory development

The approach, which is considered most appropriate for the developing countries, is the participatory development model. This model is based on the premise (supported by ample experiences) that sustainable drainage development can best be nurtured through active farmers participation in the planning and implementation of the works. Comparable participatory approaches are also being pursued in related sectors (irrigation, rural water supply, etc.).

Some initial experiences with participatory drainage development have been gained in Pakistan, India and some other countries (see INCID 2000) but no, widely tested, universally successful blueprints have yet been established. However, a number of critical preconditions have been identified as follows:

Organisation: participation can generally be applied within the context of a regular government led project. The involved government agencies need of course to be committed and sensitised to the participatory approach. Farmer's interest may be best represented by an existing respected democratic local leadership structure (preferably already dealing with water or general agriculture). If not existing, it needs to be established (see also *Operation and Maintenance* below). Facilitation by a mutually trusted competent third party (e.g., an NGO) is often helpful in cases where both parties (government and farmers) are inexperienced and/or where the farmers need assistance to define their interests and to organise their inputs.

Communication: development plans should be announced and explained to the farmers and the public as soon as possible in order to avoid false perceptions and rumours. The farmer's consultation should also be started as early as possible. This communication and consultation should be continued throughout the various project development stages.

Drainage needs/integrated development: participation has sometimes failed because the implemented measures were not addressing the most urgent real needs of the community/ farmers. Improved drainage has not always brought about the predicted benefits because yields or incomes were constrained by other shortcomings (unreliable irrigation services, poor market conditions, etc.). Planning should therefore always be preceded by a thorough diagnosis and analysis of the prevailing drainage problems, the constraints, the community/ farmers priorities and the alternative solutions (participatory rural appraisal would generally be a suitable tool for performing this type of diagnosis/analysis).

Technology and design: participatory drainage projects have also failed because of the use of unnecessarily sophisticated and locally unfamiliar and non-repairable technology. Designs have also failed because they were based on untested operational concepts and misunderstood users wishes, traditions and cultural background. It is very common to find that the location of structures and the layout and alignments of drains did not take into account farmers preferences (inconvenient locations, not following recognised boundaries) and cultural or ethnic sensitivities (to be avoided by participatory design practices, including walk-through with the farmers; the devil is often in the detail!).

Land losses: land losses should as much as possible be minimised and shared equally amongst the farmers/landowners. Differentials should be compensated for, either financially or in kind. Some mutually beneficial trading and/or consolidation of land might be possible to free up land for compensation.

Construction: making full use of farmer labour can often reduce cash costs. Engagement of local contractors and locally available hand labour and skills will help to promote local ownership of the works. This should however not be done at the cost of construction quality (subsurface pipe drains in waterlogged land can usually best be installed by machine). Before construction activities start, farmers/owners of the land affected should be informed and consulted about minimising crop damage and inconveniences to avoid last minute disputes and interruptions of the construction work (the latter may still occur and should be dealt with tactfully).

Schedules: participatory development generally proceeds more slowly than the conventional approach. The various participatory processes (project identification, communication and reaching agreement, planning and design, construction) should ideally be left to run their course without undue time pressure. Implementation schedules, budget allocations and administrative procedures need to be flexible to cope with the inherent unpredictability of progress and lack of firm deadlines.

Operation and maintenance: the institutional/organisational structure for the future O&M can best be a continuation of the participatory organisation established during the planning and construction of the works. These organisations (often called user associations) need to be formalised and registered in accordance with the prevailing laws and regulations in order to provide them with the proper legal status and the necessary empowerment (enforcing of individual maintenance tasks, levying of contributions, etc.). For irrigated areas, especially in cases where the prevailing drainage problems are largely irrigation induced, drainage associations may best be combined with the irrigation associations even when the service areas do not exactly overlap.

Role of the Government: governments need to play a major role in drainage development in the developing countries, at least comparable to the role played in drainage development in the developed countries. This applies to research, training, extension and technical assistance but also the subsidising of desirable drainage developments in cases where the farmers cannot afford to bear the full costs.

Stakeholder involvement: participants in all aspects of drainage systems can be found at all levels of society; from farmers to the general public, to government officials. Section 3.4 describes the best processes to be followed for successful and sustainable stakeholder involvement from the social engagement point of view, which complements the technical and procedural descriptions of this Chapter.

23.2.3   Management transfer

This refers to government developed and managed drainage systems which the government in future desires to be managed by the farmers (usually represented by a user association). A mixture of policy and financial considerations usually motivates such a transfer of management. Many governments have adopted policies, which pursue the devolution of present government management responsibilities in selected sectors to the direct stakeholders, the argument being that this would improve the quality of the management. An at least equally important government motivation for such a management transfer is that it allows it to shift a considerable part of the sector costs to the new management.

Management transfer has not yet been widely applied in the drainage sector. As mentioned, most of the drainage systems in the developed countries are already managed by non-governmental bodies while in the developing countries drainage is still very much in the development stage. Management transfer is however being practised on a large scale in the irrigation sector. Countries like Mexico, Pakistan, India, etc. with large government managed irrigation systems, have already transferred or are in the process of transferring large parts of these systems (especially at the tertiary and secondary levels) to user associations. The experiences gained generally also apply to the transfer of drainage management.

23.3   Public drainage organisation

Public drainage organisations are typically formed when the solution to the prevailing drainage problems is technically, financially or otherwise beyond the means of a single party and/or is to the benefit of several parties. For drainage where the disposal opportunities are mostly determined by natural conditions of the land (topography, location of the outlet), this is very common, especially in areas with small ownership. The parties would normally include the landowners but also various residential communities, industries and civil society groups. The institutional and administrative set up of these public drainage organisations has taken many forms in different countries.

In it simplest form, a public drainage organisation may consist of a group of farmers with common drainage interests which co-operates on the basis of a legally recorded agreement on one or more specific drainage measures e.g., the joint disposal of their excess water.

More commonly, however, the organisation is established by a quorum-decision of the interested parties (simple or two-thirds majority, whatever the law requires) or be set up by special act or government decree. The first applies, for example, to the drainage district organisations in the USA (established by simple majority of the interested parties), the latter to the Water Board organisations in the Netherlands (established by degree of the provincial government) and the Water Companies and the Environment Agencies in the UK (established by Act of Parliament). These more formally established organisations usually have wider tasks than simply drainage disposal and their service areas generally extend beyond the rural areas. The Water Boards in the Netherlands are, for example, also responsible for water quality control although they were originally formed for flood control and drainage disposal purposes only. In the UK, rivers are the responsibility of the "Environment Agency" (EA). The EA works very closely with and is partly funded by the Government (Department of Food and Rural Affairs). The Water Companies are responsible for the Water Supply and the Wastewater Disposal. They are private companies regulated by an independent body. Drainage at the farm level is either the responsibility of the farmers or of the appointed Internal Drainage Board. Discharge to a receiving watercourse has to have the permission of the EA.

These public drainage organisations are generally governed by a board elected by the general assembly of all interested parties. The government will often also appoint some board members. The board will usually form an executive committee for the day-to-day management of the affairs of the organisation, in which tasks this committee will be assisted by professional staff, headed by a general manager. Typically staff consists of an administrative and an engineering wing, subdivided as in the organisational chart in Figure 23.1

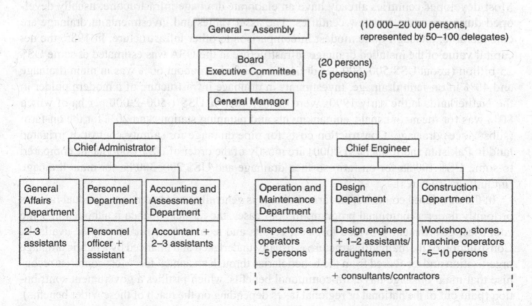

Figure 23.1 Organisational chart for a medium sized public drainage organisation

(applicable to drainage basins of about 20 000–50 000 ha with mixed farming and predominantly medium sized farms).

The functions, responsibilities and powers of the organisation would be laid down in the charter under which it was established, with the by-laws defining various aspects of the operational functioning of the organisation. For example, the by-laws would specify the maintenance obligations of the landowners. The charter and by-laws would also spell out how the expenses of the organisation are to be recovered from the beneficiaries. The assessments would usually be made based on the benefits gained from the services rendered by the organisation. For landowners this would typically be a per hectare charge, varying for example with the productivity of the land (for further details see section 23.4).

In addition to the operation and maintenance of existing main drainage systems, these organisations will often also undertake various improvement works, with the charter/by-laws specifying how the relevant decisions should be made and how the costs should be divided. The impetus for the establishment of a public drainage organisation in fact frequently arises from a communally felt desire that the drainage in the area should be improved and that an organisation through which this can be done should be established.

23.4   Financing

As is to be expected in view of the variation in development and management models, ways of financing the development and the O&M costs of drainage vary from country to country. In most countries, however, the financing is in principle based on the earlier described public/private good model and the related allocation of responsibilities. National interest, agricultural/rural development policy and farmers capacity considerations should also be taken into account, especially in the developing countries.

23.4.1   Investments

Most developed countries already have an elaborate drainage infrastructure, usually developed during over the last few centuries. The asset values and investments in drainage are considerable, although quite modest in comparison to other infrastructure. In 1985, the net capital value of the installed drainage infrastructure in the USA was estimated at some US$ 25 billion (about US$ 500–600 per drained ha), of which about 60% was in main drainage and 40% in on-farm drainage. Investments in drainage infrastructure of a modern polder in the Netherlands in the early 1990s were of the order of US$ 1 500–2 000 per ha of which 80% was for the main canals, embankments and pumping stations and 20% for the on-farm (subsurface) drainage. Construction costs for pipe drainage for salinity control of irrigated land in Pakistan and India (± yr 2000) are mostly of the order of US$ 1 000/ha, as compared to some US$ 100/ha for on-farm surface drainage and US$ 200–300/ha for main drainage (including structures).

In the developed countries, on-farm drainage is generally undertaken by individual farmers or jointly under a communal program. In both cases, the farmers finance it although in some countries, government provided low cost credits and technical assistance may be available. Construction of the main infrastructure is usually undertaken and financed by a public drainage organisation and recovered from the beneficiaries through an annual fee. Most countries recognise that main drainage has extra-communal benefits, which justifies a government contribution (paid out of the national or regional taxes depending on the reach of these wider benefits).

In the developing countries, almost all drainage development is undertaken in the form of government led and financed projects or programs. These projects/programs may not be restricted to drainage but be part of a broader rural development. They will usually include considerable training and institutional building (to provide for future management). The project/program costs are almost always fully borne by the government treasury. This also applies to cases where the participatory drainage development model is being followed (farmers may, however, provide some labour input in the on-farm development).

23.4.2 Operation and maintenance costs

In most developed countries, farmers are fully responsible for the on-farm O&M while for the main drains that responsibility rests with a public drainage organisation or, less commonly, with the local government. In 1994, the total annual costs of drainage (excluding the on-farms costs) in the Netherlands were estimated to amount to some US$ 100–150 per drained ha. A large part of these costs is charged to residential, industrial and other built-up areas or are justifiably paid out of the national or provincial budgets. The part charged through the Water Boards to the farmers is of the order of US$ 30–50 per ha. Adding some US$ 5–10 per ha for the on-farm O&M, the total farmers costs come to about 10–15% of the net farm income. Due to its lowland location and related intensive and complicated drainage infrastructure, drainage costs in the Netherlands are on the high side (the relative costs in other countries in Europe and North America are mostly in the 5–10% range).

In the developing countries, responsibility for the O&M of the drainage systems usually rests with a government department (e.g., the Irrigation Department of the Agricultural Department). Most governments have policies for charging the beneficiaries for these services (see below) but the charges collected almost never cover the full costs. Although in most developing countries, on-farm drainage is in principle the responsibility of the farmers, the distinction between main and on-farm drainage is often not relevant as very little on-farm drainage exists and where it does exist, it has often been installed and is being maintained by the government (applies to almost all subsurface drainage of irrigated land). Although many countries have now adopted stricter cost recovery policies, these have proven to be difficult to enforce in the prevailing government driven/dependent organisations.

In cases where the participatory development model or where management transfer has been applied, conditions for cost recovery are generally more conducive. Improved O&M and cost recovery is in fact almost always one of the most important tasks of the newly established organisations (local/regional users association, semi-public organisations, etc.). Initial experiences indicate that the farmer's willingness to contribute can be enhanced and that farmer's financial capacities have often been misunderstood. Valid socio-economic constraints remain, and it is widely perceived that subsistence farmers should not be asked to contribute more than 4–5% of their net income. In many cases, therefore, drainage development in the developing countries will for the time being continue to depend on government's willingness and capacity to accept not only the development costs but also part of the O&M costs.

23.4.3 Fee systems

Various systems are used to determine and to collect the fees, charges or other forms of contribution to be made by the beneficiaries (EEA 2001). The discussion here will mostly be based on the public/private good model whereby public drainage organisations need to collect fees for the financing of the main drainage infrastructure. Much of the discussion will, however, also be applicable/relevant to the participatory development model. Provided specifics mostly relate to principles and practices followed by the Water Boards in the Netherlands (for more details on Dutch Water Boards see Dolfing and Snellen 1999).

Costs and budgets: most public drainage organisations are by law required to recover costs made on their behalf from the beneficiaries. This refers to the O&M costs as well as the capital costs. In some organisations (e.g., some drainage districts in the USA), new developments need the specific approval of the board while the involved capital costs are recovered by a specific surcharge. Most organisations with long-term financial planning include capital costs of small investments in the regular (multi-year) budgets and annual fees. Renewal and replacement investments are normally financed by the depreciation and interest components in the regular budgets.

Liable contributors: the area served by a drainage association/organisation usually constitutes a natural drainage (sub) basin and under the Dutch Law on Taxes, all parties who have some form of interest in the quality of the drainage in this area (residents, owners, businesses, industries, etc.) are legally required to contribute to the costs. Separate groups of liable contributors may be identified for different services (flood protection, water quality control, etc.).

Cost recovery: the cost recovery system is usually needs based, meaning that the required funds are determined by the costs of the proper performance of the assigned functions. The simplest needs-based cost recovery system is the flat land tax, determined by dividing the total annual costs by the served area. This is a suitable system for areas where all beneficiaries have more or less the same levels of benefit (applies e.g., to a homogeneous farmland area). As the land use becomes increasingly diverse, flat rates are mostly replaced by differentiated rates. Rates may be justifiably differentiated on the basis of variation of the derived benefits and/or the costs incurred. The first approach argues for higher rates for naturally poorly drained land and in wet years, the second for higher rates for lower lying land which require pumped disposal. For reasons of financial stability, weather related rate variations are seldom applied.

Rate setting: the applied rates should in principle be proportional to the derived benefits. Dutch Water Boards distinguish between *general interests* and *specific interests*. The first apply to all who have, in one way or other, an interest in the existence and in the proper functioning of the organisation. This generally includes all residents and businesses in the service area of the organisation. The general interest is usually assessed by comparison with the 'without case' and may cover up to 25–30% of the total costs.

Most Dutch Water Boards differentiate the specific interest rates on the basis of recognised interest categories. The categories may vary per area but usually include the following:

* residents (usually the heads of registered households)
* owners of non-built up land (farmland, recreational land, nature reserves, etc.)
* owners of residential buildings
* owners of business and industrial estates.

The rate setting is then usually done in two steps. First the costs are percentage wise divided over the interest categories on the basis of the assessed relative differences in levels of benefit. In the second step, the individual contributions within each category are fixed on the basis of the assessed economic value of the involved land, site or building (the same assessment base as used for the municipal property tax). Farmers are usually assessed on the basis of a unit rate per ha.

More refined rate settings not only take into account the differential benefits but also the differential costs to be incurred by the organisation to meet the drainage requirements of a particular interest group. The rapid discharge of rain from built-up, paved and greenhouse areas for example necessitates costly higher disposal capacities, which justifies higher charges.

Fee collection: the fees may be collected directly by the drainage organisation itself or indirectly as part of some general tax. Direct collection general contributes to more transparent financing and more public accountability. It is followed by the Dutch Water Boards, although for reasons of cost saving, the fee collection may be combined with those for other public services roughly covering the same or overlapping areas (sewerage charges, water supply fees, etc.).

Good fee collection requires a very accurate and up-to-date registration of the assessment basis of the various categories of beneficiaries and an equally capable administration of the billing and payments. Most Water Boards have computerised databases linked to other public databases (land registration, general taxes, etc.) and automated billing systems. Collection percentages in the developing countries are often unacceptable low due to the lack of adequate registration and administration, lack of self-interest and vigilance (applies especially to indirect collection), but also due to corruption and fraud.

Irrigated areas: in these areas, the managing organisation will often provide both irrigation and drainage services and the question arises whether the fee systems should be separate or be combined. It might be argued that separate fee systems enhance transparency and possibly lead to better service. Irrigation and drainage are however often strongly interdependent (with over-irrigation contributing to waterlogging and over-drainage contributing to drought). Also considering collection costs, a combined fee system is often to be preferred.

Water quality: the Dutch Water Boards are also responsible for the control of the water quality in their service area, the costs of which are recovered separately from the polluters. The applied cost recovery system is again needs-based, and each polluter is charged in proportion to its share of the pollution of the Board's waters (the polluter pays principle). This share is expressed by the assigned number of pollution units, whereby one unit equals the pollution caused by one person. The average household is generally assessed as three pollution units. Small offices, businesses, industries etc. are also charged on the basis of assessed pollution units but all large polluters are charged on the basis of the actual pollution as laid down in the issued wastewater disposal permit.

Chapter 24

Maintenance

Constructed drainage systems are usually subject to conditions which in time affects their functioning. Measures (called maintenance) therefore need to be taken to safeguard the level of performance for which they were designed. The discussion here will focus on the maintenance of open drains and subsurface pipe systems. The poor maintenance of drainage systems in the developing countries is of special concern, warranting separate analysis and discussion (section 24.5).

24.1 Classification

A distinction can usually be made between the following maintenance categories:

Annual/minor maintenance: maintenance measures known to be needed at least once per year (weed clearance in open drains, debris clearance from culverts and bridges, greasing and lubrication of moving parts of weirs, etc.).

Periodic/major maintenance: preventive maintenance undertaken at pre-determined intervals or on the basis of maintenance needs established during periodic inspections (desilting and reshaping of canals, flushing of pipe drains, painting, repair and/or replacement of worn parts of structures, etc.).

Emergency maintenance: repair/replacement after break-down of system components due to accidents or unforeseen structural failure; may also apply to cases where preventive maintenance is not feasible or where the damage or temporary loss of function is acceptable. Foreseen repairs/replacements should be covered by the periodic maintenance.

Deferred maintenance: some temporary, minor maintenance neglect can generally be allowed but gross deferment of maintenance is almost always bad policy as it leads to unreliable functioning of the system and undermines the institutional discipline and user's commitment.

Rehabilitation and modernisation: rehabilitation refers to the rebuilding of (part of) systems, which have fallen badly into disrepair and cannot be kept in a proper functional state by normal maintenance. Rehabilitation will often be combined with the introduction of new designs/technology (modernisation). Well-maintained systems may also periodically need some modernisation to meet new requirements, to improve efficiencies and/or to save O&M costs. Rehabilitation and modernisation are generally not part of the regular maintenance and are financed from separate budgets.

24.2 Organisation, planning and execution

The overall framework of organisation of the maintenance of drainage systems has already been outlined in Chapter 22. This section adds some specifics on the planning and execution of the maintenance activities.

Maintenance, although essentially a technical activity, requires good planning and administration. Annual maintenance requirements are generally partly determined on the basis of the accepted maintenance frequencies, partly on the basis of the results of field surveys. These requirements are then compiled into annual and multi-annual work plans which define maintenance works to be undertaken at various places and times and which form the basis for the staff and equipment planning, costs estimate and budgets.

Such planning requires accurate and up-to-date geographical and technical information on the systems and its various components (designations, alignments, locations, accessibility, soil conditions, environmental values, use of adjoining land, site drawings, design and actual longitudinal profiles and cross-sections and water levels, functional, dimensional and structural information of structures, history/experiences of previous maintenance activities). The actual state of functioning of the various system components should be kept up to date by regular surveys. This information used to be compiled in various maps, drawings, technical files and other conventional information systems but these are increasingly being replaced by computer-based data filing/processing systems.

In most drainage organisations, maintenance is organised as a separate, specialised activity (Figure 23.1) with the maintenance division having its own staff, equipment and budget. Large organisations will usually have regional centres/workshops for minor maintenance while the execution of periodic maintenance may be more centralised. Most periodic maintenance lends itself to contracting and this mode of execution is widely applied by most organisations. Some organisations are experimenting with long-term performance contracts for canal maintenance, but most organisations do their minor maintenance themselves as it is often difficult to specify the required work in advance.

24.3 Maintenance of open drainage canals

Drained land generally has a rather dense system of open type of drains, the maintenance of which poses a heavy physical as well as a financial burden. The length of the communally maintained open drains varies from some 40–60 m per ha in the polders of the Netherlands, 30–50 m/ha in the Mekong delta of Vietnam, 7–10 m/ha in the Nile valley of Egypt, 5–7 m/ha in the Imperial Irrigation District, California, USA to some 2–5 m/ha in a typical irrigation command in India/Pakistan. The total canal maintenance costs in the Netherlands in 1985 averaged US\$ 1.00–1.50 per m (see also Table 24.1).

24.3.1 Problems

Almost all maintenance needs of drainage canals derive from the occurrence of one or more of the following three flow restricting processes.

Weed growth: canals generally provide a conducive environment for plant growth (water, nutrients, sunlight). The occurrence of various types of aquatic plants in a cross-section is shown in Figure 24.1. Some plant growth on the side slopes is generally welcome as it provides stability but most of the abundant growth of aquatic plants is unwelcome as it hinders the flow of water.

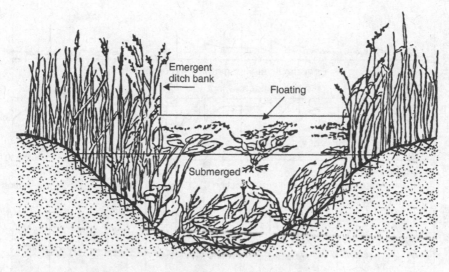

Figure 24.1 Aquatic vegetation commonly found in drainage canals

The composition of the vegetation varies per climatic zone and depends on the specific ecological conditions in the canal. The growth cycle of the canal vegetation is usually quite similar to that of the field vegetation. In mid-season, the underwater biomass may easily occupy some 20–50% of the wet cross-section and proportionally reduce the transport capacity. The vegetation also traps much sediment and provides habitat for some waterborne diseases (malaria, mosquitoes, bilharzia, snails). Decomposing biomass and detritus sediments may adversely affect fishery and other aquatic life.

Siltation: sediment may enter the canals with the drainage water (especially with the overland flow drainage water) but much of it may also be generated by internal morphological processes (erosion, scouring, sloughing of side slopes, and banks, etc.). Even properly designed canals will frequently operate under non-regime flow and/or bed conditions under which sediments are deposited and/or picked up. Bed deposits in drains often also include various debris, detritus, rotting biomass and tree remains.

Sloughing/erosion of side slopes/banks: sloughing/erosion may be caused by lack of sufficient vegetative and/or mechanical bank protection, irregular flow regimes, uncontrolled inflow of runoff water, unstable layers, excessive bank loading, inflow by seepage and piping, etc. Risks are most acute for fresh cuts before the side slopes have become vegetated and have stabilised.

24.3.2   Requirements

Drainage systems are generally designed to meet certain performance targets (usually a combination of water levels and discharges) when the canals are in a good state of maintenance (section 13.1.1). As the canal capacity is directly proportional to both the cross-section and the roughness, canals which are not in the assumed good state of maintenance, can only carry

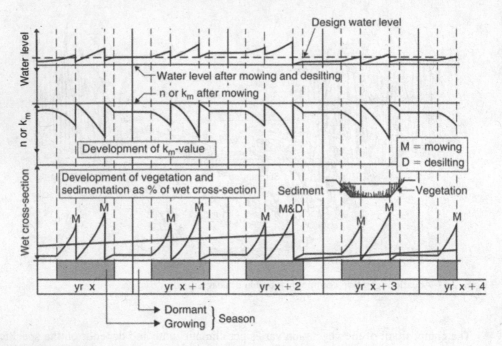

Figure 24.2 Impact of vegetation and siltation on canal water level

the design discharge at a higher water level. High canal levels hinder the proper functioning of the field drainage systems and may even lead to flooding. Figure 24.2 shows how the canal water levels (at design discharge) may be expected to vary, depending on the timing of the maintenance activities.

For siltation and sloughing, the impact is straightforwardly due to the reduction of flow section. The impact of dense vegetation, although hydraulically most correctly evaluated through the k_m-value (n-value, Manning/Strickler see section 13.1), may in practice often also be seen as a restriction in available flow section as the flow velocities in dense vegetation are close to nil (Querner 1995).

Canal maintenance requirements should rationally be based on a comparison of the costs of controlling the maintenance induced high water levels and the damage caused by these high levels (see also methodologies as used in section 2.1). In practice this is almost never done although it may generally be assumed that the applied maintenance frequencies and other empirical rules are implicitly based on the above considerations.

Applied maintenance frequencies and practices are also much more based on experience than on scientific research. Study of the factors and processes which cause the drainage systems to become less functional can, however, greatly help to make the maintenance more effective. Understanding of the anatomical and physiological characteristics of the weeds and the origin and transport of the sediments can help to make the canal maintenance measures better timed and targeted.

Some indicative information on applied frequencies have been compiled in Table 24.1. The weed clearance timings should be attuned to the cycles of the weed growth and functional

Table 24.1 Maintenance frequencies and costs for years indicated

Country	Costs for year indicated
Egypt (main drainage canals in Fayoum, 1990)	
• mechanical weed clearance: 1 × per yr	US $ 0.50 per m for small canals[1]
	US $ 0.75 per m for large canals
India (small open drains in Chambal command, 1995)	
• manual weed clearance (slashing): 1 × per 2 yrs	US $ 0.10 per m
• manual desilting/resectioning: 1 × per 5 yrs	US $ 0.50–1.00 per m
India (Uttar Pradesh 1999)	
• small drains (Q < 5 m³/s)	US $ 0.2 per m
• medium drains (Q = 5–15 m³/s)	US $ 0.4 per m
• large drains (Q > 15 m³/s)	US $ 0.8 per m
Netherlands (main drainage canals, 1985)	
• mechanical weed clearance: 1–2 × per yr	US $ 0.3–0.5 per m for small canals
	US $ 0.4–0.6 per m for large canals
• desilting/resectioning: 1 × per 5–10 yrs	US $ 0.6–1.0 per m for small canals
	US $ 1.5–2.0 per m for large canals
USA (main canals in the Imperial Irrigation District, 1992)	
• mechanical weed clearance: 1 × per yr	US $ 0.5 per m
Rule of thumb	O&M ~ 2% of the construction costs

(1) Note: top width < 7 m for small drains and > 7 m for large drains

demands of the canals. In temperate climates, growth is heavy in spring and in early summer. As high discharges occur mostly in autumn and winter, it is most critical that the canals are well cleared before the onset of autumn. Maintenance during summer can, however, not be neglected as otherwise the canals may become too choked to cope adequately with the occasional high summer rains. In tropical climates with pronounced dry and rainy seasons, drains should be cleared before the arrival of the rainy season. The weed control requirements during the dry season are quite minimal as drainage canals dry up and hardly any growth and discharge occur. In irrigated areas, drains should be in a well cleared state at the start of the main irrigation season. Regardless of the season the clearance activities can have a marked effect on watertable heights in adjacent fields and reduce the risk of waterlogging.

Silt clearance (desilting) and re-sectioning (reshaping) requirements are more variable and specific. Siltation may be minimised by designing for non-eroding flow velocities and providing for smooth curves, energy dissipation, bank protection, silt traps, etc. Weeds may reduce flow velocities to the extent that fines that would otherwise remain suspended become deposited. Mature canal sections may need very little maintenance, but new sections normally remain unstable for two to three years until a good sod has become established. Sloughing of side slopes is often due to seepage inflow and/or a combination of saturation of banks and steep slopes and can best be solved by appropriate interception drainage. Siphons in drains (e.g., to cross an irrigation canal) are notorious silt/debris traps and should wherever possible be avoided.

Design can reduce but not obviate the need for maintenance. Weed growth in canals is enhanced by light and so deep narrow shaded canals are less conducive to weed growth than wide-open shallow canals. Low flow velocities, allowing suspended load to settle and thus light to penetrate deeply, also enhances weed growth. Cuneate (wedge shaped) type low flow sections can help to dry up canal beds during periods of only trickle flow (to reduce weed growth and/or to allow for grazing). There are many different plant species growing in canals and designs which retard the development of one type, may well promote another.

24.3.3 Methods and equipment

The applied canal maintenance methods are quite diverse but may generally be grouped into the following categories.

Mechanical weed clearance

Mechanical weed clearance is by far the most widely applied method and a range of equipment has been developed to suit the range of prevailing conditions (canals of various shapes and sizes, different weed types and densities, different accessibilities and means of weed disposal, etc.). Some types of equipment clear the vegetation by uprooting the plants while others are cleared by mowing. Mowing has to be done quite frequently but is often preferable to uprooting as it is less destructive to the section. Another distinction is between self-harvesting equipment (collection of the cleared biomass) and non-harvesting equipment (leaving the cleared biomass in the canals and/or on the side slopes and banks). Harvesting may of course also be done in a separate operation.

Most popular are the cutterbar/flail mowers and the mowing-bucket type of equipment (Figure 24.3). The mowers are mostly used for the weed control of the dry section (banks and upper side slopes) while the mowing-buckets are mostly used for the wet section (bed and lower side slopes), clearing weeds as well as some silt and depositing the cleared material on the canal bank for later transport and disposal. The equipment is usually attached to and operated by a tractor (side) mounted hydraulic arm. Sturdy agricultural wheel tractors (90–115 HP) are suitable for light work but heavy-duty industrial wheel or track tractors are required for the sturdy long reach equipment used for the larger canals.

Extra heavy-duty work (reach > 6–7 m) may require the equipment to be attached to a more stable regular excavator. Suitable access roads to and service paths along the canals are essential for the effective use of all tractor-based equipment. Boat mounted equipment is an

Figure 24.3 Weed clearance machinery. Drain cleaning by tractor mounted mowing bucket and the self-contained Herder One (on the right) with rotating cabin, mowing bucket and flail mower, or crash barrier mower or the weed brusher. (from Herder BV Brochure 2019, www.herder.nl/en/herder-one/)

alternative but unfortunately its application is often seriously constrained by the prevailing obstructions.

The need for harvesting and disposal depends on the quantity of the cleared biomass, the floating biomass canal transport capacity and on environmental considerations (section 24.3.4). Spreading of floating weeds may largely be controlled by a strategic location of regularly cleaned trash racks. The transport/disposal of the collected biomass from the canal banks is often quite costly (may well cost more than the actual clearance).

Manual weed clearance

In most developed countries manual maintenance is more costly than mechanical maintenance and is generally only applied in cases where machines cannot operate or where mechanical work is environmentally undesirable. Manual work is however in many cases still the most cost-effective and most widely applied method of canal maintenance in the developing countries. It is simple to organise and does not require costly equipment, logistical support, roads/service paths. It is most suitable for community maintained small canals (< 5–6 m top width). Efficiencies may often be increased by the introduction of improved tools (hand sickles, chain sickles, rakes, portable power mowers, etc.). Long handle tools, which allow working from the banks, should be used when there are risks of contracting bilharzia or other waterborne diseases.

Burning

For canals which periodically dry out burning provides a cheap and effective method of weed control. The burning should of course be under control and not cause undue damage to structures and adjoining land.

Chemical weed control

Herbicides can be very effective for weed control in canals, but their use requires knowledge of the type and growth cycles of the prevailing vegetation and of the safe methods of herbicide application. Indiscriminate use may destroy desirable side slope vegetation, harm aquatic life and endanger the health of the maintenance personnel. For further details on environmental concerns, reference is made to section 24.3.4.

Biological weed control

A fair number of experiments have been conducted around the world on the control of aquatic weed by biological methods. Work on grass carp in California, Egypt and the Netherlands has shown successes but also that it requires elaborate measures to keep the carp population at the desired level and to keep it contained. Grazing by goats, sheep and ducks have been shown to help with the weed control on the service roads, banks and side slopes and some of the aquatic weeds. As mentioned earlier, good design can make canal conditions less conducive for weed growth. None of these methods have however yet been proven to be fully operational, robust, reliable and (cost) effective. The desired degree of weed control could generally not be achieved without some additional mechanical clearance.

Silt clearance/resectioning

This work calls for equipment which is suitable for heavy silt removal as well as for accurate re-sectioning of the beds, and side slopes. In the past, this work was mostly done by draglines and this equipment is still widely used for the desilting of large canals. Draglines may, however, leave jagged side slopes (starting point for re-growth and degradation) and for small to medium sized canals, the more manoeuvrable hydraulic excavator is often more suitable. The equipment (mostly dredging buckets) may be tractor attached/mounted but for heavier work requires regular hydraulic excavators. Small type floating suction dredgers for desilting of canals have also been developed but this equipment is mostly used for special cases only.

Manual desilting/re-sectioning is physically demanding, time consuming and not always most cost effective. Badly neglected canals may require some prior manual brush and tree clearance for machines to be able to work. Badly damaged side slopes should not only be repaired but causes should also be addressed (provision of dedicated inlets, timely stoppage of rain cuts). Banks, service roads, bunds and spoil banks should also be regularly inspected and maintained.

24.3.4   Environmental considerations

Ideally, the applied maintenance practices should meet both the land use and environmental requirements. While the former focus on the need for reliable discharge, the latter focus on conservation of an ecologically diverse canal habitat. Broader societal and various technical, operational and financial considerations need also to be taken into account. As described below, compromises satisfying all parties can often be found.

Frequency and timing of the weed clearances: the agricultural interest is to keep weeds under control and the drain functional throughout the main rainy/crop-season while the environmental interest is to have the weed control practices attuned to the demands of the habitat. Weed control should also not interfere with the recreational use of the area. Acceptable compromises can often be found by mapping the critical environmental values and adapting the scheduling of the weed clearance events accordingly.

Biomass disposal: agricultural and environmental interests largely coincide on this issue. Farmers do not like to have the cleared material on their land while the aquatic life is not burdened by oxygen consuming decomposing material. Weed control in the canals is also reduced as is the biomass with its nutrients.

The compromises generally do not represent the most efficient and cost-effective practices from an operational point of view. More staff and equipment may be needed to meet the diverse requirements and to compensate for the constraints in scheduling. The biomass disposal also involves high costs. For environmentally friendly maintenance, the involved stakeholders must generally be prepared to pay a cost.

24.4   Maintenance of pipe drains

Pipe drains need to be cleaned regularly when there is a danger of entry and deposition of fines in the pipe or when the flow becomes restricted for any other reason.

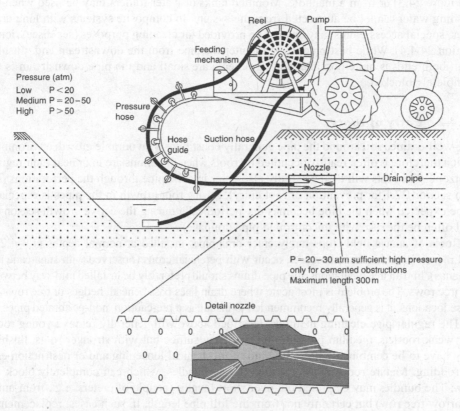

Figure 24.4 Drainpipe cleaning by means of a jetting nozzle

24.4.1   Pipe cleaning

Deposits in the pipe used to be removed by a combination of flushing and rodding (pushing a jointed bamboo rod through the pipe to loosen the deposits) but with modern flushing, using pressurised water jetted from a self-propelling nozzle (Figure 24.4) rodding is rarely needed. Water pressures at the nozzle of 10–15 atm. suffice to remove loose deposits but fail to remove deposits that have become consolidated (e.g., due to lack of regular maintenance). Such deposits may be removed by flushing at higher pressure, but this can disturb the stabilised soil around the pipe and lead to renewed entry of fines.

Entry of sediments is especially to be expected shortly after installation when the soil above and around the pipe has not had time to consolidate. Pipes therefore usually need to be cleaned soon after installation (one to two years) while from then on a frequency of five to ten years is normally sufficient. Instead of cleaning at regular intervals, it may also be done 'as required' with the latter to be established by some low level of performance monitoring (casual inspection of outflows in manholes/outlets will generally be quite sufficient). This approach is particularly appropriate for areas with minimal sedimentation risks.

Most commercially available drainpipe flushing machines have a maximum effective reach of about 250–300 m. The cleaning may be done from the pipe outlet into a ditch (as in Figure 24.3) or from a manhole. Mounted tanks or water trailers may be used when the flushing water cannot be abstracted from the system. In composite systems with long drain lines, special access points may have to be provided for cleaning purposes (see discussion in section 24.4.4). While best practice is to enter the pipe from the downstream end, flushing from both ends is quite possible when gradients are small and the pipe downstream is not completely blocked.

24.4.2   Entry of roots

The well-drained soil around the pipe generally constitutes a favourable growth environment which therefore is often readily occupied by roots when the drains are in or near their regular rootzone. The roots will then generally also enter into the pipe through the perforations (see also section 22.1 for prevention of root entry). Strong root growth in the pipes is especially to be expected when the pipe becomes dry periodically during the active growth season (as will often be the case with uncontrolled pipe drainage).

Roots of annual field crops rarely enter drainpipes installed at normal depths (1.0–1.2 m and more). Problems may, however, occur with perennial crops (observed with sugarcane and bananas). In fruit tree plantations, the pipe drains should preferably be installed mid-way between the tree rows. The problem is most acute where drain lines pass beneath hedges or tree rows. At these locations, it is generally recommended to install, as a pre-caution, non-perforated pipes.

The regular pipe cleaning methods described above will generally remove young roots. For weak rootlets, medium pressure flushing may suffice but with stronger roots, flushing may have to be combined with some kind of mechanical loosening and/or destruction e.g., by rodding. Mature roots may form thick hardy bundles, which can completely block the flow. The bundles may be pulled out over short lengths (couple of meters, e.g., from under a narrow tree row) but certainly not from the full pipe length. In such cases, replacement is often the only solution.

24.4.3   Chemical clogging (iron ochre, gypsum)

In poorly drained iron sulphide (FeS_2) holding soils, water soluble ferrous sulphate ($FeSO_4$) may form when the soils are drained and oxidised. At low pH, the ferrous sulphate may enter the drains where it is liable to oxidise to insoluble ferric hydroxide, $Fe(OH)_3$, either by direct contact with the air or by bacterial action, which shows up as an orange-brown slimy filamentous deposit in and around the drain. By drying and ageing, the ferric hydroxide eventually turns into ferric oxide deposits which clog the pipe entry points and in extreme cases may totally clog the envelope and the pipe (ochre clogging).

The problem is to be expected the first-time peat is drained, with recently reclaimed marine soils and soils containing iron pyrite in the parent material. The leaching of iron usually rises to a peak rate some two to three years after the installation of drains, subsequently falling off until eventually it ceases to be a problem. When serious problems are expected, it is advisable to start with ditches and later switch to pipes. In areas where there is only a slight risk of ochre clogging, pipes may be used immediately provided they are frequently flushed. The problem can be diagnosed on the basis of the pH of the groundwater and its divalent iron content (Eggelsmann 1981) but in most cases the problem will be clearly evident by the visible signs (occurrence of the orange-brown slimes in the ditches).

The deposition of lime and gypsum in pipe drains leading to blockage of the perforations or even to full blockage of the pipe, has also been observed in some incidental cases. Such deposition is to be expected when slowly moving lime/gypsum rich effluent is exposed to strong evaporative conditions which may occur at pipe outlets during prolonged hot and dry periods. It may be kept partly under control by introducing slowly dissolving copper sulphate blocks into the upstream end of drain lines, although full control will necessitate more rigorous measures.

24.4.4 Access facilities

The access facilities for maintenance, inspection and monitoring of pipe systems have already been described in section 8.5 (see also EPADP/RWS 2000). Figure 24.5 presents three different designs for the installation of these facilities.

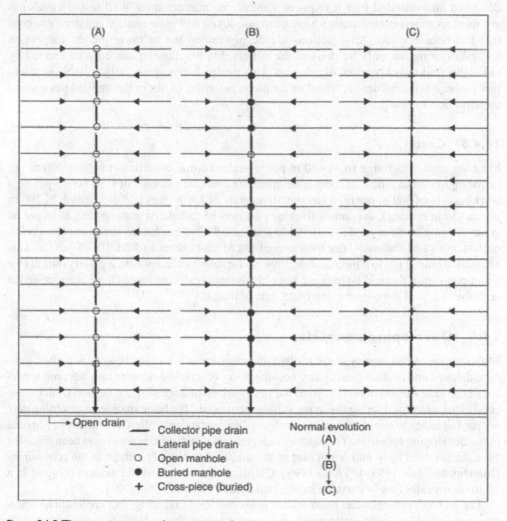

Figure 24.5 Three maintenance design approaches

Design A: in this design a dense network of manholes has been installed which provide easy access for the maintenance of the entire pipe system. It is quite costly while surface manholes are known to be sensitive to abuse. This design is suitable for situations where there is little experience with pipe drainage since construction errors can be readily diagnosed and corrected. The dense network also provides ample opportunities for inspection, monitoring, research, demonstration and learning.

Design B: this design is suitable for situations where considerable experience with pipe drainage has already been gained. Top quality installation can, however, not yet be guaranteed while some access for inspection and monitoring is also still valued. The manholes are mostly of the buried type and do not provide cleaning access for the entire pipe system. Malfunctioning pipes cannot be directly located but enough open manholes are provided for the search to be compartmentalised and the digging-up limited.

Design C: in this design it is assumed that the installed system has been so well designed and installed that it requires virtually no maintenance. Well tested envelopes are used, competent contractors have been employed and high-quality construction control has been applied. All junctions are deeply buried tee or cross pieces. Inspection and monitoring can only be done at the outlets. Malfunctioning can only be located by extensive and rather random digging up. This design is obviously only suitable in situations where full confidence, based on ample experience, exists in the applied design and construction methods.

24.4.5   Costs

Flushing capacities range from 400 m per hr under normal conditions to about 200 m per hr for highly silted pipes. At a machine operating cost (incl. labour) of US\$ 60–75 per hr, a working rate of 300 m per hr, a cleaning frequency of 1×5 years and pipe length of 200 m per ha (50 m spacing), the annual flushing cost may be calculated at some US\$ 8–10 per ha (year of price levels around 2000). Adding some US\$ 2–5 per ha for structures (manholes, outlets, etc.) and materials, the total annual O&M cost comes to US\$ 10–15 per ha. This estimate would apply to a pipe drainage system for salinity control with a gravity outfall (for a pumped sump, it would be higher). Roughly the same estimate would have been arrived at with the '~2 % of the construction costs' rule of thumb.

24.5   Developing countries

Maintenance of drainage systems is often an arduous task while the benefits are often indirect, delayed and random (hardly any benefits in dry years). The benefits only become apparent in the long run, often well beyond the time horizon of a subsistence farmer. Under these conditions, also considering the many other urgent needs, the basic maintenance philosophy of 'paying today to avoid tomorrows problems' understandably often has little ready appeal in the developing countries. This applies especially to infrastructure which has been installed by a distant third party and is not part of the traditional cultural heritage of the community (Jurriëns and Jain 1993, IPTRID 1999). Cleaning of main drains in Pakistan resulted in a 30 to 60 cm fall of the watertable in adjacent fields (Vlotman et al. 1993).

The problem of inadequate maintenance is not restricted to drainage but applies to all infrastructure (irrigation, roads, water supply, etc.). As long as the above identified generic constraints to good maintenance continue to prevail, it is almost illusory to expect high and adequate standards of maintenance. Better maintenance is realistically only to be expected when:

- the beneficiary and community appreciation of the need for maintenance has culturally been sufficiently accepted; i.e. stakeholder involvement
- the systems are planned and constructed in a more participatory manner (section 23.2.2 and section 3.4)
- the maintenance arrangements are devised in such a manner that they are within the financial means of the party responsible (see section 23.4.2).

Moreover, farmers can only be expected to pay for maintenance when they get 'value for their money' i.e., when responsible maintenance institutions are responsive to their needs and when the payers see the impact of their contribution (IPTRID 1999). As described earlier (section 4.4), fee payments should also not exceed a reasonable percentage of their net income.

Chapter 25

Performance assessment and benchmarking

Performance Assessment (PA) and benchmarking involve the assessment of the current state of performance of a (drainage) system and the evaluation of the findings against widely accepted good performance standards (the benchmarks). PA is different from monitoring and evaluation (M&E) which is a more generic assessment methodology used to assess impacts of interventions rather than performance of systems. PA provides system managers with insight into the system shortcomings on the basis of which remedial actions can be taken. A key role in this process is played by the indicators which are the (sets of) parameters which provide the required information on the system (further described in section 25.2). A distinction may be made between:

- *Operational Performance Assessment*: this concerns the performance of the regular drainage systems at the project/scheme level. The PA may be applied to entire systems or to specific system components (sub-units). Generally, these systems were designed to achieve rather specific and explicitly formulated goals, which can serve as the benchmarks against which the performance can be assessed
- *Strategic Performance Assessment*: here the focus is on the contribution of drainage to the general socio-economic development of the region/country and on the justification for the use of the involved financial, natural and other resources. The benchmarks for this assessment may be derived from the relevant regional and national plans but, by their nature, these are less specific and explicit than the operational benchmarks.

The description here focuses on the operational type of PA. Typically these take the form of periodic assessments of the operational status and performance of the system but may also include one-time special purpose assessments. When the results of project/scheme PA are compared with the performance of other similar schemes this is referred to as *benchmarking*, a term often used in (private) business and management for similar activities.

25.1 Drainage design and performance

Land drainage systems are designed to control the occurrence of excess water and salts in the root zone and as such, directly or indirectly, create favourable conditions for crop growth and farm operations. In some cases, land drainage has also non-agricultural benefits such as improved public health and environmental conditions. The expected performance of a drainage system is expressed in the basic design criteria which generally state to what degree the system is expected to control the excess water/salts. Full control may not be feasible and most designs provide for only a partial (but significant) control which takes into account the involved costs-benefit relationships. Some typical examples of basic drainage design criteria are (see also sections 2.1 and 20.2):

- *Subsurface (field) drainage*: the system should under normal conditions be able to control the watertable at 70–100 cm depth over at least 90% of the project area during at least 90% of the cropping season. The system should at any time be able to lower the watertable to 30 cm depth within two days after the ending of the recharge event. The system should (in conjunction with the water supply system) prevent salt built up in the rootzone. The system should mobilise as few solutes as possible to minimise downstream impacts
- *Surface (field) drainage*: the system should be able, in the case of a 1x5 yrs rainfall event, to remove the excess water from the land within two days after the ending of the rainfall while during the entire disposal period, there should be no erosion of the land and no overtopping of the canals and embankments
- *Main drainage*: the system should be able to convey the inflows to the outlet point(s) while maintaining water levels which allow unconstrained discharge of the field system and flow velocities which cause no harmful scouring or sedimentation
- *Controlled drainage*: the system should balance removal and conservation of water i.e., the maximum amount of water should be stored in surface and subsurface facilities, and solute mobilisation should be as low as is reasonably achievable (ALARA principle).

For standard operational PA work, the observed state of performance may normally be evaluated on the basis of compliance with the standards for which the systems were designed. However, not all of these standards may have been explicitly expressed in the design criteria. This applies, for instance, to the control of soil salinity, which is clearly a major drainage objective but is, generally, not an explicit design criterion. Design criteria stipulate the required removal of excess water/salts but not the beneficial effects of this removal. Although such objectives are not captured in the design criteria, they should of course be included in the PA. In addition, water quality aspects and environmental consideration are to be considered.

25.2   Indicators

Indicators play a crucial role in PA. It is precisely the use of indicators which distinguishes PA from other comparable assessment efforts like M&E programmes. Bos et al. (1994) provides an overview of the desirable attributes of PA indicators while Vincent et al. (2003) present a long list of potential indicators for drainage purposes. For operational PA of drainage systems, indicators may be classified as follows:

- *Primary indicators*: these indicators capture the goals which are directly addressed by the design. These goals would either be explicitly mentioned in the design criteria or if not, they would be directly linked to these criteria (the mentioned case of soil salinity). Examples are:

Watertable Regime

   ○   average seasonal or annual watertable depth (suitable for assessing waterlogging and partially suitable for assessing salinity control)

- o   rate of draw-down (suitable for assessing aeration and workability control)
- o   watertable fluctuations over time (for assessing trends and periods of high watertables and over drainage)

Flooding, Ponding and Erosion

- o   frequency and duration of excess water on the land
- o   occurrence of perched watertables
- o   seasonal and annual soil loss by erosion

Soil Salinity

- o   soil salinity conditions in the root zone during the critical growth stages of the crops
- o   annual salt balances at the root zone, field, catchment or basin level

Effluent Quality

- o   solute concentration and composition to assess quantity and type of mobilisation of solutes other than salts
- o   seasonal balance of minerals to assess danger of eutrophication etc. (e.g., the mandatory mineral bookkeeping in the Netherlands, MINAS programme, started in 1989 and succeeded by successive programs, van Grinsven et al., 2016).

- *Secondary indicators*: the purpose of these indicators is to provide additional information on the state of performance, principally to help the identification of under-performance causes. They would only be applied when the causes of under-performance are not clear. Examples: hydraulic indicators (water levels in manholes, drain discharges), maintenance indicators (sediment in pipes and open drains, and weed growth in the canals), etc.
- *Special indicators*: these indicators would be used to assess impacts on the system performance of agricultural, environmental, water management, socio-economic and institutional and other relevant conditions.

Tuohy et al. (2018) used the following indicators to assess the performance of nine drainage systems (5 ground water systems with drain depths between 1–2 m; i.e., systems tapping directly in the ground water, vs 4 shallow drainage systems with pipe drains at < 1m depth supplemented by gravel backfill followed by mole drainage (Mulqueen 1985, Tuohy et al. 2016b) or subsoiling at spacings of 1–2 m) for grassland production and utilisation of dairy farms in Ireland:

- Drainage flow start (in hours)
- Drainage flow peak occurrence and magnitude (hours and mm/h)
- Drainage flow lag time (hrs)
- Drainage peak flow rate (mm/h)
- Flashiness index (describing variations in drainage discharge), and
- Total drainage discharge and drainage discharge hydrographs compared with rainfall hydrographs.

They assessed the efficiency of the drainage systems, which is a measure of its ability to respond to rainfall event and discharge appropriate volumes of water. The ground water systems had better response time than the shallow systems and higher total discharges. The Flashiness Index is calculated for the event as the sum of the difference between quarter-hourly discharges divided by the sum of the average quarter-hourly discharge. Deelstra (2015) used a similar method to analyse subsurface drainage runoff in Norway.

25.3 Performance assessment procedure

Figure 25.1 shows as an example, the PA procedure applied to the rehabilitation of an existing drainage system. It involves sequential steps with each subsequent step only undertaken when the previous step has confirmed its necessity and, therefore, the PA process may end after each step.

The procedure may for example be stopped when the complaint assessment indicates that the complaints require no further action (and it would be resumed when the complaints persist or when other new information becomes available). Each step requires a set of indicators which will generally become more specific and detailed as the procedure enters the next step.

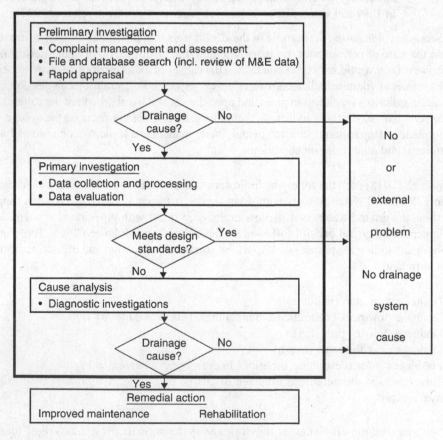

Figure 25.1 Standard performance assessment procedure

25.3.1   Preliminary investigations (first step)

This first step is proposed to include the following four activities:

1   Complaint management
2   File/database search: this includes the age of the project together with the applied tech-
     nology (materials and construction methods); the applied quality control; the contract
     documents; and other indications may be included, and each indication may have value
     singularly or in combination with others
3   Agricultural data search: crop productivity and cropping pattern
4   Rapid appraisal: a short field survey to assess the drainage conditions.

During this first step, the need for the second step is assessed. The second step requires considerable field work and expenditure and should only be undertaken when the preliminary
investigations have confirmed that there are sound indications that there are indeed waterlogging and/or salinity problems in the area or in a considerable part of the area, and that
these problems are most probably due to a malfunctioning of the drainage systems.

25.3.2   Primary investigation (second step)

This step is followed when there is a major waterlogging and/or salinity problem in the area
and these problems are due to a malfunctioning of the drainage system. In this step, this
assumption is confirmed or rejected by collecting indicator data (e.g., watertable depth and
soil salinity) and comparing these to the accepted standards or good performance. This step
may be broken down into two sub-steps:

1   Data collection and processing: monitoring the selected indicator parameters followed
     by some form of processing to facilitate the use of the collected data. For instance the
     Coefficient of Variation (CV = standard deviation/average) indicates the representative-
     ness of a data set as follows (Vlotman et al. 2000):

     CV < 0.25 average is representative
     CV = 0.25–0.5 average is moderately representative
     CV = 0.5–0.75 average is poorly representing the data set
     CV > 0.75 average is meaningless for investigation purpose

2   Data evaluation: comparing the collected indicator data with the accepted standards on
     the basis of which judgments can be made on the performance of the drainage systems.

It is of course possible that these Primary Investigations reveal that there is no real waterlogging and salinity in the area or that the prevailing conditions are not due to malfunctioning
of the drainage systems. In this case the observed problems are properly reported and the
performance assessment is ended.

25.3.3   Cause analysis (third step)

This phase is entered when the Main Investigations have confirmed that the performance of
the installed drainage system does not meet the expected standards. The remaining task is

then to identify the cause(s) of the under-performance of the system(s). An example of such a cause analysis in presented in section 25.4.

25.4　Performance checking of pipe systems

The overall functioning of a pipe drainage system may be checked by watertable observations, individual malfunctioning drains being identified by observing discharges. The causes of malfunctioning of a pipe drain may be diagnosed and localised by measuring the flow resistances as evidenced by the head losses along the flow path as shown in Figure 25.2 (Cavelaars 1994, see also section 11.1).

Case 1: impeded infiltration or percolation, usually evident by water ponded on the surface or by the occurrence of a perched watertable at some depth. Where the flow impedance is due to a hardpan layer, ripping may solve the problem. Often the cause of impeded infiltration is soil compaction, e.g., due to field traffic under wet conditions.

Case 2: high resistance in the groundwater flow towards the drain, usually due to under-design. Either the hydraulic conductivity was over-estimated, the impermeable layer occurs higher in the profile than was assumed, the drainage coefficient was under-estimated, or the drain spacing calculation formula used was not applicable. Horizontal flow and radial flow resistances may in principle be separated by placing an additional piezometer at about 0.7 D from the drain, although this is seldom necessary (they are normally correlated, when one is high so is the other); The appropriate remedial measure is to install additional drains.

Case 3: high entry resistance, a common cause of malfunctioning where drains have been installed in a trench by unsuitable methods, under unsuitable wet conditions or by the trenchless method below the critical depth. Also, an envelope may have been omitted or may have

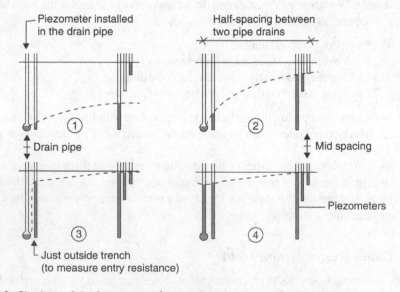

Figure 25.2 Checking of the functioning of a pipe drain by means of piezometers

become clogged. Not much can be done other than installing new drains. There is a high risk of drain failure when installation in a trench is done under wet conditions (high watertable, during rain, etc.). Puddled soil around the pipe hinders the entry of water and such installation should, as a matter of principle, be avoided.

Case 4: obstructed pipe flow due to either:

- too high a water level in the collector ditch (to be corrected by cleaning or deepening the ditch and/or by lowering the local drainage base)
- too small a pipe diameter (to be corrected by installing additional drain lines)
- misalignment of the pipes e.g., due to poor installation or due to mechanical damage to the pipe (to be corrected by repairing the line; the trouble spot may be located during cleaning by measuring the length of flushing line inserted at the outlet to clear the blockage)
- clogging of the pipe by fines or by iron compounds (ochre) or blockage of the flow by root growth in the pipe (see section 24.4). Siltation should be prevented as much as possible by providing a good filter in problem soils and should otherwise be alleviated by regular pipe cleaning. The problems and possible remedies of ochre clogging are described in section 24.4.3.

The detailed investigations can generally be restricted to a few representative sites, from which extrapolations may be made (the same causal factors generally being instrumental across a wider area). Drain failure mostly occurs or will become apparent within the first year to two years after installation. Once past this initial period, pipes can function almost permanently when maintained regularly.

25.5   Causes of under-performance of drainage systems

The performance assessment, to be really useful, should generally not be confined to the identification of the under-performance of systems but also provide enough insights in its causes to provide guidance to more detailed diagnostic investigations, or even directly to the required remedial measures. The main and most common causes of under-performance of drainage system are reviewed below.

Poor or Faulty Construction. Clearly, systems cannot perform as designed when they are not constructed as designed. Construction quality should be assured by using the correct materials, the correct machinery and the correct methods and procedures, by preparing proper construction documents and specifications, by selecting qualified contractors and generally by providing good quality control through adequate testing and supervision. The appropriate remedial measure for poor construction is partial or full rehabilitation of the malfunctioning systems or system components.

Inadequate Main Drainage. Poor performance of field drainage systems may be caused by poorly functioning main drainage systems. High water levels in open disposal drains, and non-functioning pumping stations are typical examples. The cause may be inadequate maintenance, or the problem may also be due to provision of insufficient (design) capacity.

Inadequate Maintenance. Systems generally need maintenance to keep them in a good functional state and cannot perform as designed when this maintenance is not provided. This is one of the most common causes of under-performance of systems managed by the chronically under-funded public agencies in the developing countries. It requires efficient institutional and organizational arrangements and availability of a steady stream of finance. Depending on the state of neglect, the problems may be solved by accelerated maintenance, but it may also require considerable rehabilitation.

Faulty Design. This may be due to inadequate problem diagnosis and/or inadequate field investigations or due to the use of inappropriate design methods and/or criteria. Such systems may either under-perform or perform as designed but still not solve the drainage problems. Very little can be done in such cases, except to prepare a new project design (rehabilitation), based on proper diagnosis and field investigations.

Inadequate Secondary Measures. Structural drainage systems often perform better when supported by secondary drainage measures such as chemical amendments (gypsum), (green) manuring, sub-soiling, deep ploughing, suitable crop rotations with deep rooting crops, land levelling, etc. These measures may have to be repeated from time to time.

External Constraints. Factors external to the drainage system can render them ineffective. Examples are high soil salinity due to under irrigation and non-response of farmers to the improved conditions due to lack of incentives (product prices, lack of credit, land tenure, market prices, inadequate or inappropriate government policies and regulation, institutional short comings, etc.). In the case of under-irrigation, the solution to the problem is technically straight forward but, as with the measures to overcome the non-responsiveness of farmers, only sound government policies can alleviate the external constraints.

References and further reading

* Asterix with first last name indicates not a reference but recommended for further reading.

AASHTO* 1986: Geotextile Specification prepared by a joint committee of AASHTO-AGC-ARTBA Task Force 25. American Association of State Highway and Transportation Officials (AASHTO), Washington DC, USA

Abdelraouf*, R.E. and Ragab, R. 2017: The benefit of using drainage water of fish farms for irrigation: field and modelling study using the SALTMED model. Irrig. and Drain. 66: 758–772. Wiley Online Library (wileyonlinelibrary.com) DOI: 10.1002/ird.2180.

Abdel-Shafy*, H.I and Mansour, Mona S.M. 2013: Overview on water reuse in Egypt: Present and Future. Sustainable Sanitation Practice. Issue 14, 2013.

Amlan, D. and A. Das 2000: Optimisation-based simulation and design of tile drainage systems. Journal of Irrigation and Drainage Engineering, ASCE, Reston VA, USA 126(6), 381–388.

Ankum*, P. 2001: Polders, drainage and flood control. Lecture Notes, College CT4460. Delft University (TU Delft), Delft, the Netherlands, 310 pp.

Appelo, C.A.J. and D. Postma 1999: Geochemistry, groundwater and pollution. AA Balkema, Rotterdam, the Netherlands

ASAE 1998: Design of Agricultural Drainage Pumping Plants. ASAE Standards EP369.1 Dec. 94. St. Joseph, MI, USA, 830–836.

ASAE 2003: Design, Construction and Maintenance of Subsurface Drainage in Arid and Semi-arid Areas. Engineering Standard EP 463.1, St. Joseph, MI, USA

ASCE 1969: Design and construction of sanitary and storm sewers, Manuals and Reports on Engineering Practices no 37. American Society of Civil Engineers, ASCE, Reston, VA, USA

ASTM 1996: Annual Book of ASTM Standards 04.02: Concrete and Aggregates. American Society for Testing and Materials, West Conshohocken, PA, USA, 724 pp.

Awan, Q.A, and Mahmood, S. (Eds). 2001. Second National Seminar on Drainage in Pakistan, Apr. 18–19, 2001. Univ. of Agriculture, Faisalabad, Pakistan. Nat. Drainage Program, WAPDA, Lahore. Pakistan, 417 pp.

Bahçeci, Idris, Abdullah Suat Nacar, Lui Topalhasan, Ali Fuat Tari and Henk P. Ritzema. 2018: A new pipe-envelope concept for subsurface drainage systems in irrigated agriculture Irrig. and Drain. Wiley Online Library (wileyonlinelibrary.com) DOI: 10.1002/ird.2247 (also presented at 13th International Drainage Workshop, 4 Mar 2017, Ahwaz, Iran)

Bailey, A.D. and B.D. Trafford 1978: Drainage construction techniques in England and Wales, Tenth Congress on Irrigation and Drainage, Question 34.2, International Commission on Irrigation and Drainage, New Delhi, India

Bear*, J. 1979: Hydraulics of groundwater. McGraw-Hill, New York, USA

Bear*, J., D. Zaslavsky and S. Irmay 1968: Physical principles of water percolation and seepage, Arid Zone Research XXIX. UNESCO, Paris, France, 463 pp.

Beers, van W.F.J. 1965: Some nomographs for the calculation of drain spacings, Bulletin 8. International Institute for Land Reclamation and Improvement, Wageningen, the Netherlands, 53 pp.

Beers*, van W.F.J. 1976: The augerhole method, Bulletin no 1, 4th Ed. International Institute for Land Reclamation and Improvement, Wageningen, the Netherlands, 23 pp.

Bergschenhoek BV 2003: the Netherlands at www.bergschenhoek-ct.com (last viewed Sep 2018)

Bernstein*, L 1974: Crop growth in relation to salinity. In: Drainage for agriculture, J. van Schilfgaarde (Ed.), Agronomy Monograph no 17, American Society of Agronomy, Madis on, Wisconsin, USA

Beven*, K. and P. Germann 1982: Macropores and water flow in soils. Water Resource Res., 18: 1311–1325.

Bhutta, M.H. and W. Wolters 2000: Reducing drainage requirements through interception and lining. Proc. of the 8th International Drainage Workshop. International Commission on Irrigation and Drainage, New Delhi, India, 137–152.

Biswas*, A.K. (Ed.) 1976: Systems approach to water management. McGraw-Hill, New York, USA, 429 pp.

Boast, C.W. and D. Kirkham 1971: Augerhole seepage theory. Proceedings of the Soil Science Society of America, 35:3 65–374.

Baoa, A., Huanga Y, Mab Y, Guoa H, Yongqin 2017: Assessing the effect of EWDP (Ecological Water Diversion Project) on vegetation restoration by remote sensing in the lower reaches of Tarim River. Ecological Indicators. Volume 74, March 2017, Pages 261–275. http://dx.doi.org/10.1016/j.ecolind.2016.11.007.

Boehmer*, W.K. and J. Boonstra 1994: Tubewell drainage systems, In: Drainage principles and applications. Pub. 16. 2nd Ed. International Institute for Land Reclamation and Improvement, Wageningen, the Netherlands, 931–964.

Boonstra, J. and N.A. de Ridder 1990: Numerical modelling of groundwater basins; a user-oriented manual. Pub. 29 (2nd Ed.) International Institute for Land Reclamation and Improvement, Wageningen, the Netherlands, 250 pp. (http://archive.is/ITA7).

Boonstra, J., M.N. Bhutta and S.A. Rizvi 1997: Drainable surplus assessment in irrigated Agriculture: Field application of the groundwater approach. Irrigation and Drainage Systems 11: 41–60.

Bos, M.G., D.H. Murray-Rust, D.J. Merrey, H.G. Johnson and W.B. Snellen. 1994. Methodologies for assessing performance of irrigation and drainage management. Irrigation and Drainage Systems 7: 231–261.

Bouma*, J. 1977: Soil survey and the study of water in the unsaturated zone, Soil Survey Paper no 13. Netherlands Soil Survey Institute, Wageningen, the Netherlands

Boumans, J. 1979: Drainage Calculations in Stratified Soil Using the Anisotropic Soil Model to Simulate Conductivity Condition, Proc. of the International Drainage Workshop, Publ. 25. International Institute for Land Reclamation and Improvement, Wageningen, the Netherlands, 108–122, 730 pp.

Boumans, J.H. 1987: Drainage in the arid regions. Proceeding of the Symposium 25th International Land Drainage Course. Pub. 42. International Institute for Land Reclamation and Improvement, Wageningen, the Netherlands, 22–41.

Bouwer*, H. 1978: Ground-water hydrology. McGraw-Hill, New York, USA, 480 pp.

Bowen*, R.L. and R.A. Young 1986: Appraising alternatives for allocating and cost recovery for irrigation water in Egypt. Agricultural Economics, 1(1), 35–52.

Bowler*, D.G. 1980: The Drainage of Wet Soils. Hodder and Stoughton, Auckland, New Zealand

Bressler*, E., B.L. McNeal and D.L. Carter 1982: Saline and sodic soils: principles, diagnosis and modelling. Advances in Agricultural Science 10, Springer Verlag, Berlin, Germany

Brown*, R.H. et al. (Eds.) 1972: Groundwater Studies (loose leaf publication) latest supplement 1977). UNESCO, Paris, France

Bureau of Reclamation* 1974: Design of Small Dams. US Department of Interior, Washington DC, USA

Bureau of Reclamation* 1977: Groundwater manual. United States Department of the Interior, Washington DC, USA

Cai, X., D.C. McKinney and L.S. Lasdon 2001: Solving non-linear water management models using a combined genetic algorithm and linear programming approach. Advances in Water Resources, 24(6), 667–676.

Campbell, D. 1994: Design and Operation of Smallholder Irrigation in South Asia. Technical Paper 256, Irrigation and Drainage Series, The World Bank, Washington DC, USA

Cavelaars, J.C., W.F. Vlotman and G. Spoor 1994: Subsurface drainage systems, Chapter 21. In: H.P. Ritzema (Ed.). Drainage Principles and Applications, Publ. 16, 2nd Ed. International Institute for Land Reclamation and Improvement, Wageningen, the Netherlands, 827–930, 1125 pp.

Cedergren*, H.R. 1967: Seepage, drainage and flow nets. 2nd Ed. Wiley, New York, USA, 534 pp.

CEOS 2018. Feasibility Study for an Aquatic Ecosystem Earth Observation System. Dekker A.G. and Pinnel N. (eds). Committee on Earth Observation Satellites (CEOS, www.ceos.org). Version 2.0, March 2018, 195pp.

Chadwick*, A. and J. Morfett 2004: Hydraulics in Civil and Environmental Engineering. 4th Ed., Routledge, NY, USA, 648 pp.

Chow*, Ven T. (Ed.) 1964: Handbook of Applied Hydrology. McGraw-Hill, New York, USA

Chow, Ven T. 1983: Open channel hydraulics. McGraw-Hill, New York, USA

Christen, E.W. and J.E. Ayars 2001: Subsurface drainage system design and management in irrigated agriculture; Best management practices for reducing drainage volume and salt load. TR 38/01. CSIRO Land and Water, Griffith NSW, Australia, 130 pp. (www.clw.csiro.au/publications/technical2001/tr38-01.pdf)

Christen*, E.W., J.E. Ayars and J.W. Hornbuckle 2001: Subsurface drainage design and management in irrigated areas in Australia. Irrigation Science 21; 35–43.

Clemmens, A.J., T.L. Wahl, M.G. Bos, and J.A. Reprogle 2002: Water measurement with flumes and weirs. ILRI Pub. 58, Wageningen, the Netherlands, 381 pp.

Concaret*, J. et al. 1981: Drainage Agricole. Chambre Regionale d'Agriculture de Bourgogne, Dijon, France (in French)

Corey, A.T. 1981: Pumped Outlets for Drainage Systems. Transactions of the ASAE, St. Joseph, MI, USA, 24 (6) 1504–1507.

CV 1988/2000: Cultuurtechnisch vademecum. Handboek voor inrichting en beheer van land, water en milieu. Elsevier, Doetinchem, the Netherlands, 270–287, 1085 pp. (in Dutch).

Darley, J. 2017: Low-Flow Bypass Systems, John Darley MLC, South Australian Upper House. darleyoffice@parliament.sa.gov.au.

Dasputa*, A.K. and D.W. Pearce 1978: Cost-Benefit analysis. MacMillan, London, UK

Declstra, J. 2015: Subsurface drainage runoff behaviour in Norway. Zeverte-Rivza, S. (Ed.) Proceedings of the 25th NJF Congress: Nordic View to Sustainable Rural Development 186–191.

de Ridder*, N.A. 1994: Groundwater investigations. In Ritzema, H.P. (Ed.): Drainage principles and applications, Pub.16. 2nd Ed. International Institute for Land Reclamation and Improvement, Wageningen, the Netherlands, 33–76.

DID Malaysia* 1961: Manual of the department of irrigation and drainage. Min. of Agriculture, Govt. of Malaysia, Kuala Lumpur, Malaysia

DID Malaysia* 1977: Hydrological aspects of agricultural planning and irrigation design, hydrological procedures No 20, Department of Irrigation and Drainage, Min. of Agriculture, Govt. of Malaysia, Kuala Lumpur, Malaysia

DID/LAWOO 1996: Western Johore integrated agricultural development project. Peat soil management. Final Report. Dept. of Irrigation and Drainage, Min. of Agriculture (MOA)/Netherlands Land and Water Research Group (LAWOO), Kuala Lumpur/Wageningen, Malaysia/the Netherlands, 171 pp.

Dieleman*, P.J. (Ed.) 1963: Reclamation of salt affected soils in Iraq. Pub. 11. International Institute for Land Reclamation and Improvement, Wageningen, the Netherlands, 175 pp.

Divis*, J. and Dvorak J. (Eds.) 1966: Proceedings of the Symposium on Hydrological and Technical Problems of Land Drainage. Czechoslovak Scientific and Technical Society, Agriculture and Forestry Section, Prague, Czech Republic

Dolfing, B. and W.B. Snellen 1999: Sustainability of Dutch Water Boards: Appropriate design characteristics for self-governing water management organisations, ILRI Special Report no 9, Wageningen, the Netherlands, 45 pp (http://archive.is/ITA7).

DRI staff 2001: Proceedings Final Technical Workshop on Technical Research Studies of DRP & DRP2 Projects. Ain Sukhna, Egypt, Jun 22–24, 2001. Drainage Research Institute, Kanater, Egypt, ILRI, the Netherlands, 172 pp.

DRP/DRI 2001: Drainage Research Project I & II. Final Report, Dec 1994 – June 2001. Drainage Research Institute (DRI) and International Institute for Land Reclamation and Improvement (ILRI), Cairo, Egypt. Jun. 2001, 172 pp.

Dukhovny, V.A. and P.D. Umarov 2000: The land salinisation and drainage of Central Asia. Volume I. Proceedings of 8th ICID International Drainage Workshop, New Delhi, India, 67–88.

Dumm*, L.D. 1960: Validity and use of the transient flow concept in subsurface drainage, Proc. of the ASAE Winter Meeting, Memphis Tennessee, December 1960. American Society of Agricultural Engineers, St. Joseph, MI, USA

Dunne*, T. and L.B. Leopold 1978: Water in environmental planning. W.H. Freeman, San Francisco, USA

Durnford*, D.S., B.J. Gutwein and T.H. Podmore 1987: Computerized drainage design for arid, irrigated areas. Proceedings Fifth National Drainage Symposium, Chicago, USA. 14–15 Dec. 1987. American Society of Agricultural Engineers; St. Joseph, Michigan; USA

Durnford, D.S., T.H. Podmore, and E.V. Richardson 1984: Optimal drain design for arid, irrigated areas. Transactions of the ASAE American Society of Agricultural Engineers, St. Joseph, MI, USA, 27(4), 1100–1105.

Eaton, F.M. 1950: Significance of carbonates in irrigation water. Soil Science 69; 123–133.

ECAFE* 1972: Design criteria for drainage at the farm level in Asia and the Far East. Water Resources Development Division, ECAFE Secretariat, Bangkok, Thailand

Edelman*, J.H. 1972: Groundwater hydraulics of extensive aquifers, Technical Bulletin 13. International Institute for Land Reclamation and Improvement, Wageningen, the Netherlands, 216 pp.

EEA 2001: Sustainable Water Use in Europe, Part 2: Demand Management. Environmental Issue Report No 19. European Environmental Agency (EEA), Copenhagen, Denmark. 94 pp.

EEA* 2003: Europe's Water: An indicator-based assessment. Summary. European Environmental Agency (EEA), Copenhagen, Denmark. (Luxembourg; Office for Official Publications of the European Communities), 24 pp.

Eggelsmann, R. 1981: Dränanleitung. Verlag Wasser und Boden, Alex Lindow & Co., Hamburg (also available in English: Subsurface Drainage Instructions 1978), Germany

Ehteshami, M., L.S. Willardson, and W.R. Walker 1988: Use of a surface irrigation model in subsurface drainage design. Paper No. 88–2587, American Society of Agricultural Engineers, St. Joseph, Michigan; USA

EPADP/RWS 2000: Drainage along the River Nile. Nijland, H.J. (Ed.). Egyptian Public Authority for Drainage Projects (EPADP), Min. of Water Resources and Irrigation, Egypt and Directorate General of Public Works and Water Management (RWS), the Netherlands, 323 pp.

Ernst, L.F. 1962: Grondwaterstroming in de verzadigde zone en hun berekeningen bij aanwezigheid van horizontale evenwijdige open leidingen. Verslagen van Landbouwkundige Onderzoekingen 67 (15), the Hague, the Netherlands (in Dutch).

Evans, R.O., J.W. Gilliam and R.W. Skaggs 1996: Controlled drainage management guidelines for improving drainage water quality. North Carolina Coop. Ext. Service Pub. AG-433 (1991 first issue, 1996 latest electronic revision)

FAO* Generic page with Irrigation and Drainage Papers last accessed June 2019.

FAO* 1971: Salinity seminar Baghdad, Irrigation and Drainage Paper no 7. Food and Agriculture Organisation, Rome, Italy, 279 pp.

FAO* 1972: Farm water management seminar. Irrigation and Drainage Paper no 12. Food and Agriculture Organisation, Rome, Italy, 342 pp.

FAO* 1973: Drainage machinery. Irrigation and Drainage Paper no 15, Food and Agriculture Organisation of the United Nations, Rome, Italy, 117 pp.

FAO* 1974: Soil survey in irrigation investigations. Soils Bulletin. Food, and Agriculture Organisation, Rome, Italy, 198 pp.

FAO 1976: Drainage Testing. Irrigation and Drainage Paper no 28. Food and Agriculture Organisation of the United Nations, Rome, Italy, 185 pp.

FAO 1977: Crop Water Requirements. by J. Doorenbos and W.O. Pruitt. Irrigation and Drainage Paper no 24. Food and Agriculture Organisation, Rome, Italy

FAO 1980: Drainage Design Factors. Irrigation and Drainage Paper 38. Food and Agriculture Organisation of the United Nations, Rome, Italy, 52 pp.

FAO 1985: Water quality for irrigation. Irrigation and Drainage Paper no 29 rev1. Food and Agriculture Organisation, Rome, Italy, 174 pp.

FAO 1992a: Wastewater treatment and use in agriculture, by M.B. Pescod. Irrigation and Drainage Paper no 47. Food and Agriculture Organisation, Rome, Italy, 124 pp.

FAO* 1992b: The use of saline waters for crop production. by Rhoades, J.D., Kandiah, A. and Mashali, A.M. FAO irrigation and drainage paper no. 48. Rome, Italy, 133 pp.

FAO 1994: Water quality for agriculture, Irrigation and Drainage Paper no 29 Rev1. Food and Agriculture Organisation, Rome, Italy, 90 pp.

FAO 1995a: Environmental impact assessment of irrigation and drainage projects. Irrigation and Drainage Paper No. 53. Food and Agriculture Organisation, Rome, Italy, 85 pp.

FAO* 1995b: Prospects for the Drainage of Clay Soils, Irrigation and Drainage Paper no 51. Food and Agriculture Organisation, Rome, Italy, 145 pp.

FAO 1996a: Control of water pollution from agriculture. Irrigation and Drainage Paper no 55. Food and Agriculture Organisation, Rome, Italy, 126 pp.

FAO* 1996b: Guidelines for planning irrigation and drainage investment projects. FAO Investment Centre Technical Paper no 11, Food and Agriculture Organisation, Rome, Italy, 189 pp.

FAO 1997: Management of agricultural drainage water quality, by Madramootoo, Ch. A., W.R. Johnston and L.S. Willardson (Eds.) Water Reports no 13. Food and Agriculture Organisation, Rome, Italy, 107 pp.

FAO 1999: Soil salinity assessment; methods and interpretation of electrical conductivity measurements. by Rhoades J.D., F. Chanduvi And S. Lesch (Eds.). Irrigation and Drainage Paper No 57. Food and Agriculture Organisation, Rome, Italy, 166 pp.

FAO 1999a: Transfer of Irrigation Management Services. Irrigation and Drainage Paper 58. Food and Agriculture Organisation, Rome, Italy, 110 pp.

FAO 2002: Agricultural drainage water management in arid and semi-arid areas, by K.K. Tanji and N.C. Kielen. Irrigation and Drainage Paper no 61. Food and Agriculture Organisation, Rome, Italy, 204 pp.

FAO 2005: Materials for subsurface land drainage systems. by L.C.P.M. Stuyt, W. Dierickx, and J. Martinez Beltran. Irrigation and Drainage Paper no 60 Rev1. Food and Agriculture Organisation, Rome, Italy, 200 pp.

FAO 2007: Guidelines and computer programs for the planning and design of land drainage systems. W. H. van der Molen, J. Martínez Beltrán and W. J. Ochs. FAO Irrigation and Drainage Paper no 62, Rome, Italy

FAO/UNESCO* 1973: Irrigation, drainage and salinity: an international source book. Hutchinson, London, UK

Feddes, R.A. 1988: Effects of drainage on crops and farm management. Agricultural Water Management 14: 3–18.

Gallichand, J., D. Marcotte, S.O. Prasher and R.S. Broughton 1992: Optimal sampling density of hydraulic conductivity for subsurface drainage in the Nile delta. Agricultural Water Management, 20 (1992), 299–312.

Gee, G.W. and J.W. Bauder, 1986: Chapter 15: Particle size analysis. In: A. Klute (Ed.) Methods of soil analysis, Part 1. Physical and mineralogical methods. Agronomy Monograph No. 9 (2nd Ed.). American Society of Agronomy-Soil Science Society of America, Madison, WI, USA, 383–411.

Ghassemi*, F., A.J. Jakeman and H.A. Nix 1995: Salinisation of land and water resources: human causes, extent, management and case studies. CAB International, Wallingford, UK

Giroud, J.P. and H. Plusquellec 2020: Canal Lining Practice and the Use of Geomembranes for Saving Water Resources. Book in Publication by Taylor & Francis. Presented by Nathalie TOUZE, IGS Vice-President, Head, Irstea Antony Regional Center, France, ICID 3rd World Water Forum, 1–7 Sep. 2019, Bali, Indonesia.

Glopper*, de R.J. and H.P. Ritzema 1994: Ch 13 Land Subsidence. In: H.P. Ritzema (Ed.) , Pub. 16 (2nd Ed.). International Institute of Land Reclamation and Improvement, Wageningen (ILRI), the Netherlands, 477–512.

Goewie, E.A. and M.M. Duqqah 2002: Closing nutrient cycles by reusing sewage water in Jordan. Presented at workshop: Use of appropriate treated domestic wastewater in irrigated agriculture. Wageningen, the Netherlands, Mar. 2002, Water Intelligence Online, pdf file, 17 pp.

Government of Pakistan* 1993: Surface Drainage Manual for Pakistan. Irrigation System Management Project (USAID no. 391–0467), Office of the Chief Engineering Advisor, Ministry of Water and Power, Pakistan, 112–139, 168 pp.

Grant*, E.L., W.G. Ireson and R.S. Leavenworth 1976: Principles of engineering economy. Ronald, New York, USA

Gray*, D.M. (Ed.) 1970: Handbook of the Principles of Hydrology. Canadian Committee for the International Hydrological Decade, Ottawa, Canada

Gregory*, K.J. and D.E. Walling 1973: Drainage basin, form and process. Edward Arnold, London, UK, 458 pp.

Gupta, S.K., R.K. Singh and R.S. Pandey 1992: Surface drainage requirements of crops: application of a piecewise linear model for evaluating submergence tolerance. Irrigation and Drainage Systems 6, p 249–261.

Haie, N. and R.W. Irwin 1987: Diagnostic expert systems for land drainage decisions. Irrigation and Drainage System 1(1).

Hall, M.J. and D.L. Hockin 1980: Guide to the design of storage ponds for flood control in partly urbanised catchment areas. Technical Note 100. Construction Industry Research and Information Centre, London, UK

Hall, N. 2001: Linear and quadratic models of the southern Murray-Darling basin. Environment International, 27(2–3), 219–223.

Hall*, W.A. and J.A. Dracup 1970: Water resources system engineering. McGraw-Hill, New York

Hartge*, K. 1978: Einführung in die Bodenphysik. Ferdinand Enke, Stuttgart, Germany (in German)

Heuperman, A.F., A.S. Kapoor and H. W. Denecke. 2002: Biodrainage, principles, experience and applications. International Programme for Technology and Research in Irrigation and Drainage (IPTRID), FAO, Rome, Italy

Hiler*, E.A. 1976: Drainage requirements of crops. Proceedings of the Third National Drainage Symposium. American Society of Agricultural Engineers, St Joseph, Michigan, USA

Hillel*, D. 1980: Fundamentals of soil physics. Academic Press, New York, USA, 413 pp.

Hooghoudt, S.B. 1936: Bijdragen tot de kennis van enige natuurkundige grootheden van de grond. Verslagen Landbouwkundig Onderzoek 42 (13) B: 449–541, The Hague, the Netherlands (in Dutch)

Hooghoudt*, S.B. 1940: Bijdrage tot de kennis van enige natuurkundige grootheden van de grond. Verslagen van Landbouwkundige Onderzoekingen 46 (7): 515–707, the Hague, the Netherlands (in Dutch)

Hornbuckle, J.W., Christen, E.W. and Faulkner, R.D. 2007: Evaluating a multi-level subsurface drainage system for improved drainage water quality. agricultural water management 89 (2007) 208–216.

Hornbuckle, J.W., Christen, E.W. and Faulkner, R.D. 2012: Analytical solution for drain flows from bi-level multiple-drain subsurface drainage systems. Journal of Irrigation and Drainage Engineering © ASCE/American Society of Civil Engineers. DOI: 10.1061/(ASCE) IR.1943–4774.0000438.

Hudson*, A.W. et al 1962: The draining of farmlands, Bulletin 18. Massey Agricultural College, Palmerston, New Zealand

Huisman*, P., Cramer W., van Ee, G. Hooghart, J.C., Salz, H. and Zuidema, F.C (Eds). 1998: Water in the Netherlands. Netherlands Hydrological Society (NHV), Delft, the Netherlands, 186 pp.

Heimhuber, V., M.G. Tulbure, M. Broich, Z. Xiec, M. Hurriyet. 2019: The role of GRACE total water storage anomalies, streamflow and rainfall in stream salinity trends across Australia's Murray-Darling Basin during and post the Millennium Drought. International Journal of Applied Earth Observation and Geoinformation Volume 83, November 2019 (https://doi.org/10.1016/j.jag.2019.101927).

ICID* 1978: Drainage construction techniques for vertical/tubewell drainage. International Commission on Irrigation and Drainage, New Dehli, India

ICID/World Bank* 1998: Planning the management, operation and maintenance of irrigation and drainage systems, Technical Paper 389. World Bank, Washington DC, USA, 137 pp.

IDW13 2017: Capiphon drainage, Capi – Capillary action, Phon – Siphon action. For more informa-tion see: www.greenability.com.au.

ILACO/Euroconsult* 1985: Agricultural compendium for rural development in the tropics and sub-tropics. Elsevier, Amsterdam, the Netherlands, 739 pp.

ILRI: All IRLI publications see http://archive.is/ITA7 (last accessed September 2018).

ILRI 1963: Reclamation of salt affected soils in Iraq. Pub. 11. International Institute for Land Reclama-tion and Improvement, Wageningen, the Netherlands, 175 pp.

ILRI 1964: Code of practice for the design of open watercourses and ancillary structures, Bulletin 7. International Institute for Land Reclamation and Improvement, Wageningen, the Netherlands, 82 pp.

ILRI 1972: Field book for Land and Water Management Experts. Provisional Edition. A. Volker (Ed.). International Institute for Land Reclamation and Improvement, Wageningen, the Netherlands, 672 pp.

ILRI 1973*: Acid sulphate soils. Proceedings of an international symposium, Bulletin 18, International Institute of Land Reclamation and Improvement, Wageningen, the Netherlands, 450 pp.

ILRI 1974: Drainage principles and applications, Volume IV, Pub. 16, International Institute for Land Reclamation and Improvement, Wageningen, the Netherlands, 476 pp.

ILRI 1979*: Proceedings of the international drainage workshop. J. Wesseling (Ed.), Pub. no 25, Inter-national Institute for Land Reclamation and Improvement, Wageningen, the Netherlands, 730 pp.

ILRI 1982*: Proceedings of the Bangkok symposium on acid sulphate soils, Pub. 31. International Institute for Land Reclamation and Improvement, Wageningen, the Netherlands, 450 pp.

ILRI 1982*: Proceedings of the international symposium polders of the world. International Institute for Land Reclamation and Improvement, Wageningen, the Netherlands, three volumes: 721, 726, 82 pp. and final report in 1983, 443 pp.

ILRI 1987: Proc. of the Symposium 25th International Course on Land Drainage Pub. 42. International Institute of Land Reclamation and Improvement, Wageningen, the Netherlands, 353 pp.

ILRI 1993*: Selected papers of the Ho Chi Minh City symposium on acid sulphate soils, Pub. 53. International Institute of Land Reclamation and Improvement, Wageningen, the Netherlands, 425 pp.

ILRI 1994: Drainage principles and applications, H.P. Ritzema (Ed.). Pub.16, 2nd Ed. International Institute for Land Reclamation and Improvement, Wageningen, the Netherlands, 1125 pp.

INCID 2000: Proceedings of the 8th ICID International Drainage Workshop. Volume III: Socio-economic issues, managerial and participatory aspects of drainage. Indian National Committee on Irrigation and Drainage, Ministry of Water Resources, New Delhi, India, 376 pp. (see www.icid.org, Text Delivery Service)

IOCCG* 2018: Earth Observations in Support of Global Water Quality Monitoring. Greb S., Dekker A. and Blinding, C. (Eds.) IOCCG Report Series, No. 17, International Ocean Colour Coordinating Group, Dartmouth, Canada. 125 pp.

IPTRID 1999: Realizing the value of irrigation system maintenance. Issue paper no 2, International program for Technology and Research in Irrigation and Drainage. Food and Agriculture Organisa-tion, Rome, Italy

IRRI* 1985: Soil physics and rice. International Rice Research Institute, Philippines

ISO 2003*: Plastics piping systems for non-pressure underground drainage and sewerage – Unplas-ticized poly (vinyl chloride) (PVC-U). International Standard Organisation. ISO 4435:2003. This standard was last reviewed and confirmed in 2015. www.iso.org/standard/36994.html?browse=tc.

ISO 2020*: Design using geosynthetic – part 9: barriers. Presented by Nathalie TOUZE, IGS Vice-President, Head, Irstea Antony Regional Center, France, ICID 3rd World Water Forum, 1–7 Sep. 2019, Bali, Indonesia.

Itoh, T., H. Ishii and T. Nanseki 2003: A model of crop planning under uncertainty in agricultural man-agement. International Journal of Production Economics, 81–82, 555–558.

IWMI 2000: Irrigation Induced River Salinisation, Five Major Irrigated Basins in the Arid Zone. Inter-national Water Management Institute, Colombo, Sri Lanka

James*, L.D. and R.R. Lee 1971: Economics of water resource planning. McGraw-Hill, New York, USA, 615 pp.

Janert*, H. 1961: Lehrbuch der Bodenamelioration, Band II. VEB Verlag für Bauwesen, Berlin, Germany (in German)

JNCID* 2000: Proceedings of the Asian regional workshop on sustainable development of irrigation and drainage for rice paddy fields. Tokyo July 2000. Japanese National Committee on Irrigation and Drainage, Japan

Jones, S.B., J.M. Wraith and D. Or 2002: Time domain reflectometry measurement principles and applications. Hydrological Processes, 16, 141–153.

Jurinak, J.J. and Suarez, D.L. 1990: Chapter 2, Overview: Diagnosis of salinity problems and selection of control practices. In: ASCE Manuals and Reports on Engineering Practice; No 71, New York, USA, pp. 43–63.

Jurriëns, M. and K.P. Jain 1993: Maintenance of irrigation and drainage systems; practical experiences in India and in the Netherlands. International Institute for Land Reclamation and Improvement, Wageningen, the Netherlands, 251 pp.

Jury*, W.A., W.R. Gardner and W.H. Gardner 1991: Soil physics. 5th Ed. John Wiley, New York, USA, 328 pp.

Kamra*, S.K., K.V.G.K. Rao 1994: Modelling long-term impacts of sub-surface drainage in India. GRID – IPTRID Network Magazine Issue 4, HR Wallingford, Wallingford, UK, 8–9.

Karassik*, I. J. et al. (Eds.) 1976: Pump handbook, McGraw-Hill, New York, USA

Kinori*, B. Z. 1970: Manual of surface drainage engineering. Elsevier, Amsterdam, the Netherlands

Kirkham, D. 1945: Proposed method for field measurement of permeability of soil below the watertable. Proceedings of the Soil Science Society of America, 11: 93–99.

Kirkham, D. and C.H.M. van Bavel 1948: Theory of seepage into auger holes. Proceedings of the Soil Science Society of America, 13: 75–81.

Kirkham, D. 1949: Flow of ponded water into drain tubes in soil overlying an impervious layer. Trans. Am. Geophys. Union, 30(3), 369–385.

Kirkham, D. 1957: The ponded water case, theory of land drainage. Chapter 3, Drainage of agricultural lands, J. N. Luthin, ed., American Society of Agronomy, Madison, WI.

Kirkham, D., van der Ploeg, R. R., and Horton, R. 1997: Potential theory for dual-depth subsurface drainage of ponded land. Water Resource Res., 33(7), 1643–1654.

Kirpich, P.Z. 1940: Time of Concentration of Small Agricultural Watersheds. Civil Eng. 10, 362.

KNMI 2018: Monthly weather data. Koninklijk Nederlands Meteorologisch Instituut (KNMI), Royal Netherlands Meteorological Institute (accessed August 2018).

Koerner, R.M. 1994: Designing with geosynthetics. 3rd edition, Prentice-Hall, Englewood Cliffs, New Jersey, USA, pp. 1–313, 783.

Kraijenhoff van der Leur, D.A. 1973: Ch. 15, Rainfall-runoff relations and computational models. In: Drainage principles and applications. Volume II, publ. 16. International Institute for Land Reclamation and Improvement, Wageningen, the Netherlands, 245–320.

Kroes, J.G. and J.C. van Dam (Eds.), 2003: Reference manual SWAP version 3.0.3, Alterra, Green World Research. Alterra-report 773. Wageningen, the Netherlands, 211 pp.

Lallana, C., W. Krinner, T. Estrela, S. Nixon, J. Leonard, J.M. Berland and T.J. Lack 2002: Sustainable water use in Europe, Part 2; Demand management. Environmental Issue Report No 19. European Environment Agency (EEA), Copenhagen, Denmark, 94 pp. (www.eea.eu.int)

Liggett*, J.A and D.A. Caughey. 1998: Fluid Mechanics: An Interactive Text (CD-ROM or hard copy). American Society of Civil Engineers (ASCE), www.asce.org), Reston VA, USA

Linsley*, R.K. and J.B. Franzini 1979: Water resources and environmental engineering. McGraw-Hill, New York, USA, 716 pp.

Linsley*, R.K., J.B. Franzini, D.L. Freyberg and G. Tchobangolous 1992: Water resources engineering. McGraw-Hill, New York, USA, 841 pp.

Loucks*, D.P., J.R. Stedinger and D.A. Haith: 1981. Water resources systems planning and analysis. Prentice Hall, Englewood Cliff, NY, USA, 559 pp.

Loveday*, J. (Ed.) 1974: Methods for analysis of irrigated soils, Technical Communication 54. Commonwealth Bureau of Soils, Harpenden, UK

Luthin*, J.N. (Ed.) 1957: Drainage of agricultural lands. Agronomy Monograph no 7. American Society of Agronomy, Madison, USA, 620 pp.

Luthin*, J.N. 1966: Drainage Engineering, Wiley, New York, USA, 281 pp.

Luthin*, J.N. and J.C. Guitjens 1967: Transient solutions for drainage of sloping lands. Journal of Irrigation and Drainage Div., Proc. of the ASCE 93 (IR3): 43–51.

MAFF*. 1975: Drainage and the economy of the farm, Technical Management Note no 23. Ministry of Agriculture, Fisheries and Food, London, UK, 30 pp.

Mangelsdorf*, J.K. Scheurman and F.H. Woiszi 1989: River morphology. Science Physical Environment no. 7. Springer Verlag, Germany

Marino*, M.A. and J.N. Luthin 1982: Seepage and groundwater. Developments in water science: 13, Elsevier, Amsterdam, the Netherlands, 489 pp.

Marshall*, G.R. and Jones, R.E. 1997: Significance of supply response for estimating agricultural costs of soil salinity. Agricultural Systems, 53(2–3), 231–252.

Marshall*, T.J. and J.W. Holmes 1979: Soil physics. Cambridge University Press, London/Amsterdam, UK/the Netherlands, 343 pp.

McDonald*, M.G. and A.W. Harbaugh 1988: A modular three-dimensional finite-difference ground water flow model, techniques of water resources investigations, Book 6. US Geological Survey

McKinney, D.C., X. Cai, Mark W. Rosegrant, C. Ringler and Ch. A. Scott 1999: Modelling water resources management at the basin level: Review and future directions. SWIM Paper 6. International Water Management Institute, Colombo, Sri Lanka, 71pp.

McNeal, B.L. 1974: Soil salts and their effect on water movement. In: Drainage for agriculture, J. van Schilfgaarde (Ed.), Agronomy Monograph 17, American Society of Agronomy, Madison, Wisconsin, USA, 409–432.

MDBA 2015: Tips for Developing an Engagement Plan. Murray-Darling Basin Authority (MDBA), Canberra, Aug 2015.

MDBC 1995*: Murray Darling Basin initiative. Brochure. Murray-Darling Basin Commission, Canberra, Australia. 9 pp.

MDBC 2001: Basin salinity management strategy 2001–2015. Pdf file from www.mdba.gov.au. Murray-Darling Basin Ministerial Council, Rep No I&D 6719. Canberra, Australia. 26 pp.

Minhas*, P.S. and R.K. Gupta 1992: Quality of irrigation water: assessment and management. Central Soil Salinity Research Institute, India Council of Agricultural Research, India

Mirhahar*, M.B and Sipraw, A. M. 2000: On-farm tile drainage with farmers participation: past experience and future strategies. In: Proceedings National Seminar on Drainage in Pakistan, August 16–18, 2000, Muet, Jamshoro, Pakistan. Nat. Drainage Program, WAPDA, Lahore. Pakistan, 1–14.

Mishan*, E.J. 1972: Cost-Benefit analysis. George Allen and Unwin, London, UK, 364 pp.

Moormann*, F.R. and N. van Breemen 1978: Rice: soil, water and land. International, Rice Research Institute, Los Banos, Philippines

Mulqueen, J. 1985: The development of gravel mole drainage. J. Agric. Eng. Res. 32, 143–151.

Natural Resource Conservation Service (NRCS): www.nrcs.usda.gov; drainage publications Accessed September 2018.

Nazir, A. 1979: Tubewells, theory and practise. Monograph 4. Pakistan Academy of Science, Islamabad, Pakistan

NEDECO 1991: Mekong Delta master plan. Working Paper no 3. United Nations Development Program

Nijland, H.J., F.W. Croon and H.P. Ritzema 2005: Subsurface Drainage Practices – Guidelines for the implementation, operation and maintenance of subsurface pipe drainage systems, 2005, 607 pp (See http://archive.is/ITA7)

NRCS 2001: Water management (Drainage), Part 650 Engineering Field Handbook, National Engineering Handbook Chapter14

NRCS 2001: Part 624 Drainage, Chapter 10, watertable control, National Engineering Handbook, Apr. 2001, 104 pp. Available online in pdf format from www.wcc.nrcs.usda.gov/wetdrain/wetdrain-handbooks.html

NRCS 2013 and 2015: WinTR-55, Version 1.00.10 DB (last updated 4/20/2015) Windows 7 for 32 and 64 bit PCs. See also SCS 2003.

NRCS* 2018: Handbooks available on-line:
NEH Section 16 – Drainage of Agricultural Land
Chapter 14 - Water Management (Drainage)
Part 624, Chapter 10 - Watertable Control
Part 650 - Engineering Field Handbook

Ochs*, W.J., L.S. Willardson, C.R. Camp Jr., W.W. Donnan, R.J. Winger Jr. and W.R. Johnston 1980: Drainage requirements and systems. In: Design and Operation of Farm Irrigation Systems, M.E. Jensen (Ed.), Monograph 3. American Society of Agricultural Engineers, ASAE, St. Joseph, MI, USA, 235–280.

Ogino*, Y. and K. Murashima 1991: Planning and design of subsurface drainage for paddies in Japan. In: W.F. Vlotman (Ed.). Proc. of the 5th International Drainage Workshop, Lahore, Pakistan. Feb 8–15, 1992. WAPDA/ICID, 4.1–4.9.

Ogino, Y. and K. Murashima 1993: Subsurface drainage systems of large size paddies for crop diversification in Japan. Proceedings of the 15th Congress of the International Commission on Irrigation and Drainage, New Delhi, India, Vol. 1A-1D, 1461–1468.

Olsthoorn, T. N. 1985: The power of the electronic worksheet: modelling without special programs. Groundwater 23 (3): 381–390.

Olsthoorn, T. N. 1998: Groundwater modelling: calibration and the use of spreadsheets. Thesis (doctoral). Delft University Press, 296 pp. https://trove.nla.gov.au/version/44162969

Oosterbaan, R.J. 2000: SALTMOD: Description of principles, user manual and examples of application, Special Report. International Institute for Land Reclamation and Improvement, Wageningen, the Netherlands. See other ILRI publications on http://archive.is/ITA7.

Parsons*, J.E., R.W. Skaggs and C.W. Doty 1990: Simulation of controlled drainage in open-ditch drainage systems. Agric. Water Management. Vol. 18: 301–316.

Poiree*, M. et C. Ollier, 1978: Assainissement agricole. Eyrolles. Paris, France (in French)

Polubarinova-Kochina*, P. Y. 1962: Theory of groundwater movement. Princeton University Press, Princeton, USA

Pons, L.J. and I.S. Zonneveld 1965: Soil ripening and soil classification, Pub. 13. International Institute for Land Reclamation and Improvement (ILRI), Wageningen, the Netherlands, 128 pp.

Prosida* 1978: Notes on coastal drainage design. DG Water Resources, Ministry of Public Works, Jakarta, Indonesia

Querner, E.P. 1995: A model to estimate timing of aquatic weed control in drainage canals; flow resistance and hydraulic capacity of water courses with aquatic weeds. Irrigation and Drainage Systems, 11:157–184.

RAJAD* 1998: Operation and maintenance manual for subsurface drainage system. Rajad Agricultural Drainage Research Project, Kota, Rajasthan, India, Canadian International Development Agency (CIDA), Hull, Quebec, Canada

Rakoczi*, L. and K. Szesztay 1968: Surface water hydrology. Lecture Notes of the International Post-Graduate Course on Hydrological Methods for Developing Water Resources Management. Research Institute for Water Resources Development, VITUKI, Budapest, Hungary

Reddy*, J.M. 1994: Optimisation of furrow irrigation system design parameters considering drainage and runoff water quality constraints. Irrig Sci. Berlin, W. Germany, Springer Verlag. 15(2/3): 123–136.

Reddy*, J.M. and V. Martinez 1993: Optimisation of furrow irrigation systems design considering drainage and runoff water quality. Proceedings of an International Conference on design and management of irrigation systems as related to sustainable land use, 14–17 Sep. 1992 Leuven, Belgium

Reeve*, R.C. and N.R. Fausey 1974: Drainage and timeliness of farming operations. In: Drainage for agriculture, J. van Schilfgaarde (Ed.), Agronomy Monograph no 17. American Society of Agronomy, Madison, Wisconsin, USA, 55–66.

Reich*, B.M. 1963: Short duration rainfall-intensity estimates and other design aids for regions of sparse data. Journal of Hydrology, 3:3–28.

Rhoades, J.D. 1974: Drainage for salinity control. In: Drainage for agriculture, J. van Schilfgaarde (Ed.), Agronomy Monograph no 17. American Society of Agronomy, Madison, Wisconsin, USA, 433–470.

Rishel*, J. 2002: Water Pumps and Pumping Systems. McGraw Hill. ISBN: 0-07-137491-4, 912 pp.

Ritzema, H.P. (Ed.) 1994: Drainage principles and applications. ILRI Pub. 16, 2nd Ed. International Institute for Land Reclamation and Improvement, Wageningen, the Netherlands, 1125 pp. See other ILRI publications on http://archive.is/ITA7.

Roche*, M.,1963: Hydrologie de Surface. Gauthiers-Villars, Paris, France (in French)

Rodier*, J.A. and C. Auvray 1965: Flood computations in West Africa. Proceedings of the Symposium of Budapest. International Association of Scientific Hydrology, Pub. no 66 (1): 12–38.

Russo*, D. and G. Dagan (Eds.) 1993: Water flow and solute transport in soils, Advanced Series in Agricultural Sciences no 20. Springer Verlag, Berlin, Germany

Ruthenberg*, H. 1980: Farming systems in the tropics. 3rd Ed. Clarendon Press, Oxford, UK (with contributions by J.D. MacArthur, H.D. Zanstra and M.P. Collins).

Saadat*, S. Bowling, L. and Frankenberger, J. 2017: Effects of controlled drainage on watertable recession rate. Trans. ASABE 60 (3), 2115–10032.

Saadat, S. Bowling, L. Frankenberger, J. and Klaivko, E. 2018: Estimating drain flow from measured watertable depth in layered soils under free and controlled drainage. Journal of Hydrology 556 (2018) 339–348.

Salamin, P. 1957: Relations entre les irrigations et lévacuation de léau. Proceedings of the Third ICID Congress R7-Q10. International Commission on Irrigation and Drainage, New Delhi, India

Schothorst*, C.J. 1982: Drainage and behaviour of peat soils. In: Proceedings of the International Symposium on Peat Soils Below Sea Level, Pub. 30. International Institute for Land Reclamation and Improvement (ILRI), Wageningen, the Netherlands, 130–163.

Schroeder*, G. 1968: Landwirtschaftlicher Wasserbau. Springer Verlag, Berlin, Germany (in German)

Schulte Karing*, H. 1968: Die Unterbodenmelioration. Landes Lehr und Versuchsstation, Ahrweiler, Germany (in German)

Schultz*, E. 2000: Irrigation, Drainage and Flood Protection in a Rapidly Changing World. Irrigation and Drainage 50: 261–277.

Schultz*, E. and W.S. de Vries 1993: Some aspects of maintenance of drainage systems in flat areas, Proceedings of the 15th ICID Congress, the Hague, Netherlands. International Commission on Irrigation and Drainage, New Delhi, India

SCS 1971: Drainage of agricultural land. Section 16, National engineering handbook. USDA-SCS, Washington DC, USA

SCS 1972: Hydrology. Section 4, National engineering handbook. US Department of Agriculture, Washington DC, USA

SCS 1985: National Engineering Handbook, Section 4. Soil Conservation Service, US Department of Agriculture, USA

SCS* 1986: Guide for determination the gradation of sand and gravel filters. Soil Mechanics Note No 1, 210-VI, Revised Jan 1986 (corrected Feb 1986), USDA-SCS, Engineering Division, Washington DC, USA, 6+ pp.

SCS 1988: Standards and specifications for conservation practices (SSCP). Standard on Subsurface Drains No. 606, USDA-SCS, Washington DC, USA, 606/1–606/7.

SCS 1991: US Dep. of Agriculture, National engineering field handbook, Sub Chapter C, Sect. 650.1428 (b), see also SCS 1988.

SCS* 1994: Gradation design of sand and gravel filters. NEH Part 633–26. National Engineering Publications from the US SCS. Water Resources Publications, Highlands Ranch, CO, USA, 37+ pp.

SCS 2003: Simplified flood peak and hydrograph development for small watersheds. Technical Release no 55, TR-55, Natural Resources Conservation Service (formerly SCS), US Department of Agriculture, Washington DC, USA (WinTR-55)

Segeren*, W.A. and H. Smits 1974: Drainage of newly reclaimed marine clayey sediments, peat soils and acid sulphate soils. In: Drainage Principles and Applications, volume IV, Pub. No. 16, International Institute for Land, Reclamation and Improvement, Wageningen (ILRI), the Netherlands, 261–295.

Shafiq-ur-Rehman 1995: Laboratory testing of envelope materials for pipe drains. Unpublished MSc thesis Dep. Agr. And Bio Systems Eng. McGill University, Canada

Shainberg*, I. and J. Shalhevet (Eds.) 1984: Soil Salinity under Irrigation, Ecological Studies 51. Springer Verlag, Berlin, Germany

Shaw, E.M. 1983: Hydrology in practice. Van Nortrand Reinhold, Wokingham, UK.

Shen*, H.H., C.C.K. Liu, M.H. Tang, K.H. Wang and A.H.D. Cheng 2002: Environmental Fluid Mechanics: Theories and Applications. American Society of Civil Engineers (ASCE). Reston VA, USA

Simon*, A. L. 1981: Practical hydraulics. Wiley, New York, USA

Skaggs, R.W. 1987: Design and Management of Drainage Systems. Proc. Fifth National Drainage Symposium. American Society of Agricultural Engineers, St. Joseph, MI, USA

Skaggs, R.W. and J. van Schilfgaarde (Eds.) 1999: Agricultural drainage. Agronomy Monograph 38. ASA, CSSA and SSSA, Madison, WI, USA. 58–64, 695–718, 719–763, 1328 pp.

Skaggs*, R.W., M.A. Breve and J.W. Gilliam 1994: Hydrologic and water quality impacts of agricultural drainage. Critical Reviews in Environmental Science and Technology 24 (1): 1–32.

Smart*, P. and J.G. Herbertson 1992: Drainage design. Blackie, Glasgow/London and van Norstrand Reinhold, New York, USA

Smedema, L.K. 1988: Watertable control indices for drainage of agricultural land in humid climates. Agricultural Water Management 14, p 69–77.

Smedema, L.K. 1990: Natural salinity hazards in irrigation development in the (semi) arid zone. Symposium on land drainage for salinity control in the arid and semi-arid regions. Volume 1. Drainage Research Institute, Water Research Centre, Cairo, Egypt, 22–35.

Smedema, L.K. 1993: Drainage performance and soil management. Soil Technology 6, 183–189.

Smedema, L.K. 2007: Revisiting Currently Applied Drain Depths for Waterlogging and Salinity Control of Irrigated Land in the (Semi) Arid Zone. Irrig. and Drain. 56: 379–387, DOI:10.1002/ird

Smedema, L.K. and K. Shiati 2002: Irrigation and salinity: a perspective review of the salinity hazards of irrigation development in the arid zone. Irrigation and Drainage Systems 16 (2): 161–174.

Smedema, L.K., S. Abdel-Dayem and W.J. Ochs 2000: Drainage and agricultural development. Irrigation and Drainage Systems 14: 223–235.

Smith*, K.A. and C.E. Mullins (Eds.) 1991/1993: Soil analysis; physical methods. Marcel Dekker, New York, USA

Smith*, L.P. and B.D. Trafford 1975: Climate and drainage, Technical Bulletin 34. Ministry of Agriculture, Fisheries and Food, HM Stationary Office, London, UK, 119 pp.

SOBEK 2019: www.deltares.nl/en/software/sobek/ last accessed 19/04/19.

SOURCE model 2018: https://ewater.org.au/products/ewater-source/for-rivers/groundwater-surface-water-link-model/

Spoor, G. 1979: Soil disturbance with deep working field implements in field drainage situations. In: Proceedings of the International Drainage Workshop, J. Wesseling (Ed.), Pub. 25. International Institute for Land Reclamation and Improvement, Wageningen, the Netherlands

Spoor, G. 1995: Application of mole drainage in the solution of subsoil management problems. In: Jayawardane, N.S and B.A. Stewart (Eds). Subsoil management techniques. Advances in Soil Science III, Lewis Publishers. 67–107.

Stumpf, R. P., T.W. Davis, T.T. Wynne, J.L. Graham, K.A. Loftin, 2016: Harmful Algae, 2016 – Elsevier Volume 54, April 2016, Pages 160–173.

Stumpf*, R.P., L.T. Johnson, T.T. Wynne, D.B. Baker 2016: Forecasting annual cyanobacterial bloom biomass to inform management decisions in Lake Erie. J. Great Lakes Res., 42 (6) (2016), pp. 1174–1183, 10.1016/j.jglr.2016.08.006.

Strzepek, K.M. and L. Garcia 1987: Optimal design of tile drains under uncertainty in soil properties. Proceedings Fifth National Drainage Symposium, Chicago, USA. 14–15 Dec. 1987. American Society of Agricultural Engineers; St. Joseph, Michigan; USA

Surfer* 2019: www.goldensoftware.com/products/surfer.

Tan*, C.S., C.F. Drury, M. Soultani, H.Y.F. Ng, J.D. Gaynor and T.W. Welacky 1997: Effect of controlled drainage/sub-irrigation on tile drainage water quality and crop yields at the field scale. Greenhouse & Processing Crops Research Centre, Agriculture & Agri-Food Canada, Harrow, ON, Canada N0R 1G0. COESA Report No.: RES/MON-008/97, 34 pp.

Tanji, K.E. (Ed.) 1990: Agricultural salinity assessment and management, Manuals and Reports on Engineering Practices no 71. American Society of Civil Engineers, ASCE, USA

Taylor*, H.M., W.R. Jordan and T.R. Sinclair 1983: Limitations to efficient water uses in crop production. Am. Soc. of Agron, Madison, Wisc., USA

TBS Soest BV 2003: HDPE structures. Soest, the Netherlands, www.tbs.nl

Thorn*, P.B. 1959: The design of land drainage works. Butterworths, London, UK, 235 pp.

Tietenberg*, T. 2000: Environmental and natural resources economics. 5th Ed. Harper-Collins, NY, USA, 630 pp.

Tuksöz, S. and D. Kirkham 1971: Steady Drainage of Layered Soils: theory and nomographs. Journal of Irrigation and Drainage Div., ASCE, USA

Topp, G.C., J.L. Davis and A.P. Annan 1980: Electromagnetic determination of soil water content: applications of TDR to field measurements. Proc. Third Colloquium on planetary water, Oct 27–29, 1980. Niagara Falls, New York State University, Buffalo, NY, USA

Tuohy, P., J. Humphreys, N.M. Holden, J. O'Loughlin, B. Reidy, and O. Fenton 2016a: Visual Drainage Assessment: A standardised visual soil assessment method for use in land drainage design in Ireland. Irish Journal of Agriculture and Food Research. DOI: 10.1515/ijafr-2016–0003 IJAFR.55(1). 2016.AOP.

Tuohy, P., J. Humphreys, N.M. Holden, and O. Fenton 2016b: Runoff and subsurface response from mole and gravel mole drainage across episodic rainfall events. Agric. Water Manage. 169, 129–139.

Tuohy, P., J.O. O'Loughlin, D. Penton and O. Fenton 2018: The performance and behaviour of land drainage systems and their impact on field scale hydrology in an increasingly volatile climate. Agricultural Water Management 210 (2018) 96–107.

Tyagi*, N.K. and K.V.G.K. Rao 1996: Optimising water management system design for control of shallow water-tables. Proceedings of 6th International drainage workshop. Drainage and the environment, Ljubljana, Slovenia, April 21–29, 1996, ICID, New Delhi, India. 645–652.

Uhden*, O. (Ed.) 1978: Taschenbuch Landwirtschaftlicher Wasserbau. Fränkische Verlagshandlung, Stuttgart, Germany (in German)

UNEP/WHO* 1995: Guidelines for safe use of excreta and wastewater in agriculture and aquaculture. Water, Sanitation and Health Electronic Library. UNEP/WHO. www.who.int/entity/water_sanitation_health, NB. e-refs only from 2000; website is general water, sanitation and health website, but an excellent website to obtain current information; most of their e-refs have been checked the first half of 2018, including the US EPA, WHO and OECD).

USACE* 1990: HEC-1 and HEC-2 Packages, Hydrologic Engineering Centre, US Army Corps of Engineers, USACE, Davis, California, USA. See www.hec.usace.army.mil/software/

USBR* 1973: Design of Small Dams. 2nd edition. US Bureau of Reclamation. Water Res. Tech. Pub. Denver Colorado, USA (revised printing of 1974), 816 pp., 235–237.

USBR 1974: Earth manual. 2nd Ed. US Bureau of Reclamation. Water Res. Tech. Pub. Denver Colorado, USA (2nd printing in 1985), 1–64.

USBR 1978: Drainage manual. A Water Resources Technical Publication. Bureau of Reclamation, US Department of the Interior, Washington, USA, 286 pp. (new edition in 1993)

USBR 1984: Drainage manual. US Department of the Interior, Washington DC, USA, 212–218.

USBR 1993: Drainage manual. (Revised Reprint). United States Department of the Interior, Bureau of Reclamation, Denver, Colorado, USA, 212–218, 321 pp.

USBR 1999: Integrated system for agricultural drainage management on irrigated land. Final Report Grant Number 4-FG-20-11920.USBR, Westside Resources Conservation District, Five Points, CA, USA

USBR 2017: Reclamation; Managing water in the West. Quality of Water, Colorado River Basin, Progress Report No. 25. U.S. Department of the Interior, Bureau of Reclamation, Upper Colorado Region. 131 pp.

USDA 1954: Diagnosis and improvement of saline and alkali soils, Agricultural Handbook no 60. US Department of Agriculture, Washington DC, USA

USDA* 1962: Field manual for research in agricultural hydrology. Agricultural Handbook no 224. US Department of Agriculture, Washington DC, USA

USDA 1987: Farm drainage in the United States: History, status and prospects. Miscellaneous Publication 1455. Economic Research Service, United States Department of Agriculture, USA, 170 pp. (www.usda.gov).

USDA/SCS 1985: National Engineering Handbook, Section 4. Soil Conservation Service, US Department of Agriculture

van Achthoven, T., Shohan, H. and Parlin, B.W. 2000: The reclamation of waterlogged saline lands with subsurface drainage: an overview of the Haryana Operational Pilot Project (HOPP). 8th International Drainage Workshop, ICID, Volume I, New Delhi, India, 515–528.

van Bakel, P. J. Th. 2003: Controlled drainage in the Netherlands revisited? Paper No 012. Presented at the 9th International Drainage Workshop, September 10–13, 2003, Utrecht, the Netherlands. 6 pp. (www.ilri.nl)

van Dam, J.C., Huygen, J., Wesseling, J.G., Feddes, R.A., Kabat, P., van Walsum, P.E.V., Groenedijk, P. and van Diepen, C.A. 1997: Modelling Water Flow and Solute Transport for Agricultural and Environmental Management. Theory of SWAP version 2.0 Wageningen University and Research Centre. Wageningen. the Netherlands, 167 pp.

van Grinsven, H.J.M., Tiktaka, A. and Rougoor, C.W. 2016: Evaluation of the Dutch implementation of the nitrates directive, the water framework directive and the national emission ceilings directive. NJAS – Wageningen Journal of Life Sciences 78 (2016) 69–84 (http://dx.doi.org/10.1016/j.njas.2016.03.010).

van Hofwegen*, P.J.M. and B. Schultz (Eds.) 1997: Financial aspect of water management. Proceedings of the Third National ICID Day. AA Balkema Publishers, the Netherlands, 113 pp.

van Hoorn, J.W. 1974: Drainage of heavy clay soils. In: Drainage principles and applications, Pub. 16, volume IV, International Institute for Land Reclamation and Improvement, Wageningen, the Netherlands, 298–326.

van Hoorn, J.W. 2002: Some observations with respect to sodicity hazard of irrigation water. Agricultural Water Management 61:229–231.

van Hoorn, J.W. and J.G. van Alphen 1994: Salinity control. In: Drainage principles and applications, Pub. 16. International Institute for Land Reclamation and Improvement, Wageningen, the Netherlands, 533–600 (http://archive.is/ITA7).

van Schilfgaarde*, J. (Ed.) 1974: Drainage for Agriculture, Agronomy Monograph no 17, American Society of Agronomy, Madison, Wisconsin, USA, 700 pp.

van Zeijts, T.E.J. 1992: Recommendations on the Use of Envelopes Based on Experiences in the Netherlands. In: Vlotman (Ed.) Proceedings of 5th International Drainage Workshop, Lahore, Pakistan, ICID, IWASRI, 1992, Vol. III, 5.88–5.96.

Verdier*, J. and J.L. Millo 1992: Maintenance of irrigation systems: a practical guide for managers, Paper no 40. International Commission on Irrigation and Drainage, New Delhi, India

Verruijt*, A. 1982: Theory of ground-water flow. 2nd Ed. MacMillan, London, UK, 144 pp.

Vincent, B., W.F. Vlotman and D. Zimmer 2003: From performance assessment of drainage systems towards benchmarking. Proceeding of the 54th ICID Conference, Montpellier. International Commission on Irrigation and Drainage, New Delhi, India, 25 pp.

Vlotman, W.F., Syed Rehmat Ali, Abdul Rauf Chishti, and Arshad, M. 1993: Tale of watertable fluctuations (water level recorder data 1985–1992), Fourth Drainage Project. Working Paper, IWASRI

Publication No 143, International Waterlogging and Salinity Research Institute (IWASRI), Lahore, Pakistan, 35 pp.

Vlotman, W.F., M.N. Bhutta, S. Rehmat Ali Shah, and A.K Bhatti 1994: Design, monitoring and research (Main Report) Fourth Drainage Project, Faisalabad, 1976–1994, IWASRI Pub. No. 159 (NRAP Report 53), International Waterlogging and Salinity Research Institute, Lahore, Pakistan, 223 pp.

Vlotman, W.F. 2000a: EM38 workshop proceedings, February 2000, New Delhi, India, Special Report. International Institute for Land reclamation and Improvement, Wageningen, the Netherlands (http://archive.is/ITA7).

Vlotman, W.F. 2000b: Irrigainage, rethinking irrigation and drainage design. Proc. 8th International Drainage Workshop, Jan 31 – Feb 4, 2000, New Delhi, India. Vol. II, 175–192.

Vlotman, W.F., L.S. Willardson and W. Dierickx 2000: Envelope design for subsurface drains. ILRI Pub. No 56, International Institute for Land Reclamation and Improvement, Wageningen, the Netherlands, 358 pp. See also http://archive.is/ITA7.

Vlotman*, W.F. and Jansen, H.C. 2003: Controlled drainage for integrated water management. Paper No 018. Presented at the 9th International Drainage Workshop, September 10–13, 2003, Utrecht, the Netherlands. 16 pp. (www.ilri.nl)

Vlotman, W.F. and Ballard, C. 2013: Water, Food and Energy Supply Chains for a Green Economy. First World Water Forum, 29 September – 3 October 2013, International Commission on Irrigation and Drainage (ICID), Mardin, Turkey

Vlotman, W.F. and Ballard, C. 2014: Water, Food and Energy Supply Chains for a Green Economy. Irrig. and Drain. Published online in Wiley Online Library (wileyonlinelibrary.com) Irrig. and Drain., 63: 232–240. doi: 10.1002/ird.1835.

Vlotman, W. F. and Ballard, C. 2016: Water, Energy and Food Supply Chains for a Green Economy. ICID News 2016 (First Quarter), 31 March 2016. p. 3–4.

Vlotman, W.F. 2017: Beyond Modern Land Drainage. Keynote speech and paper at thirteenth International Drainage Workshop IDW13, International Commission on Irrigation and Drainage (ICID), March 4–7, 2017. Ahwaz, Iran

Volker*, A. 1981: Reclamation and polders. Lecture Notes. International Institute for Hydraulic and Environmental Engineering, Delft, the Netherlands (see also Ankum* 2001)

Wahl,* T.L., T.B. Vermeyen, K.A. Oberg and C.A. Pugh (Eds.) 2003: Hydraulic measurements and experimental methods 2002 (CD ROM). Proceedings of the specialty conference held in Estes Park, Colorado, July 28 – August 1, 2002. American Society of Civil Engineers (ASCE), Reston, VA, USA

Walker*, R. 1972: Pump Selection. Ann Arbor Science, Ann Arbor, Michigan, USA

Wanielista, M 1978: Stormwater Management, Quantity and Quality. Ann Arbor Science, Michigan, USA

Ward*, F. A., and A. Michelsen 2002: The economic value of water in agriculture: concepts and policy applications. Water Policy, 4(5), 423–446.

Wesseling*, J. (Ed.) 1979: Proc. of the international drainage workshop. The International Institute for Land Reclamation and Improvement (ILRI), Pub. No. 25., Wageningen, the Netherlands, May 16–20, 1978, 731 pp. See also http://archive.is/ITA7.

Wesseling*, J. 1974: Crop growth and wet soils. In: Drainage for Agriculture, J. van Schilfgaarde (Ed.), Agronomy Monograph no 17. American Society of Agronomy, Madison, Wisconsin, USA, 7–38.

Wilson*, E.M. 1974: Engineering Hydrology. MacMillan, London, UK

Wilson-Fahmy, R., G. Koerner, and R. Koerner, 1996: Geotextile Filter Design Critique, in Recent Developments in Geotextile Filters and Prefabricated Drainage Geocomposites, ed. S. Bhatia and L. Suits (West Conshohocken, PA: ASTM International, 1996), 132–161. https://doi.org/10.1520/STP15598S

Winston*, W.L. 1994: Operations research, applications and algorithms. 3rd Ed. Duxbury Press, Belmont, CA, USA, 1138 pp.

Winter*, E.J. 1974: Water, soil and the plant. MacMillan, London, UK

WMO, 1974: Guide to hydrological practices. Technical Pub. no 168, 3rd Ed. World Meteorological Organisation, Geneva, Switzerland

Wolfram Mathematica* 2019: www.wolfram.com/mathematica/new-in-11/.

Workman, S.R., W.R. Skaggs, J.E. Parsons & J. Rice (Eds.), 1990: User's manual DRAINMOD. Raleigh, USA

Yaron*, B., E. Danfors and Y. Yaadia (Eds.) 1973: Arid Zone irrigation. Chapman and Hall, London, UK

Yong*, R.N. and B.P. Warkentin 1975: Soil properties and behaviour. Series: Development in geotechnical engineering. Elsevier, Amsterdam, the Netherlands, 5:449 pp.

Youngs, E.G., 1968: Shape factors for Kirkham's piezometer method for determining the hydraulic conductivity of soil in-situ for soils overlying an impermeable floor or infinity permeable stratum. Soil Science, 106: 235–237.

Zâvoianu*, I. 1985: Morphometry of drainage basins. Developments in Water Science no 20. Elsevier, Amsterdam, the Netherlands

Zeeuw, de J.W. and F. Hellinga 1960: Neerslag en afvoer. Landbouwkundig Tijdschrift, the Netherlands 70:405–421. (in Dutch)

Zeeuw, de J.W. 1973: Ch. 16, Hydrograph analysis for areas with mainly groundwater runoff. In: Drainage Principles and Applications, vol. II, Pub. 16, International Institute for Land Reclamation and Improvement, Wageningen, the Netherlands, 321–358.

Index

Note: Page numbers in *italics* indicate a figure and page numbers in **bold** indicate a table on the corresponding page.